Designing quality buildings
A BRE guide

BRE is the UK's leading centre of expertise on the built environment, construction, sustainability, energy, fire and many associated issues. Contact BRE for information about its services, or for technical advice:

BRE, Garston, Watford WD25 9XX
Tel: 01923 664000
enquiries@bre.co.uk
www.bre.co.uk

BRE publications are available from
www.ihsbrepress.com
or
IHS BRE Press
Willoughby Road
Bracknell RG12 8FB
Tel: 01344 328038
Fax: 01344 328005
brepress@ihsatp.com

Requests to copy any part of this publication should be made to the publisher:
IHS BRE Press
Garston, Watford WD25 9XX
Tel: 01923 664761
Email brepress@ihsatp.com

BR 487

© Copyright BRE 2007
First published 2007
ISBN 978-1-86081-899-8

Foreword

Today's building industry has no shortage of information: indeed, the challenge is in selecting and using what is really important. Construction technology is constantly developing, and the regulatory framework within which decisions are made about design, materials, construction, costs, maintenance and so on, is changing too, as clients become more demanding and as pressures for construction to become more sustainable increase.

I am delighted to welcome and endorse this BRE guide both personally and on behalf of BRE Trust. The Trust's mission is:

'to promote and support excellence and innovation in the built environment for the benefit of all.'

This book contributes to these aims in three ways.

- It brings together in a single volume, the most important aspects of the accumulated knowledge and experience of a team of BRE's most experienced architects, engineers, surveyors and scientists in a concise format that sets out key principles and advice.
- The guide provides a route map through the maze of standards, regulations and guidance that circumscribe our industry.

- Sustainable construction can only be achieved by applying lessons about successful performance of materials and techniques over time: this guide will enable clients and designers to make sound decisions that will contribute to sustainable construction based on a thorough understanding of the fundamental physical, chemical and environmental characteristics of buildings.

For generations of building professionals, the book *Principles of modern building* was a trusted companion to those needing authoritative guidance on the application of building science and technology. I believe that *Designing quality buildings* is a worthy successor to that classic book, and I commend it to you.

Sir Neville Simms
Chairman, BRE Trust
January 2007

Acknowledgements

This book has been produced with contributions from all parts of BRE. A special mention must be made of *Kathryn Bourke, Mike Clift, Stephen Garvin, Harry Harrison* and *Peter Trotman* for drafting and reviewing a significant part of the text, and of the following for their contributions:

Martin Aris
Paul Blackmore
Alan Ferguson
James Fisher
John Griggs
Richard Hartless
Colin Hill
Paul Littlefair
Oliver Novakovic
Richard Phillips
Barry Reeves
Dave Richardson
Keith Ross
Gerry Saunders
Chris Scivyer
John Seller
Hilary Skinner
Kristian Steele
Nigel Smithies
David Strong
Mike Wright
Tony Yates

Preface

'The highest available knowledge of pure science and the most effective methods of research are needed in building as in any other field of research. Building research as a whole, however, is concerned with the principles of an exceptionally wide range of science... ... Results of past scientific research are not at present fully utilised in building because there is no suitable bridge between the research worker and the architect or designer'.

(DSIR 1919)

Then as now! These observations from nearly a century ago highlight what has proved to be a continually recurring problem for the construction industry in the UK. In spite of the nearly instant availability of information (some might argue that the most relevant information has become buried in an avalanche of virtually useless data), there is still a need for considered overviews to aid the busy designer. That is one of the main purposes of BRE's *Digests* and other leaflets and reports, and it is to update, enhance and consolidate that effort that this book has been produced.

There are perhaps as many interpretations of the meaning of the word 'quality' as there are designers. For the purposes of this book, however, it is the technical excellence of a building that underlies our central theme, rather than its aesthetic excellence, in accordance with the BRE tradition of adopting a non-controversial approach to matters of aesthetics. Technical excellence includes sustainable construction, which means designing buildings that are fit for purpose, adaptable and durable. When short-life components are used, they should be safe

and easy to look after and replace, and at the same time have a minimal impact on the environment and be affordable.

Many of the chapters have been based on construction elements but this should not be allowed to hide the fundamental fact that an holistic approach, developed in the earlier chapters, is required to improve both the design and construction processes if we are to achieve improved quality in our finished buildings.

The scope of the book has been restricted to low-rise construction, if only in order to keep the book to a manageable size. It is also primarily about new construction, where opportunities for application of sustainability principles are perhaps greatest, but many of these principles, if not the practices set out in the following pages, apply in equal measure to refurbishment of the existing stock of buildings. It gives the construction professional a solid understanding of many key design and specification principles and a good starting point when talking to specialists.

The past few years have radically changed the national framework for the construction industry as well as the research which underpins it, and new challenges and opportunities are shaping the form and content of our buildings. Our homes, offices, schools, factories, hospitals (in fact all types of public and private building) will need to reflect and respond to:

- a changing age structure in the population,
- changing family structures and patterns of living,

- new materials technologies,
- new dispersed working patterns following IT developments,
- new control technologies (eg allowing lower energy use and increased water conservation),
- better security and fire protection,
- recycling of land and building materials,
- greater use of natural energy sources,
- better structural assessment and repair technologies,
- improved protective finishes for timber, metals and concrete,
- rapid and economical replacements for items that are life-expired,
- new insulation materials and better heating and lighting products, to cut energy consumption,
- climate change,
- a more demanding and articulate building user.

These, in summary, form the background to our chosen topics.

There are many references to building regulations in the following pages. Buildings constructed meticulously in accordance with these regulations should ensure that minimum standards of performance and safety are met. However, in many cases, higher standards than those cited in building regulations will be both justifiable and affordable when measured against long-term sustainability criteria. That is the real meaning of 'quality'.

Key points have been added to draw the reader's attention to important design information.

When selecting products, systems, installers or maintenance companies BRE strongly recommends that you choose those that have been independently approved by a third-party certification body. The performance of an excellent product can be severely undermined by poor installation or maintenance. BRE produces lists of approved products and services which can be viewed free of charge at www.RedBookLive.com.

Readers' feedback for the next edition is of course welcome. It can be addressed to:
The Publisher, IHS BRE Press, BRE, Bucknalls Lane, Watford WD25 9XX.

Mike Clift
Stephen Garvin
Harry Harrison
Peter Trotman

January 2007

Contents

1 Introduction

Good designers have always sought to provide functional, cost-effective and pleasing buildings for their clients. BRE first published *Principles of modern building* in 1938-9. In the introduction to the 3rd edition in 1959, the Director of Building Research stated the objective as:

'to help the reader to grasp principles, to sense the interconnection of requirements, and to appreciate their relative importance in particular circumstances….There is relatively little change, however, in the basic requirements that a building must satisfy; strength and durability, exclusion of the weather, admission of light and air, control of noise, and so on, at reasonable cost….What materials, put together in what ways, what methods of design, will meet these requirements?'.

The book went on to deal with the following topics:
● Strength and stability,
● Dimensional stability,
● Exclusion of water,
● Heat insulation,
● Ventilation,
● Sound insulation,
● Daylighting,
● Fire protection,
● Durability, composition and maintenance,
● Building economics,
● Principles of use of materials.

Each of these issues is still highly relevant to designers of buildings, but perhaps we would query the statement that there is relatively little change in the basic requirements. The process of design now takes place in the context of changing expectations and constraints, in a world where information is available almost instantly, but knowledgeable overviews are few and far between. The designer (of whatever discipline or background, within whatever form of contract) has the role of interpreting the client's requirements to turn them into a building that meets both the immediate need and the foreseeable future requirements. This entails translating functional issues into technical and practical designs and specifications, which can be built effectively, using available and cost-effective materials, labour and techniques.

In the process of design many legislative and regulatory requirements must be met — these changing requirements are summarised in each chapter. There is also a requirement to comply with good or best practice — key aspects of these are highlighted and sources of further guidance are listed at the end of each chapter.

We now recognise that we are living in the context of a rapidly changing and evolving social and technological age, with new concerns about the planet we live on and the sustainability of our society. This introductory chapter on sustainable design deals with the context of building design in the early 21st century. It covers some of the enduring and current issues and relates them to the basic concepts of design.

Guidance has not been explicitly tailored to 'traditional' or 'modern methods of construction' as each will be required to meet the same performance requirements. The main difference lies in the fact that where a whole system or kit is selected as the basis of design, the predetermined selections should still be tested against the basic design criteria. In principle, either can be the basis of sustainable design and specification.

Sustainable design and specification

'Sustainable construction is the set of processes by which a profitable and competitive industry delivers built assets (buildings, structures, supporting infrastructure and their immediate surroundings) which:

- enhance the quality of life and offer customer satisfaction,
- offer flexibility and the potential to cater for user changes in the future,
- provide and support desirable natural and social environments,
- maximise the efficient use of resources.'

(Source: OGC 2000)

The 'triple bottom line' of sustainability (Figure 1.1) is the successful balancing of social, economic and environmental issues. Sustainability is almost always context-dependent; what improves a problem in one area can exacerbate it in others, depending on the circumstances. It also follows that a fundamental rule of sustainable construction is that there must be an identified need for the development in the first place. The built environment has a key role to play in sustainability and appropriate building design is one of the key tools to improve sustainability in construction.

Real-world design is not about 'absolutes'; this becomes abundantly clear when trying to deal with more than one aspect of design at a time, and where interdependent alternatives might have competing or conflicting priorities. This makes the 'perfect' optimisation of a design difficult to achieve since

Office of Government Commerce (OGC). *Achieving sustainability in construction procurement.* 2000. 12 pp. Available as a pdf from www.ogc.gov.uk

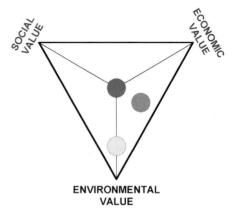

KEY

Project with perfect balance of Social, Economic and Environmental Value

Project with high Environmental Value but little Social or Economic Value

Project with equal balance of Economic and Environmental Value but little Social Value

Figure 1.1 Projects have varying balance between the triple bottom-line values of sustainability

there is often no common function against which alternatives can be measured. Different aspects of a design will be measured against different functions, with perhaps a large gain or loss in one being needed to compensate for a small loss or gain in another.

For example, increasing an area of glass to achieve an increase in the level of natural light within a building (eg to exceed the measure of a specific minimum daylight factor) may lead to significant increases in the amounts of solar gain on some elevations and an unavoidable need for mechanical cooling. In this context, the desire for natural light and the need to avoid overheating can be seen to be in conflict and to be difficult to resolve.

However, the issue can be resolved if considered at a higher level, one where the questions of 'how much glass' and 'what temperature' can be balanced against each other. Examining the cost of installing and running a cooling system as opposed to the cost of artificial lighting would be one way of doing this, as well as considering other factors. The designer needs to be mindful of functionality. For example, people generally prefer daylight to artificial light and are thought to be more productive and perform better in daylight. This should make it more worthwhile investing in shading devices, although it is not easy to demonstrate payback.

All this illustrates the crucial importance of the notion of hierarchy when dealing with the many uncertainties and conflicts that surround the types of problems that designers have to deal with all the time, and which lie at the heart of the concept of any design and especially sustainable design. Otherwise, it would be impossible to make design decisions that are both reasonable and, within limits, objective.

In any project, the designer(s) will need to recognise the balance sought by their clients between cost, environmental value and social value. Generally, not all aspects will have exactly equal value: it is relatively rare for social or economic values to be sought at any price. It is, however, part of the expertise of the designer to seek to strike an optimum balance within the constraints of the budget. Maximising sustainability is best achieved by considering it from the outset of the design process.

Box 1.1 gives a brief overview of each of the key values of sustainability. More detail is given in the following sections when they are described in

---

Box 1.1 Key values of sustainability

Environmental concerns and principles

Environmental sustainability (in relation to the built environment) considers the local, national and global environment and the impacts of development on them.
This means:

- Using land wisely and protecting areas of natural beauty, scientific interest, etc.
- Using the least amount of energy and finding more environmentally friendly forms of energy, with less damaging emissions to the environment and people.
- Limiting the amount of water treated for human consumption and increasing the use of environmentally friendly (grey) water supply and drainage systems.
- Reducing the amount of road traffic to alleviate congestion, reducing air pollution and limiting the land required for roads and car parks.
- Reducing the amount of raw materials used for construction, and considering appropriate means of extraction and/or processing for materials that are plentiful.
- Encouraging local sourcing of materials, thereby reducing transport costs and impacts.
- Providing safe disposal of used materials that cannot be re-used or recycled.
- Protecting and enhancing wildlife and biodiversity.

Social concerns and principles

Social sustainability in relation to the built environment entails providing a healthy, attractive and desirable place for people to live and work.
Consider, wherever possible:

- A high quality built environment (one that the majority of people find attractive, safe and comfortable).
- A mix of housing (types and tenures) and land uses (housing, employment, health, education and leisure).

- An appropriate density of buildings for the type of area.
- Provision of facilities locally (shops, schools, chemists, health centres).
- High accessibility throughout the area with good public transport and provision for walkers and cyclists, and recognition of different individuals' changing mobility.
- A reduction in the domination of the car, particularly in residential areas.
- Measures to improve air quality.
- Provision of a high standard of urban design with sufficient public green space and areas of beauty.
- Designs that reduce the opportunities for and fear of crime.
- Designs that reduce noise nuisance and provide some quiet spaces.

Economic concerns and principles

Economic sustainability varies depending on the nature of the community, but it is linked to the economic health of the surrounding region.
Within the urban situation some of the factors needed to make an area economically viable for the future include:

- Providing employment sites to meet projected needs.
- Providing an appropriate intensity of land use to ensure viability of local business.
- Providing good infrastructure links to key trading centres by both public and private transport.
- Supporting local trades and businesses during construction and regeneration activities.
- Ensuring that owners and occupiers will be able to afford running costs, foreseeable maintenance and repair activity, and anticipation of likely timing and cost.
- Ensuring that buildings have some flexibility to meet changing requirements without sacrificing cost effectiveness in the short term.

relation to specific issues, such as climate change, together with an overview of some key tools for balancing the competing values of sustainability.

The principles of space planning

What do buildings provide for us? At the most basic level they provide space protected from the elements so that we can carry out our intended activities. This can be developed into more detailed descriptions of:
● the type of spaces,
● the services,
● the interrelationship of spaces, and
● the relationship to the environment outside,
but essentially they should meet the needs of people and organisations.

Clearly, some buildings make a greater contribution than others to the efficiency and profitability of the business operations carried out within them, ie their functional performance.

Functional performance has a natural life cycle. It should typically be at a maximum on or shortly after completion of a new building. It will then decline as the building fabric starts to wear and the occupants' requirements change.

In each sector for which the building industry constructs buildings, there are different drivers for the construction and different functional requirements placed on the building. In some, the need to measure functional performance is more obvious than in others.

For some other buildings that are more closely integrated into a manufacturing or logistics process, the functional performance is even more critical. For example, the floor of a high rack automated warehouse has to perform to a high standard to ensure the racks are able to function. Passenger interchange buildings not only need to provide reliable mechanical equipment (escalators, lifts, baggage handling, etc.) but the planning must avoid cross-flows on main thoroughfares. The building may need to operate 24 hours a day throughout the whole year.

Constructing and running office buildings is such a small proportion of the cost of employing the staff over the design life of the building, that it is clearly important to understand and optimise the contribution of the building to their activities that take place within it. In commercial buildings there is clearly a link between how well a building supports the activities taking place within it and the profitability of that operation. The exact relationship of these factors is not simple because so many of the factors which determine profitability have nothing to do with the building(s) housing the operation. Separating out the non-building factors would require considerable research.

There is, however, a factor which links function and profitability and that is productivity. Some industry sectors and some existing research have addressed this. In the manufacturing sector, productivity is usually expressed as output per person (because even here people costs are such a large proportion of overall costs). However, it can be expressed by building. In other building types though, the concentration has been on the impact of indoor environmental factors. Productivity is quite difficult to measure, even in environments such as call centres because of the impact of non-building factors.

Away from commercial buildings decision-making is usually based on other factors. In the case of houses, most residents make decisions based on their own functional performance assessment as only they can.

The functionality of every form of built facility, including roads, railways, retail outlets, houses, hospitals, schools, offices, can be expressed in terms of the flow of added value. Figure 1.2 shows how objects, information or people flow in, something is done within the physical realm of the building and objects, information and people flow out with value added to them. Smooth flow equates with high functionality and high added value, while interrupted flow equates with low functionality and waste.

For instance, in a railway station, the continuous flow of passengers to and from trains adds value to journeys for as long as people can move freely. The continuous progress made by passengers in their journeys is a 'smooth flow'. On a motorway, the flow of traffic adds value to journeys even if

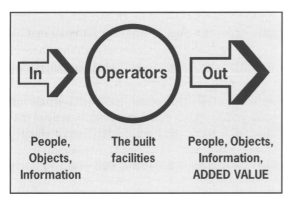

Figure 1.2 Flow chart of added value through a building

movement is slow, but any unplanned stoppage causes delay, frustration and waste. This is an example of interrupted flow, where progress towards the goal is not continuous. Time passes without value being added to the journey.

Smooth flow does not necessarily require continuous physical movement of people and objects themselves. Smooth flow is about continuously adding value or progressing towards a goal. For instance, in a shop, people who stand still to examine goods on a shelf, continue to progress smoothly towards their goal so long as they are:

● able to look at the goods and decide whether or not to buy, and,
● they are not in the way of other shoppers or staff.

As soon as shoppers find that they have been interrupted through being unable to see or reach goods, waste is generated. Time passes without value being added, and worse (for the business), the shoppers may be sufficiently frustrated that they leave the shop to make their purchase elsewhere.

Guidance and toolkits

There are a number of tools that help establish how well a building is going to perform and provide guidance on good practice design. Box 1.2 describes some of the available toolkits.

Principles and application of environmental sustainability in design

In the UK, construction activity is responsible for around:

● 72 million tonnes of waste from sites and demolition (17% of the total waste produced in the UK),
● 10 million tonnes of unused materials,
● 260 million tonnes of quarried materials,
● 20% of water pollution incidents,
● 60% of timber used in the UK.

(Source: Sustainable Construction Task Group 2002)

Whether at site, building or component level of design and specification, taking account of environmental issues can change typical or default behaviour and have a long-term impact on issues of national or global importance. Over the last decade, climate change has emerged as one of the most pressing environmental issues. Concerns about its wide implications are now leading to changes in both legislation and building regulations in the UK. In the next section on *BRE sustainability tools*, some of the tools to measure the environmental impact of buildings are described. In the section on climate change, the changes to construction practice required as a result of climate change are reviewed.

SCTG. *Reputation, risk and reward.* Report prepared by BRE and Environment Agency. 2002

Box 1.2 Toolkits for space planning in buildings

Housing Quality Indicators

http://www.housingcorp.gov.uk/

Housing Quality Indicators (HQI) were developed in response to a perceived fall of standards in housing association developments and changing housing needs that affected the definition of housing quality. It is managed by the Housing Corporation.

The HQI system evaluates housing schemes on the basis of quality rather than simply on cost. Housing Associations complete an HQI on line as part of the normal funding process.

HQI provides an assessment of quality of key features of a housing project using three main categories: location, design and performance. These three categories produce 10 quality indicators that make up the HQI system:

● location,
● visual impact, layout and landscaping,
● open space,
● routes and movement,
● size,
● layout,
● noise, light and services,
● accessibility,
● energy, green and sustainability issues,
● performance in use.

The Housing Corporation provides supporting guidance with its Scheme Design Standards setting out its required design and quality issues for housing projects receiving a Social Housing Grant.

Design Quality Indicators

http://www.dqi.org.uk/

Design Quality Indicators (DQI) is a tool that assists a building's procurement team to check the evolution of design quality at key stages in the process. There are currently two versions of the DQI:

● the generic tool that can be applied to any building project, and
● the DQI for Schools, a version specially adapted to meet the needs of schools. This version was developed by the Construction Industry Council (CIC) and the DfES.

The main tool in the DQI is the DQI questionnaire, containing a series of nearly 100 statements that measure success factors applicable to most buildings. The statements are organised into three main sections and 10 sub-sections:

● *Functionality* looks at the way the building is designed to be useful. The statements are split into: use, access and space
● *Build Quality* looks at the materials, and the different systems and conditions inside the building. The statements are split into performance, engineering services and construction.
● *Impact* refers to the building's effect on the local community and environment. The statements are split into: character and innovation, form and materials, within the school and the school in its community.

The questionnaire is completed either online or on paper in a facilitated workshop setting, attended by about 12–15 participants to enable free discussion of the issues. The DQI questionnaire provides a structured framework for stakeholder discussions, and enables issues to be raised and noted that could otherwise have been overlooked. The DQI has been designed to be used at four stages of a project cycle:

● *Briefing:* to allow project aspirations to be set, addressing the opinions of all stakeholders and defining what aspects are most important to the success of the completed building. The DQI helps to set priorities and answer questions such as, 'What do we want?', 'Where do we want to spend the money?'.
● *Mid-design:* to allow the client and design teams to check whether early aspirations have been met and make adjustments accordingly. The DQI can be used throughout the design phase
● *Ready for occupation:* to check whether the brief/original intent has been achieved immediately at occupation.
● *In use:* to obtain feedback from the project team and building users and to suggest improvements to be addressed for this project and for consideration in any subsequent projects being undertaken by the project team.

Cont'd....

Box 1.2 Toolkits for space planning in buildings (cont'd)

Healthcare
http://www.dh.gov.uk/
The National Health Service has developed 'OnDesign', a web-based collection of information that aims to assist the design process in healthcare design, and to encourage networking and the sharing of knowledge and best practice between healthcare design schemes.

It caters for all stages of a scheme from concept to post-project evaluation and all building types from the smallest GP surgery to major acute hospitals. It embraces all stakeholders associated with a scheme, from Trust and architects through to a patient user group.

Healthcare buildings involve complicated design concepts that are difficult to evaluate using conventional tools such as DQI. The Achieving Excellence Design Evaluation Toolkit (AEDET) is designed to help the NHS and its Trusts manage their design requirements from initial proposals through to post-project evaluation. It forms part of the guidance for ProCure21, PFI, LIFT initiatives and conventionally funded schemes.

Commission for the built environment (CABE)
http://www.cabe.org.uk/
CABE believes that well-designed homes, streets, parks, work-places, schools and hospitals are the fundamental right of everyone. It works for a higher quality of life for people and communities, with particular concern for those living in deprived areas. It does this by making the case for change, gathering hard evidence, providing education opportunities and through direct help on individual programmes and projects.

CABE carries out design reviews for proposals which will have a significant impact on their environment. The design review panel offers free advice to planning authorities and others on the design of selected development projects in England. It is particularly interested in strategic projects in their early stages.

CABE publishes a number of guidance documents aiming to help create better buildings and places. It also publishes policy documents commenting on the latest government planning and regeneration initiatives and case studies.

British Council of Offices (BCO)
http://www.bco.org.uk
BCO produces best practice guidance for the design and specification of office buildings. In particular it recognises the importance of productivity, noting that staff costs can be as much as 80% of the operational costs of the building. It emphasises the importance of:
● the quality of the internal environment as this has an impact on staff comfort, satisfaction and productivity,
● flexible spaces to support team working,
● not reducing staff performance by maximising the use of space.

It provides guidance to the procurement team and in particular the client on:
● site issues,
● building form,
● engineering,
● finishes and fit-out, and
● specification choices that commonly need to be made.

BRE Sustainability tools
In this section, some key BRE sustainability tools are described (Figure 1.3). Other tools (and some case studies illustrating the use of them) are outlined in the BRE guide *Achieving whole life value*.

One route to optimising sustainability is to use the BRE publication *A sustainability checklist for developments* as a guide to specific aspects to

consider in the early design stages. This tool is applicable at planning stage for both detailed estates and sites. It is intended as a common framework for both planners and developers.

BREEAM (Building Research Establishment Environmental Assessment Methodology)
BREEAM is an environmental assessment method for buildings. It is a consensus-based standard and was originally developed by BRE in 1990 for application to the design of new office buildings (Yates et al 1998). Versions for other buildings quickly followed, and now it can be applied to any building type. BREEAM has also been adopted

Bourke K et al. *Achieving whole life value in infrastructure and buildings.* BR 476. Bracknell, IHS BRE Press, 2005
Brownhill D & Rao S. *A sustainability checklist for developments: A common framework for developers and local authorities.* BR 436. Bracknell, IHS BRE Press, 2002

SCHEMES   TOOLS

breeam:ecohomes   **breeam:developments**
                                 targeting & benchmarking tool

breeam:retail   **breeam:ecohomesXB**
                                 stock management tool

breeam:industrial   **breeam:envest**
                                 whole building LCA

breeam:office   **breeam:specification**
                                 the green guide

breeam:schools   **breeam:profiles**
                                 environmental lca

breeam:bespoke   **breeam:smartwaste**
                                 benchmarking + solutions

Figure 1.3 An overview of BRE design support tools

internationally in Australia, New Zealand, Hong Kong and Canada and forms the basis of environmental assessment methods in South Africa, Norway and the United States (LEED), among others.

The organisational and operating structures of BREEAM are designed to allow flexibility and wide participation. BRE is responsible for:
- developing BREEAM, taking account of new technical and legislative developments as they occur,
- training and licensing independent assessors,
- certifying assessments made by independent assessors,
- maintaining quality control.

BREEAM assesses the performance of buildings as given in the following list.
- Management: overall management policy, commissioning site management and procedural issues, including whole life costing.
- Energy use: operational energy and carbon dioxide (CO_2) issues.
- Health and well-being: indoor and external issues affecting health and well-being, including the

acoustic environment of a building (see Appendix B).
- Pollution: air and water pollution issues.
- Transport: transport-related CO_2 and location-related factors.
- Land use: greenfield and brownfield sites.
- Ecology: ecological value conservation and enhancement of the site.
- Materials: environmental implication of building materials, including life-cycle impacts.
- Water: consumption and water efficiency.

Developers and designers are encouraged to consider these issues at the earliest opportunity to maximise their chances of achieving a high BREEAM rating.

Credits are awarded in each area according to performance. A set of environmental weightings enables the credits to be added together to produce a single overall score. The building is rated on a scale of PASS, GOOD, VERY GOOD or EXCELLENT, and a certificate awarded that can be used for promotional purposes.

EcoHomes: BREEAM for new homes
EcoHomes (introduced in 2000) is the homes version of BREEAM. It provides an authoritative rating for new, converted or renovated homes, and covers both houses and apartments.

Yates A, Baldwin R, Howard N & Rao S. BREEAM '98 for offices. BR 350. Bracknell, IHS BRE Press, 1998
For BREEAM products, visit website: www.breeam.org

EcoHomes balances environmental performance with the need for a high quality and healthy indoor environment. The issues assessed are grouped into seven categories:

- Energy,
- Water,
- Pollution,
- Materials,
- Transport,
- Ecology and land use,
- Health and well-being.

Many of the issues are optional, ensuring EcoHomes is flexible enough to be tailored to a particular development or market. Other types of accommodation such as sheltered homes or student flats, can be assessed using a bespoke version of BREEAM.

Independent assessors who are trained and licensed by BRE carry out EcoHomes assessments. The assessment is based on inputs completed by the developer/designer which can be downloaded as PDF files. EcoHomes is revised annually to ensure that it remains representative of current best practice and takes account of technical and legislative changes.

Two guides are available:

- *EcoHomes: the environmental rating for homes* describes the issues covered within EcoHomes and the background to the method.
- *The Green Guide to Housing Specification* provides guidance to designers and specifiers on the environmental impacts of the main fabric elements commonly used in housing.

'Green specification' using life-cycle assessment (LCA)

Life-cycle assessment (LCA) is a powerful tool for improving understanding of the environmental impact of building materials and designs. If applied holistically, this technique will provide a 'cradle to

---

Rao S et al. *EcoHomes: the environmental rating for homes.* BR 389. Bracknell, IHS BRE Press, 2003

Anderson J & Howard N. *The Green Guide to Housing Specification: an environmental profiling system for building materials and components used in housing.* BR 390. Bracknell, IHS BRE Press, 2003

grave' assessment of the impacts of a product over its full life-cycle. It links closely to the topic of Environmental Product Declarations (EPDs — these are based on LCA), which are likely to be incorporated within the EU Construction Products Directives at some point in the future. However, LCA is a tool most appropriately used in tandem with techniques considering the broader sustainability and environmental characteristics of the whole development (such as BREEAM and EcoHomes assessments).

ISO 14040 defines LCA as:

'a systematic set of procedures for compiling and examining the inputs and outputs of materials and energy and the associated environmental impacts directly attributable to the functioning of a product or service system throughout its life'.

In LCA it is important to ensure fair comparison between the product, system or process that is being studied. Therefore, the assessment must be based on a consistent basis of functionality, eg for a product, it may be the the function the product is designed to deliver.

The inputs of raw materials and resources that go into the study system (which may be a component or an assembly of components or even an entire building) are assessed in detail, along with the environmental outputs of the product system. This is illustrated in Figure 1.4.

A full LCA aims to describe all the interactions between the product system and the environment over the full life-cycle of the product system. However, this is often either impossible or impractical due to constraints of data availability and cost or time constraints of undertaking the study. Therefore, an LCA study should clearly define the boundaries of what has been included. These may include geographical boundaries (eg for the whole country or for one building) or temporal boundaries (the time period over which the impacts will be assessed), or process boundaries (whether to include or exclude processes such as maintenance).

Whenever LCA is employed it is important to consider the results carefully. Recommended questions to ask include:

| Inputs | Upstream | The product system | Downstream | Outputs |

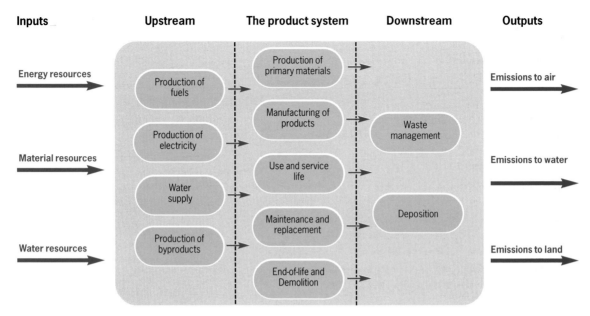

Figure 1.4 Stages of a product life-cycle

- How will the results be used?
- Is there a dominant life-cycle stage?
- What is the threshold of uncertainty?
- How precisely is the product system defined?
- How much is already known about the life-cycle stages of the product system?

BRE have produced a suite of tools based on LCA using the BREEAM environmental profiles methodology, which is compliant with the ISO 14040 series of standards. The BRE *Green Guides* all use this methodology as their basis and assess the environmental issues listed in Box 1.3.

Given the variety of environmental impacts considered, and the different methods of measuring them, the *Green Guides* have normalised and

---

Anderson J & Howard N. *The Green Guide to housing specification: an environmental profiling system for building materials and components.* BR 490. Bracknell, IHS BRE Press, 2003

BRE & Net Composites. *Green Guide to composites: an environmental profiling system for composite materials and products.* BR 475. Bracknell, IHS BRE Press, 2004

Anderson J & Shiers DE. *The Green Guide to specification: an environmental profiling system for building materials and components used in housing.* 3rd edition. Oxford, Blackwell Publishing, 2002

Action Point
Use tools like the Green Guides to inform yourself on the environmental implications of proposed specification choices.

weighted the results into an Ecopoint score. The first divides each environmental score (eg tonnes of waste) against the annual impact of one UK citizen. This dimensionless data is then multiplied by a weighting factor, derived from an extensive research study in 1999, including consultation with a wide range of groups. A great degree of consensus on relative importance between the different issues was found, and this provides the summary ratings. A score of A (best) to C (worst) is given to each product assessed depending on the level of environmental impact in Ecopoints.

A list of certificated environmental profiles can be viewed free of charge at www.GreenBookLive.com.

breeam:envest
Another useful tool for considering the environmental impact of building materials and designs is *breeam:envest*. It has been developed to allow architects and designers to assess the

Box 1.3 Environmental issues assessed in BRE's Green Guides

Climate change is associated with problems of increased climatic disturbance, rising sea levels, desertification and spread of disease. Gases recognised as contributing to climate change (greenhouse gases) include CFCs, HCFCs, HFCs, methane and carbon dioxide. Their relative global warming potential is calculated by comparing their global warming effect after 100 years to the simultaneous emission of the same mass of carbon dioxide.

Fossil fuel depletion of the limited resource of fossil fuels is measured in terms of the primary fossil fuel energy needed for each fuel.

Ozone depletion occurs as gases cause damage to the ozone layer which reduces its ability to prevent ultraviolet light entering the earth's atmosphere. This increases the amount of harmful UVB light reaching the earth's surface.

Human toxicity emissions to air and water. Some substances (such as heavy metals) have negative impacts on human health. Assessments of toxicity are based on tolerable concentrations in air, water, air quality guidelines, tolerable daily intake and unacceptable daily intake for human toxicity.

Ecotoxicity. Some substances have impacts on the ecosystem. Assessment is based on maximum tolerable concentrations in water for ecosystems.

Waste disposal. This encompasses the reduction in landfill, the noise, dust and odour from landfill sites (and other means of disposal), the gaseous emissions and leachate pollution from incineration and landfill, the loss of resources from economic use and risk of underground fires, etc. Because of a lack of data on the fate of materials sent to landfill or incineration a proxy figure for these impacts is measured by tonnes of waste produced.

Water extraction. This recognises the value of water as a resource and the depletion, disruption or pollution of aquifers or disruption or pollution of rivers and their ecosystems due to over extraction.

Acid deposition. Gases such as sulfur dioxide react with water to form acid rain. When this falls it causes ecosystem disruption often far from the original source. Gases which cause acid rain include hydrogen chloride, hydrogen fluoride, nitrogen oxides and sulfur oxides. Ammonia, though non-acidic, contributes to the long-range transport of the acidic particles.

Eutrophication is over-enrichment of water courses with excessive concentrations of nitrates and phosphates in the water. This encourages excessive growth of algae, reducing the oxygen within the water and damaging the aquatic fauna and flora within the ecosystem. Emissions of ammonia, nitrates and phosphates to air or water all have an impact.

Summer smog. In atmospheres containing nitrogen oxides and volatile organic compounds, ozone creation occurs in the presence of radiation from the sun. Although ozone in the upper part of the atmosphere is beneficial, ozone in the lower atmosphere is implicated in impacts such as crop damage and increased incidence of asthma and other respiratory complaints.

Minerals extraction. This concerns the total quantity of mineral resource extracted, both in the UK and overseas, and its impact on the local environment, eg from dust and noise. It assumes that all mineral extractions are equally disruptive of the local environment.

embodied environmental impact of different building designs. It enables the user to construct their building virtually, altering and changing the material fabric and construction elements to produce more sustainable designs. Environmental impacts are compared in Ecopoints using the environmental profiles methodology.

breeam:envest also includes operational energy demands. This allows designers to investigate trade-offs between the embodied life-cycle impacts of their building design and its operational impacts in use.

The software is web-based, allowing large design companies to store and share information in a controlled way, enabling in-house benchmarking and design comparison.

breeam:smartwaste
breeam:smartwaste is a system for evaluating waste and its generation, step-by-step. It is a web-based group of four integrated tools:
● SMARTStart™ (simple overview),
● SMARTAudit (detailed audit),

- SMARTStartLG (Local Government performance), and
- BREMAP™ (resource exchange).

It has been developed by BRE to help the construction industry to:
- identify cost savings,
- improve resource use,
- improve productivity, and
- demonstrate continuous improvement.

The issues addressed by the tools include:
- waste benchmarking,
- identifying key demolition products for re-use or recycling,
- identifying key waste products for reduction, re-use and recycling,
- sourcing local resource and waste management facilities,
- sourcing local supplies of reclaimed and recycled building products,
- demonstrating continuous improvement in resource efficiency.

breeam:smartwaste tools can be applied to any waste-generating activity, and have already been adapted and used for the construction, demolition, refurbishment, manufacturing and pharmaceutical industries. More information can be found at www.smartwaste.co.uk.

Other sources of advice
WRAP (Waste and Resource Action Programme) was established in 2001 in response to the UK Government's Waste Strategy 2000 to promote sustainable waste management. WRAP's mission is to accelerate resource efficiency by creating markets for recycled materials and products, while removing barriers to waste minimisation, re-use and recycling. WRAP produces a number of publications giving advice on how to maximise the use of recycled materials in construction. More information can be found at www.wrap.org.uk.

The Carbon Trust is an independent, not-for-profit company, set up and funded by the government to help the UK meet its international climate change obligations. It provides free, practical advice to business and public sector organisations to help reduce energy use and cut emissions of carbon dioxide, thereby reducing climate change. More information can be found at www.thecarbontrust.co.uk.

The Energy Savings Trust provides a similar service but deals mainly with the housing sector. Their web site address is www.est.org.uk.

Integrated design

'All the really important mistakes are made on the first day of the design process'
(Amory Lovins, Rocky Mountain Institute)

Few buildings are genuinely sustainable even though the basic principles are very straightforward:
- minimise artificial lighting, heating and mechanical ventilation,
- avoid air-conditioning,
- conserve water,
- use the site and materials wisely, and
- recycle where possible.

Yet, poor design, construction or building management often results in buildings having unnecessarily high environmental impact.

A fundamental requirement for success is to establish a clear client brief which places the sustainability targets alongside style, image and aesthetics in terms of importance. Putting sustainability at the core of the design brief helps to ensure that the building will be more economic, comfortable, productive, humane and more beautiful than a conventional building.

Delivering sustainable buildings requires a new design vocabulary. In particular, climate-responsive buildings that are based on the principles of bioclimatic design require the architect and engineers to work as an inter-disciplinary team. This will deliver intelligent buildings that work with natural systems to provide (for free) much of the requirement for lighting, heating, cooling and ventilation.

Holistic design

Integrating the design process enables a 'whole system' approach to be adopted. This approach often results in major capital and operational cost savings by 'tunnelling through the cost barrier' (see *Natural capitalism*). Also, substantial savings in the cost of building services equipment can be achieved by intelligent façade design and by optimising the use of free cooling and natural ventilation and by limiting summertime overheating (while also maximising daylight and useful solar heating in the winter).

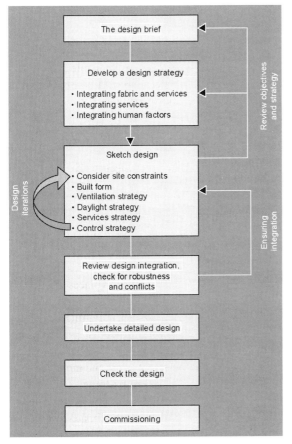

Figure 1.5 Achieving energy-efficient design through an integrated approach. Reproduced from *Energy efficiency in buildings: CIBSE Guide F* by permission of CIBSE

Hawken P, Lovins, AB & Lovins LH. *Natural capitalism: creating the next industrial revolution.* Snowmass, Colorado, Rocky Mountain Institute, 1999

These often conflicting design requirements can only be resolved through integrated design *from day one of the design process*. It is essential that the architectural concept is challenged, improved and refined in an iterative manner, with climate-responsive design solutions being incorporated as an integral part of the design.

Avoiding greenwash

Green design should be the starting point and never simply addressed by 'bolting-on' token features. Greenwash applied to the building by a marketing or PR department is no substitute for buildings whose designers have addressed the environmental, ecological and energy issues at every stage of the design process. The true credentials of a green-washed building will become immediately apparent if subjected to a BREEAM assessment and/or if an energy performance certificate is produced.

Increasingly, mandatory requirements for labelling and certification will result in buildings being assessed and differentiated in terms of their environmental and energy performance. Performance rating and ranking coupled with the growing reporting requirements associated with corporate social responsibility are likely to have a profound effect (both positive and negative) on the asset value of a building. In short, no organisation concerned about its brand equity or corporate positioning will want to occupy a badly rated building. Integrated design places the engineer (and ideally the contractor and facilities manager) alongside the architect as equal partners in the design process. Sometimes egos can get in the way of this vital engagement so expert facilitation may be required to deliver the optimum design solution (Box 1.4).

Designing for whole-life performance

The integrated whole-building design process should also consider the impact in terms of the 'whole-life' performance of the building. For example, increasing the façade from double to triple glazing may save operational energy but the embodied energy used to manufacture the glass will increase. These complex trade-offs can be assessed by using life-cycle analysis (LCA) methods.

Box 1.4 Using design charettes to deliver demonstrably better buildings

The charette is changing the way buildings are planned and delivered. Used by some of the UK's leading clients and developers, the process is promoting intelligent architecture, integrated design and innovative solutions. The charette brings client, design team and key stakeholders together at the start of a project to consider a variety of design options and create a plan that meets the needs of all parties. It involves intensive facilitated discussions during which opportunities are identified to reduce waste and environmental impact, increase efficiency and performance, and raise the comfort levels and productivity of occupants. The benefits that accrue from this approach are considerable. They support both commercial and community interests.

● **Clients:** better buildings, lower construction and whole life costs, reduced risk
● **Communities:** greater consultation, more inclusion, less disruption
● **Design teams:** fewer reworks, easier passage through planning, improved reputation
● **Occupants:** secure, healthy, comfortable and productive environments
● **The environment:** increased use of renewable resources, reduced waste and pollution, improved energy and water efficiency

The charette process starts as soon as the initial decision to procure has been taken. An expert facilitator meets with the client or developer to discuss the broad requirements of the project and to draft a brief and/or and output specification.

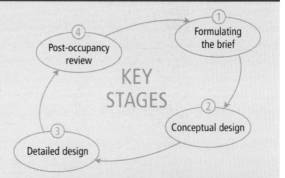

Once this has been done, a series of expertly facilitated interdisciplinary meetings with the client, design team and stakeholders are undertaken during which ideas are discussed, options debated, trade-offs agreed and targets set. This process inspires and fuels ideas. By involving key parties in this way and drawing on specialist advice, radical improvements can be achieved during the design and masterplanning process. Greater understanding and commitment develops, fresh approaches are taken, and better solutions evolve.

Learning from experience
Once a project is completed and the building has been occupied, BRE recommends carrying out a post-occupancy review to see how well the designs are working. Performance issues are assessed and occupant and user views sought. This information is analysed and fed back to both client and design teams to inform future projects and improve the operational performance of the building.

Software tools are available to assist the design team to ensure that rational and informed design decisions are made on the basis of environmental impact over the life of the building (see earlier section on *breeam:envest*).

Incentivising the design team

Current professional fee structures do not help deliver low environmental impact buildings. In particular, fees for architects and/or building services engineers that are based on a percentage of the total construction cost (or building services cost) are not helpful. An engineer who invests extra design time in working with the architect to deliver highly innovate ways of reducing the operational energy cost associated with heating, lighting, ventilation and cooling, will be penalised by

receiving a lower fee. This perverse incentive is a major barrier in delivering lower environmental impact buildings. Other fee structures should be used, ideally based on a fixed fee with an incentive element linked to the actual building performance over the first 7–10 years of operation. This has the added benefit that the design team remains motivated and incentivised to ensure that the building is correctly commissioned, operated and maintained in accordance with the original design intent. Fee structures of this type have been trialled in the USA and have delivered demonstrably better buildings, often at substantially lower cost (further details of these trials are available from the Architectural Energy Corporation, Boulder, USA, www.archenergy.com/library/general/performance)

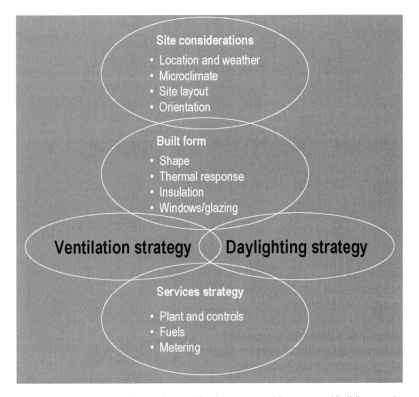

Figure 1.6 Sketch design process showing interrelationships between architecture and building services and energy efficiency strategies. Reproduced from *Energy efficiency in buildings: CIBSE Guide F* by permission of CIBSE

Designing for people

An over-arching objective of integrated sustainable design is to deliver buildings that are healthy, comfortable and productive. For example, an award-winning ultra-low energy school fails in terms of sustainability if the acoustic environment provides classrooms where the learning outcomes are poor.

Recent US studies have shown that well-designed climate-responsive buildings can increase productivity by 6–15%. Since most employers spend considerably more on salaries than they do on energy, this represents an even more attractive benefit than the savings on their fuel bills. High quality sustainable buildings have a profound and well-documented impact on productivity in workplaces, learning outcomes in schools, recovery times in hospitals and sales performance in shops (see *Greening the building and the bottom line*).

Romm J & Browning WD. *Greening the building and the bottom line.* Snowmass, Colorado, Rocky Mountain Institute, 1994

The challenge of 'green' design

Delivering sustainable architecture using climate adaptive/responsive buildings is much more challenging than designing traditional highly serviced climate-excluding buildings. Conventional buildings using energy-intensive electro-mechanical systems to provide comfortable conditions to 'fix' a badly designed building is the antithesis of intelligent sustainable design. Much greater care is also required during construction, commissioning and operation to deliver optimum performance from a climate-responsive building. For this reason, key design and specification decisions should always be undertaken by a fully integrated multidisciplinary team. This will deliver intelligent, innovative and inspired buildings that have the lowest possible environmental impact.

In summary, there is nothing stylistically prescriptive about sustainable design. It does, however, require close collaboration between all members of the design team from the early stages of a project. Even in hostile climatic conditions buildings can be designed to exploit the natural systems available *for free* to provide some or all of the ventilation, cooling, heating and daylighting.

Impact of climate change on building design and construction

Politically and scientifically, the debate over climate change has begun to change from 'Is it happening at all?' to 'What will it actually mean?'. There is a growing consensus that the climate is changing at a rate unsurpassed over the last millennium and that these changes will impact on us all. The majority of the scientific community accepts that emissions of greenhouse gases from human activities are driving climate change, and that even if production of these emissions stopped tomorrow, the climate would continue to change for many years.

Typically, we spend 90% of our lives in buildings, and rely on them to provide the infrastructure in which we live. Building stock which currently performs adequately may be compromised by future climate change. Equally, design standards need to be reviewed to minimise future problems.

> **Key Point**
> Consider climate change and its implications on your building now!

How buildings are designed to cope with climate

Buildings are designed to provide shelter and protection against the climate. They are therefore designed to cope with typical and anticipated risks of climate. There are currently three forms of performance requirement which define the way in which buildings in the UK cope with climate (Box 1.5).

These requirements have been developed based on an historical understanding of climate, materials performance and building technology. With changing climatic conditions many of the existing codes and standards are being reviewed in the light of predicted changes in climate. Complete revision may take many years.

Box 1.5 Standards for building performance

1 **Third-party certification** ensures that building components and systems are designed and manufactured to recognised industry standards. These minimum standards are set at levels that provide the specifier with some confidence that the product will achieve an effective service life within the expected exposure and climate. This third-party certification may make recommendations for additional protection to ensure that durability problems don't lead to premature failure. The considerations that are required at a component level are extensive, and these preoccupy standards committees and third-party testing authorities who are responsible for developing performance standards.

2 **Design codes** govern the required structural and performance levels of buildings by consideration of the likely statistical frequency of extreme climatic events. In broad terms, these codes set appropriate levels of performance to cope with these climatic extremes, incorporating a margin of safety.

3 **Building Regulations** set requirements that are necessary for buildings to achieve a minimum level of acceptable performance. Building Regulations typically cover health, safety, energy performance and accessibility requirements for buildings.

How UK construction is likely to change to reflect climate change

There is uncertainty about the impact of changing greenhouse gas emissions which are affected by global economic development. This has led to a number of scenarios being developed to reflect different possible future patterns of development. The timescales seem long, but reflect typical building service lives of 60 years or more, so should be considered as foreseeable. The latest climate change scenarios were published in the UK by the UK Climate Impacts Programme in 2002. These predictions include:

● UK climate will become warmer, with predictions for the average annual temperature being between 2 °C and 3.5 °C warmer by the 2080s.

● Hot summers will occur more frequently, and extreme cold winters will become less common.

- Summers will become drier and winters will become wetter.
- There will be less snowfall.
- Heavy winter rains will become more frequent.
- Average sea level is rising by about 1 mm per year.
- Higher tides will return more frequently.
- Wind speeds and gale frequencies cannot yet be predicted, but increasing frequency of severe storms seems likely.

The most immediate risks therefore are:
- increased flooding events (more severe rainfall, rising sea levels),
- weathertightness of buildings compromised (more severe winter rainfall events),
- foundation movement problems for buildings (wetter winters and drier summers),
- summer overheating.

The service life or durability of a range of building materials is likely to change. In general, materials towards the outer envelope will be at most risk. Particularly vulnerable materials and components include:
- sealants (Figure 1.7),
- jointing materials,
- plastics,
- coatings, especially paint,
- composite materials.

Flooding events

Increased flooding is anticipated in both coastal and riverside locations, but urban and flash flooding is also likely to increase (Figure 1.8). The changing flood risk has already led to a changing pattern of insurance against risk. Options currently under consideration or active development include:
- charging premiums to reflect local flooding risk and availability of flood defence measures,
- increased expense or unavailability of building insurance in areas subject to repeated flooding,
- withdrawal of mortgage availability on un-insurable buildings.

A design strategy to increase resistance to flooding involves:
- firstly, locate the building higher than predicted flood levels,
- secondly, keep flood waters away from the building, and
- thirdly, reduce the damage caused if flood waters do reach it.

Some design considerations for reducing and alleviating damage from flooding are listed in Box 1.6.

Recommendations to make a building easier to reinstate after flooding include:
- install services above anticipated flood height to avoid the need for replacement,
- install sewage backflow devices,
- provide solid ground floors or access for inspection and debris clearance from suspended floors,

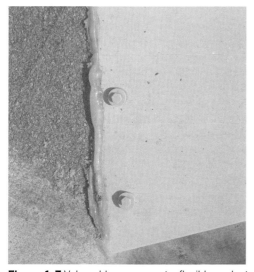

Figure 1.7 Vulnerable components: flexible sealants

Figure 1.8 Flooding caused by heavy seas at Folkestone, Kent (Courtesy of Max Hess)

Box 1.6 Design considerations to reduce or alleviate damage from flooding

- **Constructional or technological measures** that provide protection in the form of a barrier or diversionary solution to flood waters.
- **Local geography and topography:** although low-lying coastal areas are at risk and UK planning agencies are actively restricting future flood plain developments, a significant proportion of flooding problems occur away from flood plains. In fact, problems with flash flooding often occur in valleys or hilly locations.
- **Local and surrounding drainage infrastructure:** care should be taken to ensure that storm or rain water drainage systems or schemes are capable of accommodating new development. It is important to evaluate the potential risk of flooding, identify the most likely causes and predict the likely duration of the flood.
- **Use of resilient forms of construction** to reduce the level of damage, especially in areas of frequent flooding.

- provide masonry walls or, in lightweight construction, consider how to alleviate differential movement and corrosion/rot following wetting.

Further guidance is available in Garvin et al 2005 and the Flood Repairs Forum's *Repairing flooded buildings*.

Exposure to wind-driven rain

Future weather patterns are likely to include rainfall events becoming more intense, although data are not fully available. On the basis of current scenarios from the UK Climate Impacts Programme (UKCIP), the exposure increases given in Table 1.1 are predicted for Glasgow and London.

Garvin S, Reid J & Scott M. *Standards for the repair of buildings following flooding.* C623. London, CIRIA, 2005
Flood Repairs Forum. *Repairing flooded buildings: an insurance guide to investigation and repair.* EP 69. Bracknell, IHS BRE Press, 2006
UK Climate Impacts Programme (UKCIP). *Climate change scenarios for the United Kingdom.* The UKCIP02 Briefing Report. Norwich, The Tyndall Centre for Climate Change Research, April 2002. www.tyndall.ac.uk
Lacy RE. *Climate and building in Britain.* BRE Report. London, HMSO, 1977

Table 1.1 Change in exposure to winter driving rain based on current UKCIP scenarios

	Glasgow (%)	London (%)
2050's winter	+7 to +11	+9 to +15
2080's winter	+10 to +19	+14 to +25

Current scenarios offer only meagre guidance about the occurrence of wind and precipitation together (eg Figure 5.3). Box 1.7 gives a method of estimating future exposure. More sophisticated procedures are needed to achieve accurate exposure calculations, but in their absence, this method allows designers to account for future driving rain exposure.

Building features that provide extra protection against wind-driven rain are:
- recessed window and door reveals and sills with a drip that overhangs the external leaf,
- use of render finishes, external insulation systems, or rain-screen cladding,

Box 1.7 How to estimate future exposure

A first step towards evaluating the change in driving rain can be made using UKCIP scenarios in conjunction with the following relationship given by Lacy in *Climate and building in Britain*.

Driving rain = 0.222 × Wind speed × (Rainfall rate)$^{0.88}$

This relationship assumes the same pattern of wind and rain that has been traditionally experienced, and as such provides a first approximation. It can be used to estimate the percentage increase in free air driving rain for any location in the UK and as a first pragmatic step this percentage increase can be applied to the results of building exposure calculations made using BS 8104 to help inform possible future design requirements.

Key Point

Consider designing the building to perform to a higher level of exposure to the local wind-driven rain index or adopt traditional construction practice from highly exposed areas predominantly on the west coast.

- increased eaves overhangs to protect the wall head,
- use of partial fill cavity construction techniques, avoiding fully filled cavity construction,
- additional laps and fixings to roof coverings.

Foundation movement
Climate change is anticipated to lead to hotter, drier summers in the UK. The number of subsidence claims each year is closely linked to hot dry summers. Sites most affected are likely to be those on shrinkable clays with high or very high volume change potential (plasticity) (Table 1.2). These are primarily located in South-East England. As well as considering foundation depths, the position and location of trees will affect the risk.

Trees known to be associated with significant subsidence damage include: Oak, Ash, Willow, Poplar, Beech.

Further guidance on the risk from trees can be found in BRE's *Digest 298* and *Good Repair Guide 2* and NHBC Standards.

Summer overheating
As well as reacting to predictable climate changes, designers can contribute to reducing the risk of climate change by designing the building to reduce summer overheating, which in France in 2003 contributed to an additional 15,000 deaths (a 60–70% increase in death rates).

The latest predictions for Central England include:

- Today we expect one day per summer to reach 31 °C,
- By the 2080s there may be 10 days per summer reaching or exceeding 31 °C,
- By the 2080s one day per summer may reach 38.5 °C.

Some commentators see that an increased demand for air-conditioning is inevitable, but this could itself significantly contribute to emissions of greenhouse gases and the expense may lead to a further social divide between those who can afford it and those who cannot.

More details of how to design for passive ventilation and natural cooling are given in chapter 3, but the design concepts include the following.

- Design structures, building orientation and internal spaces to get the maximum benefit from natural ventilation.
- Use thermal mass in the structure to store heat during the day and release it overnight for winter conditions.
- Use thermal mass in the structure to store coolness during the night and release it during the day for summer conditions.
- Design buildings to function as mixed mode, so they can be naturally ventilated and only if necessary mechanically cooled.
- Use shading devices on windows to reduce excessive solar gains.

We can also expect the construction process to be affected by climate change. Wetter, stormier winters will prevent or delay construction. Site flooding is likely to become more of an issue. Drought conditions could cause water supply problems on

Table 1.2 Volume change potential of different clays			
Clay type	Plasticity index, Ip (%)	Modified plasticity index, I'p (%)	Volume change potential
Lower Lias	38	38	Medium
Oxford	44	42	High
Weald	43	41	High
Gault	77	77	Very high
	54	54	High
London	63	63	Very high
	46	46	High
Glacial Till: E Anglia	20	14	Low
Glacial Till: Scotland	21	15	Low

BRE
Low-rise building foundations: the influence of trees on clay soils. Digest 298. Revised 1999
Damage to buildings caused by trees. Good Repair Guide 2, 1996
NHBC. *NHBC Standards.* Chapter 4.2: Building near trees. www.nhbcbuilder.co.uk

site. Hotter weather will also make working conditions less comfortable in the summer. On the plus side, increased temperatures in winter should reduce the risk of frost to curing and drying of wet processes.

BRE's *Good Building Guide 63* gives more guidance on the impact of climate change on building design and construction.

Principles and application of social sustainability in design

An emerging body of research suggests that the way in which buildings have been designed and constructed can have significant impact on behaviour for those who occupy them. Well-designed schools are said to have lower truancy rates and improved attendance compared with poorly designed schools, and well-designed hospitals have lower bed-occupancy time and help to retain skilled staff. Increasingly, commercial organisations are coming to recognise that well-designed offices and work environments are important factors in successful recruitment and retention of highly skilled staff in competitive recruitment markets.

Health and safety also is a vital issue for the construction industry. For example, the National Audit Office's report *Improving health and safety in the construction industry* noted that across all industries poor health and safety, aside from the human cost, costs an estimated £18.1 billion per year (2.6 % of gross domestic product), and that in construction a rate of four fatal injuries per 100,000 employees is five times the average for all industries.

Social sustainability starts with the construction process itself, and this is also one of the key safety aspects of the designer's role. The safety of neighbours and operatives, as well as occupiers and maintainers, is one of the key Construction (Design

and Management) Regulations (CDM) responsibilities.

Guidance on improving the site construction stage and on improving safety within homes is included later in this chapter.

Working with the community during the design and construction period

The construction industry and its clients are becoming increasingly aware of the importance of giving more consideration to the needs of the local community. As well as reinforcing the social dimension of sustainable construction by demonstrating a commitment to corporate social responsibility, improved relationships with the local community can bring substantial business benefits both in the short and long term, eg:

● time and cost savings resulting from a smoother planning process,
● increased goodwill towards construction projects during and post construction,
● reduction in time and money spent dealing with complaints,
● a greater sense of ownership of the project among local people, thereby reducing vandalism and improving security,
● enhanced public relations for the client and contractor,
● increased potential for repeat business in the locality,
● improved job satisfaction and motivation for the workforce.

The main impacts that people are worried about during the construction period are: noise, dust, traffic, parking, lack of communication and consideration throughout the process.

A major factor in working successfully with the community is planning: potential impacts on people and businesses should be considered from the earliest stages. This requires a positive commitment from all parties. Responsibility needs to be shared to provide a coordinated approach throughout. Although much of the following guidance will be under the control of the contractors, the designer shares responsibility for minimising the impact on the local community.

BRE. *Climate change: impact on building design and construction*. Good Building Guide 63. Bracknell, IHS BRE Press, 2004

National Audit Office. *Improving health and safety in the construction industry*. HC 531. 2003/2004. Norwich, The Stationery Office. Available as a pdf from www.nao.co.uk.

Box 1.8 summarises how the main impacts can be reduced. More advice is given in BRE's *Working with the community.*

Box 1.8 Reducing the impacts on the local community

Key tips
- Keep people informed, in appropriate language, from the earliest stage and provide updates throughout.
- Provide a key contact point and include contact phone numbers on signage.

Site hoardings display warning signs and contact details for enquiries and complaints

- Communicate with local business forums and visit businesses to discuss how to minimise disruption (eg by timing deliveries, ensuring parking availability for their customers and visitors).
- Give notice of disruption to services (eg water, electricity).
- Use local suppliers (eg for materials, catering, labour).
- Have a complaints procedure that deals efficiently and effectively with local concerns.
- Deal with complaints promptly and give feedback on action taken.
- Train public-facing operatives in the complaints procedure and in how to handle complaints.

Dust and dirt impacts can be reduced by:
- Covering piles of material.
- Employing dust and dirt control measures such as water-spraying, fitting bag filters to equipment, road surfacing.
- Minimising dry cutting and sweeping, excessive vehicle movements, crushing and screening of soils and aggregates.
- Taking measures to prevent water run-off to roads and watercourses.
- Using demolition techniques that limit the amount of dust or period of dust-generating activities.
- Putting up hoardings to act as dust and noise barriers.
- Allowing visibility but not at the expense of dust containment (eg avoid extensive mesh hoardings).

Water spraying during dust-generating activities reduces dust emission and spread

- Keeping local residents informed of noisy or dusty activities and giving advice about keeping indoors, shutting windows and doors, and protecting plants and gardens.
- Using road sweepers and wheel washing.

Vibration and noise impacts can be reduced by:
- Limiting noise from human activity (eg loading of lorries, loud radios).
- Confining working hours (eg normal working day in residential areas, weekends or night work in commercial areas).
- Limiting hours of noisy work (eg pile driving, hammering, drilling, deliveries by large lorries).
- Liaising on pile driving times.
- Using silenced or quiet plant or quieter processes (eg hydraulic piling rather than percussive pile driving).
- Using barriers to reduce site noise and lining chutes and skips with noise-attenuating materials.
- Providing respite areas and times.

Hydraulic piling reduces levels of noise and vibration

Cont'd....

Box 1.8 Reducing the key impacts on the local community (cont'd)

Pollution impacts can be reduced by:
- Being aware of likely specific concerns (eg asbestos, diesel oil).
- Taking samples to allay the concerns of local residents.
- Using an incinerator (approved under the Clean Air Act 1993) rather than site bonfires.
- Controlling disposal of polluted water.
- Controlling emissions from site traffic (eg not leaving engines running unnecessarily, using low emission fuels, maintaining engines and exhaust systems).

Other traffic impacts can be reduced by:
- Locating site entrances away from busy roads and junctions. Consider multiple entrances to spread the volume of traffic.
- Operating a traffic management scheme.
- Denying access to, or minimising construction site traffic during busy times (eg school drop-off and pick-up times).
- Timing deliveries to minimise queues.
- Using local suppliers and labour where possible.
- Minimising road closures.
- Co-ordinating utilities work.

Don't leave plant running longer than necessary to reduce emissions of diesel exhaust fumes

Consider the location of site entrances and exits in the early stages of site planning to avoid traffic congestion

Parking impacts can be reduced by:
- Providing car parking on site as far as possible.
- Ensuring site workers are aware of impact of parking on road safety/congestion (eg where visibility for vehicles or pedestrians would be restricted).
- Providing appropriate locations for parking and prohibiting parking where it would impede emergency vehicle access or resident access or egress.
- Employing local labour to minimise transport issues.
- Ensuring that parking and transport routes are well signposted.
- Liaising with local police and the local authority.

Impacts on pedestrians can be minimised by:
- Providing safe alternative footpaths with handrails.
- Being proactive in repairing broken pavements, potholes, etc.
- Ensuring adequate coverage of holes, lighting and crossing points.
- Considering escorts for children, manned traffic lights, accommodating the needs of the elderly and the handicapped and those with pushchairs.
- Making sure that materials do not block pavements and entrances to local businesses, and ensuring emergency exits are kept free.

Hadi M, Rao S, Sargant H & Rathouse K. *Working with the community: a good practice guide for the construction industry.* BR 472. Bracknell, IHS BRE Press, 2004

Designing for safety

About 30% of all accidents in UK homes are from slips, trips and falls, of which:
- nearly 200,000 are on stairs,
- about 55,000 are in kitchens and bathrooms,
- about 60,000 are in sitting or living rooms.

There are about 276,000 kitchen accidents and 86,000 bathroom accidents every year. 40,000 accidents involve windows, while 22,000 involve glass or glazed doors. The following design recommendations could substantially reduce these figures. Many of these recommendations are in the accompanying documents of the national building regulations, for example, in England & Wales: Approved Document (AD):
● M (access) or
● K (stairs, etc.) in respect of avoidance of falls, but are not always applied in practice. Others are requirements of AD:
● N (glazing),
● B (fire safety) or
● P (electrical safety),
or they form part of
● the IEE Regulations (BS 7671).

Some of the recommendations also fall within the duties of designers under:
● the CDM Regulations.

Building Regulations in the UK are discussed in more detail at the end of this chapter.
    Box 1.9 summarises some important considerations when designing for safety.

Principles and application of economic sustainability in design

BS ISO 15686-1 defines Whole Life Costing (WLC) as an:

'economic assessment considering all agreed projected significant and relevant cost flows over a period of analysis expressed in monetary value. The projected costs are those needed to achieve defined levels of performance, including reliability, safety and availability'.

It is a process that aims to look at the total costs associated with the procurement, use during service life, and disposal at the end of life (Table 1.3). The objective of WLC is to make investment decisions with a full understanding of the cost consequences of different initial options. The technique is not new

British Standards Institution. BS 7671: 2001 *Requirements for electrical installations. IEE wiring Regulations.* 16th edition

but contrasts significantly with a lowest capital cost approach. It is the first step towards measuring value, rather than simply cost.

'…lowest capital cost is not a reliable measure of value for money, since it takes no account of how well buildings perform….Value for money in construction is about more than delivering a project to time and cost…A good building project must also contribute to the environment in which it is located, deliver a range of wider social and economic benefits and be adaptable to accommodate future uses… The ultimate aim is to deliver construction projects that meet the requirements of the business and all stakeholders, particularly the end users.'

(Source: NAO/CABE 2004))

'It is sometimes claimed that good design is an unaffordable luxury. Too often, an assumption is made that cheapest cost signifies best value, while good design is an optional extra which results in higher prices. Neither is true. The key to good design is that it takes account both of fitness for purpose and of the whole-life costs of a building to meet the user's needs in delivering value for money. Typically the cost of design is only about 1% of the lifetime cost of a building, and may be dwarfed by long-run energy and maintenance costs. Good design can cut maintenance costs and help to ensure that there is a higher residual value.'

(Source: DCMS 2000)

Simpler uses of WLC limit the word 'costing' to financial impacts. More complex uses extend 'costing' to include some external effects, eg impacts on the environment, users and society in general, by representing these in financial terms.

> **Key Point**
> Consider the whole life costs iteratively throughout the design process.

National Audit Office/CABE. *Getting value for money from construction projects through design: how auditors can help.* Norwich, The Stationery Office. 2004
Department for Culture, Media and Sport (DCMS). *Better public buildings.* 2000. Available as a pdf from www.culture.gov.uk

Box 1.9 Designing for safety

Outside
- Provide level access or gentle slopes to buildings.
- Ensure paths are level and easy to maintain.
- Make steps consistent, with at least three steps, and a going between 300 mm and 450 mm. Exterior ramps are covered in chapter 12.
- Consider handrails and lighting and guard changes in level with a stable barrier.
- Consider soft areas and avoid water features in areas where children play.
- Gates and fences in family homes should be at least 1200 mm high. Avoid openings of greater than 100 mm and avoid toeholds to discourage climbing.
- Provide RCD protection for garden mains-powered tools.

Thresholds and doors
- Provide level thresholds, both internal and external.
- Provide a clear opening width of 775 mm minimum for doors in dwellings. For buildings other than dwellings this minimum rises to 800 mm.
- Use safety glass in and around doors.
- Make glass panels running down to floor level visible with permanent marking or patterning.
- Use door handles that are easy to operate; spring-return lever handles tend to be easier.
- Avoid external doors that open straight onto areas used by cars.
- Provide illumination to the main entrance to a building, and, ideally, provide a cover.

Corridors
- Provide clear unobstructed width that is a minimum of 900 mm for dwellings and 1200 mm for other buildings.
- If the door cannot be approached head-on, widen corridors or doors to provide easier access.
- Provide adequate storage for shoes, etc.
- Avoid doors opening out onto corridors wherever possible but if required ensure they do not open against pedestrian flow.

Internal stairs
- Provide continuous handrails 900–1000 mm above the pitch line, on both sides wherever possible. At least one side should have a handrail. Check with AD K for wider stairs, particularly where used by the public.
- Handrails on dwelling stairs should extend 300 mm beyond the top and bottom nosings. There should be a clear distance between wall and handrail of between 60 mm and 90 mm.
- The acceptable size of a circular section handrail is between 35 mm and 45 mm in diameter. 'Pig's ear' handrails are an unacceptable form of handrail design.
- Use closed risers on dwelling stairs.

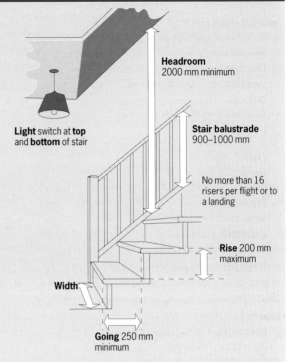

Headroom 2000 mm minimum

Light switch at **top** and **bottom** of stair

Stair balustrade 900–1000 mm

No more than 16 risers per flight or to a landing

Rise 200 mm maximum

Width

Going 250 mm minimum

Critical stairs details for dwellings

- If carpets are provided, ensure the carpet is in good condition and secure, and make nosings clearly visible.
- Prevent falls by providing barriers with openings of maximum 100 mm in diameter and without toeholds. Barriers should be at least 1100 mm high, or 900 mm where the handrail forms part of the barrier.
- Make steps consistent throughout the flight, with equal goings of 250–400 mm and a consistent rise of between 150 mm and 200 mm. For public stairs (where a substantial number of people will gather, or where the likely number of users is greater than 20 per day) the minimum going should be 300 mm and the rise should be between 130 mm and 180 mm.
- Owing to the importance of inter-step variability, there should be a maximum difference between successive goings or a flight, of 10 mm if the goings are less than 300 mm, and 15 mm if the goings are 300 mm or larger.
- In non-domestic buildings, consider stair nosings to provide a visible contrast to the rest of the tread. Nosings can also reduce the risk of a slip (see BRE's *Information Paper IP 15/03*).

Cont'd....

Roys M & Wright M. *Proprietary nosings for non-domestic stairs*. Information Paper IP 15/03. Bracknell, IHS BRE Press, 2003

Box 1.9 Designing for safety (cont'd)

Internal stairs (cont'd)
- Provide adequate artificial lighting at top and bottom, with two-way switching.
- Provide natural light where possible and remember that cross-lighting is better than head-on or at the top or bottom.
- Avoid windows that come to landing level unless well protected.
- Position windows or rooflights where they can be opened, closed, cleaned and curtains used without introducing a falling hazard.
- Provide 2000 mm minimum clear headroom throughout.
- Provide a clear landing, at least as long as the stairs is wide, at the top and bottom of all flights. Avoid doors encroaching over landing space.
- Avoid glazed doors within 2 m of the foot of flights.

So far as possible these recommendations should also apply to attic rooms or occasional-use spaces. AD K provides alternative stairs details applicable to such rooms.

Floors
- Floors should be level and stable: small changes of level are a tripping hazard.
- Avoid single steps. Where present, mark clearly and provide a handrail and adequate lighting.
- Flooring materials should be durable and secured firmly to the floor.
- Flooring which becomes slippery when wet (eg laminates or rubber) is particularly dangerous in kitchens, bathrooms and near external doors.
- Mats and rugs, if specified, should be backed to avoid slips and trips. Flooring should be easy to clean and maintain.

Windows, glass and ventilators
- Where impact is likely (ie around doors and low-level windows below 800 mm) fit safety glass to BS 6262-4.
- Make openable lights easy to operate to provide rapid ventilation.
- Where access is restricted, provide a pull-cord to trickle ventilators.
- In family homes or where children or elderly will be unsupervised, ensure windows are lockable and/or restrict opening to a maximum of 100 mm, particularly at first floor or above.
- Make windows easy to clean. Tilt and turn windows should be balanced, and capable of being locked in a reverse position for cleaning.
- Avoid fixtures such as kitchen cupboards and baths in front of openable windows. Where present, provide means of opening them without climbing on the units.

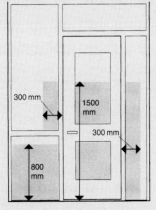

Windows, glass, ventilators:
- Consider safety glass
- Easy to clean and maintain
- Avoid collisions
- Avoid large sills
- Easy to operate
- Restricted for children

Safety glazing requirements

- Windows should not open onto circulation routes.
- Avoid window sills that will be easy for children to climb.
- Avoid glazed doors or glass panes within 2000 mm of the foot of a flight of stairs.

Lighting, heating and electric supply
- Place switches near doors and position them between 750 mm and 1200 mm above the floor. Position sockets between 450 mm and 1000 mm above the floor. Outlets, switches and controls should be at least 350 mm from room corners.
- Avoid excess lighting or shadowy areas. Provide lighting at door thresholds.
- Consider two-way switching for all circulation areas.
- Consider compact fluorescent lamps as they do not get hot in use and need changing less frequently than tungsten filament lamps.
- Provide daylighting wherever possible, but avoid solar glare, in particular in kitchen or circulation spaces.
- Consider low-temperature heating methods or protective covers for radiators.
- Fit thermostatic controls to mixer taps and showers (set to 44 °C).
- Avoid the need for portable heating devices.
- Provide RCD protection to ring main circuits.
- Ensure that fuses, MCBs and equipotential bonding provide adequate protection against electric shock.
- Provide sufficient sockets to avoid the use of extension cables (see chapter 9, *Building services*).
- Do not position sockets in wet areas (eg near kitchen sinks or in bathrooms) unless adequately protected. 13 amp sockets and switches are not permitted in bathrooms.

Cont'd....

Box 1.9 Designing for safety (cont'd)

Kitchens
- Provide access for rubbish removal and bringing in goods.
- Provide a good working triangle (the sum of the distance between cooker, sink and fridge should be between 3.3 m and 6.6 m).
- Avoid circulation routes that cross the working triangle.
- Do not place cookers near doors or windows to avoid collisions, reaching over hot plates, draughts on gas burners extinguishing the flame on cookers or curtains being set on fire.
- Avoid cupboards over the hob unit and provide at least 300 mm of counter top on each side at the same level as the hob.
- Provide level access between cooking and dining areas.
- Provide a lockable cupboard for cleaning chemicals and medical supplies.
- Fit a suitable third-party approved smoke detector in an adjacent room (see www.RedBookLive.com).

- Avoid unprotected sharp edges at head heights of adults or children. This effectively means between finished floor level (for crawling infants) and 2 m for adults.

Bathrooms
- Make shower doors and partitions of safety glass or other suitable material (EN 12600 Class 1).
- Provide good side access to the bath, preferably 700 mm wide by a minimum length of 1100 mm from the end of the taps.
- Avoid sharp corners.
- Provide grab rails for baths.
- Consider fitting a shower in addition to a bath, with a slip-resistant finish.
- Make locks operable from the outside so access can be gained after an accident or where toddlers have locked themselves in.
- Provide a lockable medicine cabinet.

Table 1.3 Costs included in WLC (Data from CCF 1999)

Acquisition	Use and maintenance	Disposal
Costs	**Costs**	**Costs**
● Acquisition by construction	● Operational costs	● Cost of disposal (eg demolition, sale)
● Site costs (eg purchase, clearance and groundwork)	● Routine maintenance (eg cleaning, energy, utilities, facilities management including landscape maintenance	● Site clean-up
● Design	● Major maintenance, commissioning and/or fitting out (eg repairs, churn costs)	
● Preliminaries		
● Construction commissioning and/or fitting out		
● Fees		
● In-house administration		
Purchase/leasing	**Income**	**Income**
● Purchase price	● Income generated through ownership of asset (eg rents from surplus space less loss of income during refurbishment or during failure of facilities	● Sale or interest in asset
● Cost of purchase/adaptation		● Sale of materials for reuse/recycling
● Fees		
● In-house administration		

CCF. *Whole life costing: A client's guide.* London, Confederation of Construction Clients. 1999

WLC can be undertaken at any stage in the life of an asset (new build or maintenance or refurbishment) and also by all organisations involved in the stages of the supply chain. At each stage, activities such as maintenance and replacements are identified, together with their probable costing and timing. WLC is used to select between competing design options on the basis of which represents lower whole life costs.

The cash flows resulting from the analysis are brought back to a common basis of measurement, normally by using discounted cash-flow techniques. This means using a discount rate to reflect the present value of the whole cash-flow analysis. Net Present Value (NPV), represents a single figure, which takes account of all relevant future incomes and expenditures for that option over the period of analysis. The discounted cash-flow analysis gives a basis for comparison between the WLC of each option.

Normally, this calculation will be undertaken on either a spreadsheet program (such as MS Excel) or on a specialist WLC program. Specialist programs will normally integrate functions such as sensitivity analysis of the inputs and/or default or benchmark data inputs.

Typical decisions informed by WLC analysis include:

● Choices between alternative components, all of which have acceptable performance (component-level WLC analysis).

● Choices between alternative designs for the whole, or part, of a constructed asset (assembly or whole asset level WLC analysis).

● Evaluation of different investment scenarios (eg to adapt and redevelop an existing facility, or to provide a totally new facility).

● Estimation of future costs for budgets or the evaluation of the acceptability of an investment on the basis of cost of ownership.

● Comparison and/or benchmarking analysis of previous investment decisions, which may be at the level of individual cost headings (eg energy costs, cleaning costs, or at a strategic level, eg open-plan versus cellular office accommodation).

Air permeability

Buildings need to be ventilated to provide a healthy environment for the occupants by supplying fresh air, getting rid of excess heat, smells and other pollutants. Many existing buildings are excessively leaky and this leads to uncontrolled heat loss,

draughts, interference with the performance of designed mechanical and natural ventilation systems, uncontrolled polluted air entering the building and reduction in the performance of the building's insulation.

Air permeability is defined in Box 1.10. CIBSE's Technical Memoranda *Testing buildings for air leakage* recommends air leakage indices and air permeabilities for some buildings types. Tests are carried out by pressurising or depressurising the building using a fan fixed to an opening such as the front entrance (see chapter 9). Tracking down the cause of the leaks is not easy but using smoke pencils or puffers can help. The building needs to be practically complete before it can be tested so fixing the problem afterwards can be costly.

However, there are currently no regulatory requirements controlling airtightness of the outer envelope of buildings in the UK, although it is anticipated that revisions to regulations will soon include a requirement. There are already standards for windows.

Some countries have standards for the entire envelope. It is likely that post-construction testing of air leakage rates will be introduced. Generally, poor performance in this respect is associated with adventitious ventilation through:

● gaps between walls and windows/doors,

● poorly fitting windows/doors,

Box 1.10 Definition of air permeability

Air permeability is the measure for airtightness of the building fabric. It measures the resistance of the building envelope to inward or outward air leakage.

It is defined as the average volume of air (in cubic metres per hour) that passes through unit area of the structure of the building envelope (in square metres) when subject to an internal or external pressure difference of 50 Pa.

The envelope area of the building is defined as the total area of the external floor, walls and roof, ie the conditioned or heated space.

CIBSE. *Testing buildings for air leakage*. Technical Memoranda TM23. London, CIBSE, 2000

- lack of draught-proofing,
- gaps round service pipes, and
- gaps where joists are built into the inner leaf of cavity walls.

Airtightness is now an important issue for the construction industry. Designers are responsible for designing an airtight building; contractors are responsible for building an airtight building, arranging for an airtightness test to be carried out after completion, and also for any necessary rectification and re-testing. There is more information on airtightness in Box 9.4 as there are crucial implications for energy conservation.

It is important that operatives are instructed on appropriate details, since in most situations remedial work in order to rectify hidden deficiencies exposed by post-completion testing will be much more expensive than getting it right in the first place.

Fire safety and means of escape

The safety of building occupants and passers-by is the key objective of fire safety. Buildings should be designed and constructed in such a way that the risk of fire is reduced. However, if a fire does occur, the design should include features to restrict its growth and the spread of smoke so that the occupants can escape safely and fire-fighters can deal with rescue and put the fire out safely and effectively.

The Regulatory Reform (Fire Safety) Order 2005, which requires a written risk assessment approach to be used for controlling fire safety in all occupied buildings employing five or more people, came into force in October 2006.

Key Point

There may be up to four stages to escaping from fire that the designer needs to consider:

Room escape from the room or area of the origin of the fire (or from the fire where only one direction is possible).

Compartment escape from the compartment of the origin.

Floor escape from the floor of origin to a protected zone and escape stair.

Building escape from the building to a place of safety at ground level.

Everyone inside a building must be provided with at least one protected means of escape to a place of safety outside the building. The route should be as short as possible. In a number of instances, an alternative route will be required, for example, where the travel distance is excessive. It will also mean that the occupants can move away from the fire to an escape in the other direction.

In the event of an outbreak of fire, it is important that the occupants are warned as soon as possible, particularly where people may be asleep, are unfamiliar with the building's layout or where there is a particularly high risk.

Designers should take into account the personal characteristics of the occupiers when considering means of escape and fire compartmentation:

- *Awareness:* whether occupants are normally awake or may be sleeping (eg in hotels). Higher standards of safety are needed for buildings where there is sleeping accommodation.
- *Familiarity:* whether the majority of the occupants are likely to be familiar (their workplace) or unfamiliar (visitor) with the layout of the building and its escape routes. Directional arrangements become more critical with unfamiliarity.
- *Ability:* the extent to which occupants are able to respond to a need to escape, or whether they are likely to rely on others because of age or some physical impairment such as deafness (response ability). It also refers to whether or not the occupants are able to use horizontal and vertical escape routes without assistance from others in the building (locomotive ability).
- *Disabled persons:* legislation requires buildings to be accessible to people with disabilities, wherever it is reasonable and practicable, and will result in most buildings (other than domestic) needing to incorporate some additional arrangements that will aid their escape.

The building regulations are concerned with the protection and health and safety of people from the dangers inherent in buildings, rather than protecting the owners of buildings from any economic loss that might occur. Therefore, it is important for designers and owners of buildings to understand that the regulations will not necessarily prevent the ultimate destruction of the building from a fire.

Fire prevention and control of its spread will not only save lives, but will reduce environmental pollution.

Fire protection

The structure must be able to resist the impact of a fire for a sufficient period of time to enable the occupants to escape to a place of safety outside the building. It must also allow fire fighters to safely gain access to rescue trapped people and to fight the fire.

Structural fire protection

The purpose of structural fire protection is to:
- minimise the risk to the occupants, some of whom may not evacuate the building immediately,
- reduce the risk to fire-fighters carrying out fire-fighting or rescue operations,
- reduce the risk to people in the vicinity of the building or in adjoining buildings from fire spread and collapse of the structure,
- resist the transfer of excessive heat, for example, between floors.

During a fire, elements of structure should continue to function and remain capable of supporting and retaining the fire protection to floors, escape routes and fire access routes, until all occupants have escaped. Elements of the structure include the main structural frames, floors and load-bearing walls and compartment walls, even though the latter may not have load-bearing functions. Conversely, load-bearing elements do not necessarily have fire-separating functions.

Fire compartmentation

The aim of compartmentation is to control the spread of fire within the building by reducing the fuel available in the initial stages of a fire. This will limit the severity and spread of the fire and enable occupants to escape from the building. It will also assist fire service personnel with fire-fighting and rescue operations. The principle of fire compartmentation is to divide the building into a series of fire-tight boxes termed compartments that

will form a barrier to the products of combustion (smoke, heat and toxic gases) and the spread of the fire.

The maximum size of compartments depends on:
- *the use of a building and its fire loading:* this affects the ease of evacuation, the potential for fires and their severity,
- *the height to the topmost floor:* this has implications for the ease of evacuation and the ability of the fire service to intervene,
- *whether the building is sprinklered:* this slows the rate of growth of a fire and may even put it out.

Semi-detached and terraced houses and individual flats and maisonettes need to be separated by compartment walls. Flats and maisonettes additionally need to be separated by compartment floors, although not within a maisonette. Institutional buildings (that here include health care) have upper limits for the size of compartment. Non-residential buildings are each treated separately when sizing the area of the compartment.

Buildings or parts of a building in different occupation (eg flats) and mixed usage pose particular problems in terms of fire safety. This is because one occupier usually does not have any control over the activities or working practices of their neighbouring occupiers.

Stairs and ducts that pass through and connect compartments need to have the same ability as the walls to restrict the spread of fire.

Cavities and voids

From a fire safety viewpoint, a cavity is a concealed space enclosed by elements of a building (including a suspended ceiling) or as part of a building element (such as cavity wall), a roof space, a service riser or any other space used to run services around the building.

Fire and smoke can spread quickly in these concealed spaces and hence throughout a building and remain undetected by the occupants of the building or by fire service personnel. Ventilated cavities generally promote more rapid fire spread around the building than unventilated cavities due to the plentiful supply of replacement air. Buildings containing sleeping accommodation pose an even greater risk to life and demand a higher level of fire precautions. For these reasons, it is important to control the size of cavities using fire-stopping and controlling the type of material in the cavity.

Internal surface spread of fire

The building contents are likely to be the first items ignited in a fire. Materials used in walls and ceilings can, however, significantly affect the spread of fire and its rate of growth. Fire spread on internal linings in circulation spaces and escape routes is particularly important because rapid fire spread in protected and unprotected zones could hinder or prevent the occupants from escaping.

Spread of fire to adjacent buildings

Buildings should be isolated from neighbouring buildings by either construction of the external wall, by spacing them apart or a combination of both. The distance between a building and its relevant boundary is dictated by the amount of heat that is likely to be generated in the event of fire. Openings in external walls may need to be limited to minimise the amount of radiation that can pass through the wall, taking into account boundary distances.

Key Point

For further information on how to comply with the Regulatory Reform (Fire Safety) Order 2005, including the selection of approved products and services,
see www.RedBookLive.com.

Building Regulations and Construction (Design and Management) Regulations

The Building Regulations (and accompanying documents) require that health and safety and energy conservation are taken fully into account in the design of new and refurbished buildings. They are, however, the minimum performance requirements; achieving full sustainable design is likely to require performance significantly in excess of the requirements of the Building Regulations. It is likely that the Sustainable Buildings Act will increasingly bring about additional regulatory requirements in respect of sustainability.

The Public Health Act of 1875 gave local authorities the power to enact by-laws with respect to the structure of walls, foundations, roofs and chimneys of new buildings for the purposes of stability, fire prevention and health. While these by-laws served a useful function, it was appreciated that they could be a barrier to innovation and in 1936 the Building Research Board commented that:

'There is too frequently in byelaws a tendency to specify particular forms of construction, forms usually based on particular materials .... The ideal is undoubtedly that the byelaws should state the performance required leaving open the methods by which the requisite standard is to be attained.'

In the 1960s, building regulations were introduced applicable to the whole country; the Scottish Building Regulations were issued in 1963 and those for England and Wales in 1964. The Building Research Board was pleased to note that:

'The form of the new regulations ... stems from the principles first put forward by this Board in 1936 and amplified in later years. Progressively the building byelaws, and now the building regulations which are replacing them, have reflected more and more the philosophy enunciated by our predecessors.'

The building regulations are operated independently of the planning system and plans may be approved before, in parallel with, or after determination of a planning application.

With certain specified exceptions, all building work, as defined in the regulations in England and Wales is governed by building regulations. Separate systems of building control apply in Scotland and in Northern Ireland. The Building Act 1984 consolidated most of the earlier primary legislation relating to buildings.

The following sections briefly describe two key regulations that need to be taken into account when designing and constructing buildings:
● the national building regulations:
  ❑ The Building Regulations (England & Wales) 2000,
  ❑ The Building (Scotland) Regulations 2004,
  ❑ The Building Regulations (Northern Ireland) 2000.

- the Construction (Design and Management) Regulations 1994.

Both building and CDM regulations are to do with the health and safety of the users and persons nearby and those who are going to build and maintain the building.

The Building Regulations (England & Wales)

The Building Regulations are made under powers provided in the Building Act 1984, and apply in England and Wales. There was a major revision of the building control system in England and Wales in 1985. The current edition of the regulations is The Building Regulations 2000 (as amended) and the majority of building projects are required to comply with them. They exist to ensure the health and safety of people in and around all types of buildings (ie domestic, commercial and industrial). They also provide for energy conservation and for access and facilities for disabled people.

The Schedule 1 Requirements were written in general terms requiring reasonable standards of health and safety. Reference was made to supporting Approved Documents, which could be complied with in various ways. Builders and developers are required by law to obtain building control approval in the form of an independent check that the building regulations have been complied with.

The technical requirements in Schedule 1 are mostly set out in a functional form for building design and construction and they are presented in 15 parts. Each part of Schedule 1 is supported by an Approved Document (AD), which gives guidance on how to comply with the requirements in some of the more common building situations.

The series of ADs give some technical detail and refer to British Standards and guidance material such as BRE publications. Compliance must be shown with the Building Regulations, but not necessarily the contents of the ADs, which give practical guidance on meeting the requirements of the regulations. There is no obligation to adopt a particular solution contained in an AD if meeting the relevant requirement in some other way is preferred.

The Building Regulations Division of the Office of the Deputy Prime Minister is responsible for the formulation of the regulations and prioritising and programming reviews of the regulations.

The Building Regulations Advisory Committee (BRAC), which is categorised as an 'advisory non-departmental public body,' advises the Secretary of State on the exercise of his power to make Building Regulations. In practice, BRAC is consulted on all matters concerned with changes to the Building Regulations, both technical and procedural.

The main work of BRAC is conducted in five committee meetings per year; while advice on more technical aspects is developed and carried forward through working parties comprising BRAC members, co-opted members where appropriate, and officials of the Building Regulations Division. The main BRAC meetings are attended by observers from other government departments and advisory organisations including BRE Trust, of which BRE is a wholly owned subsidiary.

Ideas for new regulations or changes to existing ones are handled by the Building Regulations Division.

Builders, developers and owners are required by law to obtain building control approval, an independent check that the Building Regulations have been complied with. There are two types of building control providers: the Local Authority and Approved (private) Inspectors.

Approved Documents (ADs)

Practical guidance on ways to comply with the functional requirements in the Building Regulations is contained in the series of ADs.

Each AD contains:

- general guidance on the performance expected of materials and building work in order to comply with each of the requirements of the Building Regulations,
- practical examples and solutions on how to achieve compliance for some of the more common building situations, ie deemed to satisfy.

The ADs are grouped as given in Table 1.4. At the time of publication, Parts F and L governing ventilation and energy efficiency and aiming to reduce CO_2 emissions are being revised. They are expected to come into effect in April 2006.

Table 1.4 Content of the Approved Documents of the Building Regulations 2000 (England & Wales)	
Approved Document (AD)	**Content**
A	Structure
B	Fire safety
C	Site preparation and resistance to moisture
D	Toxic substances
E	Resistance to passage of sound
F	Ventilation
G	Hygiene
H	Drainage and waste disposal
J	Combustion appliances and fuel storage systems
K	Protection from falling collision and impact
L1 and L2	Conservation of fuel and power
M	Access to and use of buildings
N	Glazing
P	Electrical safety
AD to support Regulation 7	Workmanship

The guidance in the ADs does not amount to a set of statutory requirements and does not have to be followed if the designer or builder wants to carry it out in some other way, providing it can be shown that it still complies with the relevant requirements. The guidance will be taken into account when the Building Control Service is considering whether the plans of the proposed work, or work in progress, comply with particular requirements.

However, there is a legal presumption that if the guidance has been followed, then this is evidence that the work has complied with the Building Regulations. It is still the job of the Building Control Service to consider whether the plans and work comply with the relevant requirements in Schedule 1 to the Building Regulations, not whether they necessarily follow the specific guidance or a specific example in an AD.

'Building Work' is defined in Regulation 3 of the Building Regulations to include the following types of work:

- erection or extension of a building,
- installation or extension of a service or fitting which is controlled under the Regulations,
- alteration involving work which will be relevant to the continuing compliance of the building, service or fitting with the requirements relating to structure, fire, or access and facilities for disabled people,
- insertion of insulation into a cavity wall,
- underpinning of the foundations of a building.

The Building (Scotland) Regulations 2004

The Building (Scotland) Act 2003 gives Scottish Ministers the power to make Building Regulations to do the following:

- secure the health, safety, welfare and convenience of persons in or about buildings and of others who may be affected by buildings or matters connected with buildings,
- further the conservation of fuel and power,
- further the achievement of sustainable development.

The Building (Scotland) Regulations 2004 set out the standards that should be achieved by buildings. The standards are given in Schedules 5 to Regulation 9 and are in the form of expanded functional standards. These standards are compulsory and must be complied with by building owners. However, the means of complying with the standards is at the discretion of the building owner and approval by the verifier (building control authority). The building regulations are approved by the Scottish Parliament, however, they must not represent a barrier to trade and are therefore scrutinised by the European Commission.

There are two sets of Technical Handbooks, dealing with Domestic and Non-Domestic buildings respectively, that give technical advice on meeting the requirements of the functional standards. Section 0 of the Technical Handbooks provides an introduction and also sets out the Building Regulations. To meet the requirements of the Construction Products Directive, materials and construction methods must be described by use of a suitable European Standard where such exists. The Technical Handbooks are designed to be readily updated, with the intention that any update is made once per year.

The building regulations are mandatory; however, the choice of how to comply with them lies with the building owner. The Technical Handbooks provide guidance with respect to achieving the building regulations. If the guidance is followed in full then this should be accepted by the verifier as indicating that the building regulations have been met. It is, however, acceptable to use other methods of compliance provided that they fully satisfy the regulations. Using the guidance in the Technical Handbooks is therefore likely to be the main way of complying with the regulations. However, a designer may put forward other ways of meeting the regulations. In the future, other documents may be put forward to provide guidance as the system develops.

If designers choose to use alternative means of compliance, they need to demonstrate that the alternative is at least as good. Designers may look to use Approved Documents for the England & Wales building regulations, however, they would need to ensure that the guidance in those documents meets the detail of the Scottish Technical Handbook

guidance. This is best illustrated by an example; the thermal insulation standards for Scotland have been more stringent than England & Wales since the 2002 changes. It would therefore be unacceptable to use thermal insulation values from the England & Wales Approved Documents to show compliance with the energy requirements of the regulations.

The difference between the previous Scottish Building Standards System (1959) and the new system is marked significantly by the arrangement of the technical guidance in the Technical Handbooks. However, at present all the information that was contained in the old technical standards has been transferred into the Technical Handbooks. The arrangement of the Technical Handbooks directly relates to the Essential Requirements of the Construction Products Directive, and is as follows:

- **Section 0:** Introduction
- **Section 1:** Structure (EC – Mechanical resistance and stability),
- **Section 2:** Fire (EC – Safety in case of fire),
- **Section 3:** Environment (EC – Hygiene, health and the environment),
- **Section 4:** Safety (EC – Safety in use),
- **Section 5:** Noise (EC – Protection against noise),

Table 1.5 The Building (Scotland) Regulations: Comparison of new and old technical information

Old Technical Standards	New Technical Handbooks
Part A General	**Section 0** General
	(Regulations 1–7 and 9–12)
Part B Materials and Workmanship	**Section 0** General (Regulation 8)
Part C Structure	**Section 1** Structure
Part D Structural fire precautions	**Section 2** Fire
Part E Means of escape from fire	**Section 2** Fire
Part F Combustion appliances	**Section 3** Environment and **Section 4** Safety
Part G Preparation of sites	**Section 3** Environment
Part H Resistance to transmission of sound	**Section 5** Noise
Part J Conservation of fuel and power	**Section 6** Energy
Part K Ventilation of buildings	**Section 3** Environment
Part M Drainage and sanitary facilities	**Section 3** Environment
Part N Electrical installations	**Section 4** Safety
Part P Miscellaneous hazards	**Section 4** Safety
Part Q Facilities for dwellings	**Section 3** Environment and **Section 4** Safety
Part R Storage of waste	**Section 3** Environment
Part S Access to and movement	**Section 4** Safety
Building operations regulations	**Section 0** (Regulations 13–15)

- **Section 6:** Energy (EC – Energy economy and heat retention).
- **Appendices:**
  - ❐ **A** defines the terms;
  - ❐ **B** gives a list of standards and other publications;
  - ❐ **C** gives cross-references to the 6th amendment of the technical standards).

Table 1.5 compares the technical guidance in the old standards with the new system.

The Building Regulations (Northern Ireland)

The 1979 Order is the primary legislation in Northern Ireland for building regulations. It sets out the powers, duties, responsibilities and rights of the Department of Finance and Personnel, district councils and applicants in relation to building regulation matters, including giving the Department of Finance and Personnel responsibility for writing regulations in the form of secondary or subordinate legislation. These regulations currently are the Building Regulations (Northern Ireland) 2000 and the Building (Prescribed Fees) Regulations (Northern Ireland) 1997.

The building regulations set out requirements that can reasonably be expected to be attained in buildings, having regard to the need for securing the health, safety, welfare and convenience of persons in or about buildings and of others who may be affected by buildings or matters connected to buildings. They also cover the furthering of the conservation of fuel and power.

The Office Estates and Building Standards Division of the Department of Finance and Personnel has overall responsibility for making, and overseeing the application of, Building Regulations in Northern Ireland. The work is undertaken by the Building Regulations Unit. The Unit calls upon additional specialist support from within the Department as required.

Its main tasks are to:
- ensure that the underpinning primary legislation provides a satisfactory basis for a set of regulations reflecting modern building practice,
- draft, amend and make the Building Regulations,
- maintain the Technical Booklets in support of the Regulations,
- draft, amend and make the Prescribed Fees Regulations,
- consider appeals lodged with the Department against decisions made by District Councils in their role as enforcers of the Building Regulations,
- respond to queries relating to the Building Regulations.

District Councils have a statutory duty to enforce the Building Regulations (NI). A Building Regulation application is required for the following:
1 altering or extending a building,
2 erecting a building,
3 executing works or install fittings,
4 making a material change of use.

The building regulations are supported by Technical Booklets that provide guidance on how to achieve compliance with the regulations. The Technical Booklets provide methods and standards of building, which if followed will satisfy the requirements of the building regulations. There is no obligation to follow the methods or comply with the standards set out in the Technical Booklets and designers may adopt another way of meeting the requirements, but it is necessary to demonstrate compliance.

The various regulations and Technical Booklets are set out in the Table 1.6.

The Construction (Design and Management) (CDM) Regulations 1994

The CDM Regulations are aimed at improving the overall management and co-ordination of the health, safety and welfare of those involved throughout all stages of a construction project to reduce the large number of serious and fatal accidents and cases of ill health which happen every year in the construction industry. It also aims at improving the health and safety of those who need to look after the building.

The CDM Regulations apply to most construction projects and all demolition. However, there are a number of situations where the Regulations do not apply. These include:

Table 1.6 The Building Regulations (Northern Ireland)	
Regulation	**Technical Booklet**
A Interpretation and general	
B Materials and workmanship	
C Preparation of site and resistance to moisture	C Preparation of site and resistance to moisture, 1994
D Structure	D Structure, 1994
E Structural Fire Precautions	E Fire safety, 2005
EE Means of escape in case of fire	
F Thermal insulation of dwellings	F Conservation of fuel and power, 1998
FF Conservation of fuel and power in buildings other than dwellings	
G Sound insulation of dwellings	G Sound, 1990
	G1 Sound: conversions, 1990
H Stairways, ramps, balustrades and vehicle barriers	H Stairs, ramps, guarding and protection from impact, 2000
J Refuse disposal	
K Ventilation	K Ventilation, 1998
L Chimneys, flue pipes, hearths and fireplace recesses	L Heat producing appliances, 1991
M Heat-producing appliances and incinerators	
N Drainage	N Drainage, 1990
P Sanitary appliances and unvented hot water storage	P Unvented hot water storage systems, 1994
R Facilities for disabled people	R Access and facilities for disabled people, 2000
S Thermal insulation of pipes, ducts and storage vessels	
T Control of space and water heating systems	
U	
V Glazing	V Glazing, 2000

- construction work other than demolition that does not last longer than 30 days and does not involve more than four people,
- construction work for a domestic client,
- construction work carried out inside offices and shops or similar premises without interrupting the normal activities in the premises and without separating the construction activities from the other activities,
- the maintenance or removal of insulation on pipes, boilers or other parts of heating or water systems.

The CDM Regulations place duties on all those who can contribute to the health and safety of a construction project. Duties are placed upon clients, designers and contractors and the Regulations create a new duty holder: the planning supervisor. They also introduce new documents:

- health and safety plans, and
- the health and safety file.

The role of the designer

Designers are organisations or individuals carrying out design work for a construction project, including temporary works design. Under the Regulations, the term designer includes architects, engineers, quantity surveyors, chartered surveyors, technicians, specifiers, principal contractors and specialist contractors. Design includes drawings, design details, specifications and bills of quantity.

Designers play a key role within the construction project in ensuring that the health and safety of those who are to construct, maintain or repair a structure or building are considered during the design process. They have to consider the potential effect of their designs on the health and safety of those carrying out the construction work and others affected by the work. This means there will be a need to assess the risks of the design which can reasonably be foreseen. In the main, this will include risks to those persons building, maintaining or repairing the structure as well as those who might be affected by this work.

To ensure that risks to health and safety are fully considered, a risk assessment should be carried out to:

- identify the significant health and safety hazards likely to be associated with the design and how it may be constructed and maintained,
- consider the risk from the hazards which arise as a result of the design being incorporated into the project.

The design should be altered to avoid the risk, or where this is not reasonably practicable, reduce it to an acceptable and manageable level.

Other regulations

Where housing developments are proposed, the NHBC or Zurich Building Guarantees requirements may also have to be met, together with requirements for security discussed in *Secured by design*. These have specific requirements for the design and construction of dwellings, and their guidance tends to be quite specific and well illustrated.

Planning Policy Guidance Notes

These cover issues such as transport, flood risk and site development density.

Association of Chief Police Officers (ACPO). *Secured by design.* London, ACPO Crime Prevention Initiatives. 2004 www.securedbydesign.com

2 Site investigation and preparation

Site investigation

The main purpose of site investigation is to reduce and control ground-related risks, both during and after construction.

> **Key Point**
> The ground often represents the greatest risk to any construction project.

Information from a properly conducted site investigation will provide the basis for economic and safe design and construction, and for meeting tender and construction requirements. It will also save money by avoiding the following:

- redesign during construction,
- over-conservative design,
- delays and litigation costs due to unforeseen ground conditions,
- foundation problems post-construction.

Site investigation comprises all those activities that enable hazards to be identified, a hazard and/or risk assessment to be carried out and an informed choice to be made between different remedial strategies.

Effective ground investigation need not incur excessive cost but it is important that the specifier of the site investigation has specialist knowledge and experience, and an awareness of the hazards associated with the ground and the performance requirements imposed by the construction process and final construction.

There is increasing pressure to develop brownfield sites, where previous occupation or use has either changed the geological structure of the site (fill or made ground), left the remnants of old buildings or excavations or caused contamination of the site (Figure 2.1).

The aspect of risk posed by contamination of the ground has been the focus of recent advances in the investigation and management of brownfield land. Government policy has also moved from a 'buyer beware' position to the 'polluter pays' principle. These issues need to be addressed by any developer undertaking a site investigation for contamination.

This section on site investigation is divided into risk assessment and risk management issues. Risk assessment is a process that requires sufficient information to be collected on the site and for this information to be translated into a proper evaluation of the potential risks. Risk management follows on from risk assessment and is the process by which developers reduce the risk to:

- a new development,
- people,
- groundwater, or
- the local or wider environment.

Risk management may involve physical intervention in the ground to remove or reduce the risk from contamination, or it may involve redesign of the development to accommodate the

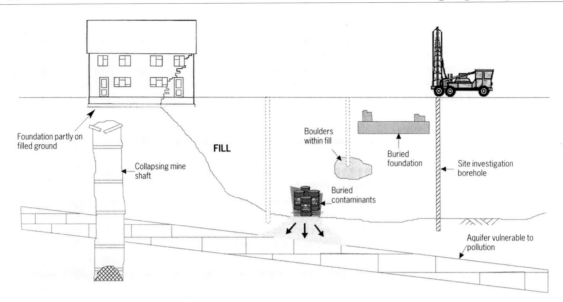

Foundation partly on
filled ground

FILL

Collapsing mine
shaft

Boulders
within fill

Buried
contaminants

Buried
foundation

Site investigation
borehole

Aquifer vulnerable to
pollution

Figure 2.1 Typical hazards on a brownfield site

contamination on site, while economically
developing the site.

A key part of the information required to design a
site investigation effectively is an understanding of
the client's requirements and the constraints
imposed by the site and any structures.

Risk assessment

This section covers the main requirements for
undertaking a risk assessment for a site. As it is
written primarily from the perspective of
contaminated land, developers should also be aware
that sites will require a geotechnical investigation
and possibly other assessments of risk (eg flooding
risk). It is necessary to seek expert advice in these
areas; note that the geotechnical and environmental
investigations can be undertaken in parallel and
often use the same consultant group or project team.
It is necessary that competent and experienced staff
are used in these specialist activities.

The site investigation is an essential component
of the risk assessment; it can involve a range of
tasks, from desk-based research through to on-site
intrusive investigations.

Table 2.1 shows a responsibility checklist for site
investigations as developed by the Association of
Geotechnical Specialists (AGS).

The contaminated land regime which is set out in
Part IIA of the Environmental Protection Act 1990,
was introduced in England on 1 April 2000, and on
1 July 2001 in Wales. A similar regime was
introduced in Scotland on 14 July 2000. The regime
is jointly regulated by local authorities and the
Environment Agency [or the Scottish Environment
Protection Agency (SEPA) in Scotland]. Local
authorities are the primary regulators and take the
lead role. Part IIA provides a risk-based approach to
the identification and clean-up (known as
remediation) of land where contamination poses an
unacceptable risk to human health or the
environment.

The legal definition of 'contaminated land' is
given in section 78A (2) of Part IIA of the
Environmental Protection Act 1990, as follows
(DETR 2000):

'... any land which appears to the local authority in whose
area it is situated to be in such a condition, by reason of
substances in, on or under the land, that:

❑ Significant Harm is being caused or there is a
significant possibility of such harm being caused;

Association of Geotechnical Specialists (AGS). *Guidelines
for combined geoenvironmental and geotechnical
investigation*. AGS, 2005. Publications available from
www.ags.org.uk

Table 2.1 Responsibility checklist (Source: AGS 2005)

Stage	Action	Responsible person(s)
1 Proposal	Identify site Form project team Appoint geotechnical adviser	Client
2 Feasibility	Carry out desk study which includes site reconnaissance and prepare a report	Geotechnical Consultant/GI Specialist (Consultant or Contractor under direction of a Geotechnical Adviser)
	Check historical records/sources of information: ● previous investigations ● previous use ● mining/landfill ● geology/groundwater ● adjacent sites ● services records Carry out hazard identification and risk assessment Prepare objectives methodology for phased GI Prepare contract documents for GI fieldwork Prepare CDM pre tender Health & Safety Plan Define supervision requirements and organisation Evaluate and recommend contractor for fieldwork or advise Term Contractor	Consultant/GI Specialist
3 Design Phased Investigation (repeat as required)	Fieldwork Laboratory testing Factual report Continuing assessment of ground conditions enhanced by adequate supervision	Consultant/GI Specialist
Interpretation	Interpretative report	Consultant/GI Specialist
Preliminary design	Risk assessment	Project team including GA
Detailed design	Geotechnical design summary reports Consider construction activities and temporary works as well as long-term maintenance requirements	Geotechnical Engineer(s) or Engineering Geologist(s) within the Project design team
	Risk assessment	Project design team including GA
4 Construction	Inspection/Trials/Monitoring Progressive modifications	Design Engineer/Main Contractor (Highly desirable to have member(s) of the Project design team involved in these activities)
	Dispute resolution	Client/Main Contractor/Designer
5 Operation	Monitoring Inspection Maintenance and renewal	Design Engineer/Main Contractor Handover required during commissioning to the Client's staff or operating Consortium's staff

GI Specialist = Ground Investigation Specialist = Firms that either operate solely as specialist ground investigation contractors and/or provide specialist consultancy advice.

GA = Geotechnical Advisor.

Consultant = Firms that offer specialist geotechnical, geological or geoenvironmental advice.

or
❏ Pollution of controlled waters is being, or is likely to be caused'.

A risk assessment aims to obtain information and data in order to identify potential sources of contamination, pathways and receptors (ie pollutant linkages) specific to an individual site. A site investigation may involve any one of a variety of contexts and will often comprise a number of stages for which different methodologies and technologies may be used.

Key Point

The concept of pollutant linkage is essential to the understanding of risk assessment for brownfield sites.

The following aspects of pollutant linkages need to be considered.

● *Sources of contamination:* the contamination may exist in either solid, liquid or gas form. Contaminants are primarily associated with previous industrial use or from the deposition of waste materials on a site. Contaminants can form part of the solid materials in the ground or be dissolved or transported in groundwater.
● *Pathways* are any potential route by which the contaminants are transported to a receptor. This can be through the soil itself or through the air by airborne dust or vapours.
● *Receptor:* the Environment Act recognises that there are a number of possible receptors of contamination. A receptor is any being or article that could be harmed by the contamination and includes people, groundwater, property, flora, fauna and the environment.

A similar risk-based approach can be taken for all forms of contamination and receptors.

Hazard identification

The first stage of the risk assessment is hazard identification. The hazard is typically the contaminant and this stage is intended to identify all possible contaminants that could be present on a site.

Desk study

The first activity is a desk study to identify the potential hazards on the site. It involves historic research as well as records of site development and uses.

As recommended by BS 10175, a desk study should comprise a combination of thorough documentary research and consultations with a range of stakeholders. The extent of the information obtained at this stage provides the basis for the intrusive site investigation work. The results of the desk study also provide information for the protection of construction personnel and the future occupiers of a contaminated site.

Information can be limited due to a lack of documents from the past. Often former industrial sites that have lain derelict for a number of years will have no records from the operation of the site, as site owners may have destroyed or disposed of such records. Information from those who previously owned or worked on such sites can compensate for the lack of records.

A desk study will contain the essential aspects outlined in Box 2.1, and as further described in Annex A to BS 5930 and Section 6.2 of BS 10175. Some key information sources for the desk study items set out in Box 2.1 are listed in Box 2.2.

Determining the previous use of the site and the types of contamination that are associated with different sites is important. A series of industry profiles was produced in the early 1990s by the former Department of the Environment and these contain significant and important information. Each profile contains details of contaminants associated with particular industries.

In addition to information on the nature or origin of hazards, it is necessary to collect information on the characteristics of the site and its setting (risk factors) that may enhance or reduce the potential for risks to receptors, for example, information on the following factors.

British Standards Institution
BS 10175: 2001 *Investigation of potentially contaminated sites. Code of practice*
BS 5930: 1999 *Code of practice for site investigations*

BS 10175: Section 6.2
● General
● Documentary research
● Site location and historical setting
● Site usage and contamination
● Geology, geochemistry, hydrology and hydrogeology
● Ecology and archaeology

BS 5930: Annex A
● General land survey
● Permitted use and restrictions
● Approaches and access
● Ground conditions
● Sources of material for construction
● Drainage and sewage
● Water supply
● Electricity supply
● Gas supply
● Telecommunications
● Heating
● Information related to potential contamination

● Current and historical land use maps
● Geological maps
● Local trade directories, local libraries and archives
● Local Authority records (eg planning, building control, land use, environmental health and pollution control departments)
● Valuation office records
● Fire Service records (particularly for combustible fills)
● Environment Agency/SEPA records on authorised processes including waste management activities, pollution incidents, surface and groundwater quality
● Coal Authority reports
● Previous site investigation reports (including borehole and trial pit logs) for information on likely ground conditions
● Hydrogeological and groundwater vulnerability maps

● The natural geology of the site that will tell the assessor something about the potential for the mobilisation of hazardous substances within a site and on the potential for subterranean fires to spread laterally.

● The hydrogeology and hydrology of the site since this tells the assessor something about the potential for:
  ❑ aggressive substances to move,
  ❑ combustion to commence or be sustained, and
  ❑ triggering events such as rising groundwater or flooding.
● Surface and groundwater quality since this may give the assessor additional evidence of the presence of hazardous substances or materials on the site.

Site reconnaissance
The second activity in the hazard identification process is site reconnaissance. This aims to verify the output of the desk study and supplement the information on pollutant linkages obtained from desk-based research by confirming (or not) the documentary record, and by gathering additional information about the site and its setting which is not obtainable from desk-based research (Environment Agency Report P5-035).

Various activities are conducted throughout this stage including the following (BS 10175, BS 5930):
● *detailed walk-over inspection*,
● *interviews* with site occupiers, neighbours and former workers of an industrial plant,
● limited ad hoc *sampling and field measurements* (if appropriate),
● *preparatory activities* for site work (plans and maps, permission to gain access, verification; identified hazards),
● *general information* (traverse proposed location of work to inspect, observe and record relevant information, check access, water levels, streams and so on),
● *ground information* (study and record surface features, assess and record relevant ground conditions, study and note the nature of

Environment Agency. *Assessment and management of risks to buildings, building materials and services from land contamination.* R & D Report P5-035/TR/01. London, 2001
British Standards Institution
BS 10175: 2001 *Investigation of potentially contaminated sites. Code of practice*
BS 5930: 1999 *Code of practice for site investigations*

vegetation, study embankment, building and
other structures in the vicinity),
● *site inspection* for ground investigation (inspect
and record location and condition of access to
working sites, observe and record obstructions,
locate and record activities in any areas, ascertain
and record ownership).

The walk-over survey is an integral part of the site
investigation process which should always be
carried out. Used in conjunction with a good desk
study, it provides valuable information that cannot
be obtained in other ways.

Box 2.3 shows some specific items that should be
addressed in any walk-over study.

The outcome of the site reconnaissance work will
be a development of the desk study work with
greater knowledge of the potential hazards,
pollutant linkages and development of the
conceptual model. The site reconnaissance should
also be used to develop the plan for intrusive site
investigation work.

There may be occasions when the desk study and
the site reconnaissance do not indicate that potential
hazards, pollutant linkages or risks exist on a site,
either in its current condition or during and after
redevelopment. In this case, which may be rare,
there may be no need for intrusive site
investigations. However, in most cases the site
investigation will progress to the intrusive stage.

The output from the desk study and site
reconnaissance will be an initial identification of
potential hazards on the site and development of
potential pollutant linkages. This can be wrapped
into a framework of a conceptual model for hazards
and risks at this stage (Defra and the Environment
Agency's CLEA Report CLR11).

The conceptual model is a textual or graphical
representation of the pollutant linkages that the
assessor considers relevant to the site being
assessed. In addition to summarising the assessor's

Defra and Environment Agency. *Model procedure for the
management of land contamination.* CLEA Report CLR11.
London, The Stationery Office, 2004
BRE. *Site investigation for low-rise building: the walk-over
survey.* Digest 348. Bracknell, IHS BRE Press, 1989

**Box 2.3 Walk-over survey items (adapted from BRE's
Digest 348)**

Slope angles
These can be more accurately assessed on site than from
an OS map. They can be interpreted in terms of the types of
materials underlying the site.

Instability
Record the presence of hummocky, broken or terraced
ground, or boggy, poorly drained hill slopes that may be
associated with landslips. Assess whether trees or hedges
are bent over.

Vegetation
On shrinkage clay, record the presence of all trees and
shrubs and their size, height, etc. In contaminated ground
these may be of poor quality or show signs of damage.

Made ground
Note the position of any infilling, compare with OS map
indications. On former industrial sites made ground or
contaminated fill may be present.

Structures
Examine structures in the area and on site and record signs
of damage. Assess whether structures no longer exist
compared with maps and photographs of the site.

Soil and rock
Look for exposure of soil and rock. Where soft soils are
present on site it should be possible to make shallow holes
using an auger to assess their characteristics.

Groundwater
Note positions of springs, ponds and other water and
compare with OS maps for any site-filling activity.

Mining and quarrying
Look for signs of mineral extraction in the area.

Solution features
Check land underlain by chalk or limestone as it may contain
naturally occurring voids or pipes filled with soft soils.

Access
Check the ease of access for drilling rigs or hydraulic
excavators for ground investigation work.

initial views about the likely risk profile of the site,
the conceptual model can also help the assessor to
decide if a further, more detailed, assessment is
required. Box 2.4 shows an example of a conceptual
model for risks to buildings set in a scenario in
which a variety of different receptors may be at risk.
Initially, the conceptual model is based on the

Box 2.4 Example of a conceptual model of risks to buildings, building materials and services (BBMS)
(Data from Environment Agency Report P5-035)

A former gas works is being assessed in relation to possible development for a leisure centre. The site is adjacent to a canal. It is known that the canal bank needs to be strengthened and that the civil engineering advice is to use driven corrugated steel piling. Site reconnaissance has confirmed the presence of extensive areas of made ground, oils and tars. There is evidence of chemical attack on existing concrete structures in some areas and severe corrosion of metal items present on the surface. The conceptual model for pollutant linkages relevant to buildings, building materials and services (BBM&S), assuming the proposed development goes ahead without remedial action is shown in matrix and graphical form below.

Conceptual model of pollutant linkages (matrix presentation assuming no remediation)

Linkage	Substance/condition	Pathway/event	Receptor
1	Soils containing hydrocarbons including tars, polycyclic-aromatic hydrocarbons (PAHs), oils and BTEX	Direct contact	Plastics components (eg pipes) including protective coatings (eg electrical, telecommunications cables)
2 & 4	Soils (and perched waters where applicable) containing high concentrations of sulfates, sulfides, elemental sulfur, ammonium compounds, carbon (coke etc.) and low pH	Direct contact or contact with contaminated pore water	Concrete foundations, service pipes (metal and plastics), steel piling
3	Extensive areas of made ground, including sub-surface structures	Settlement	Buildings and services

Conceptual model of pollutant linkages (graphical representation assuming no remediation)

Environment Agency. *Assessment and management of risks to buildings, building materials and services from land contamination.* R & D Report P5-035/TR/01. London, 2001

findings of desk-based research, site reconnaissance and (where carried out) ad-hoc sampling. However, the objective is to refine the conceptual model if and when more information about the site and its setting becomes available. The conceptual model can therefore be used to explore a range of different construction scenarios and associated risks.

Hazard assessment

The purpose of the hazard assessment is to obtain a better understanding of the risks that may be associated with the site. This information will be gained usually through additional desk-based research and exploratory investigation of ground conditions. The findings are used to refine the view of the risk profile of the site and hence the conceptual model of pollutant linkages developed during hazard identification.

In the context of risks, the designer should seek to confirm the following:

● whether contaminants or conditions are present and, if so, their characteristics,
● whether the characteristics of the site or its setting are such that various factors (risk factors) that may enhance or trigger risks to receptors are likely to apply,
● the nature of any existing structures (eg in the case of buildings 'as-built' foundation design and materials used in construction) or planned structures (eg houses, offices, bridges, sewers, roads, etc.) and any other works planned for the site (eg landscaping).

Exploratory investigation

Exploratory investigations should concentrate on suspect substances, materials or areas of the site. The investigation may include surface sampling and the use of boreholes, trial pits and similar intrusive techniques. In most cases, targeted sampling is likely to predominate although systematic sampling may be appropriate for some areas of a site (eg where there is no obvious basis for targeting sampling locations).

Risk estimation

This part of risk assessment is concerned with estimating the probability that adverse effects on specified receptors will occur under defined conditions. There are two main activities.

● Carrying out a sufficiently detailed main investigation to support risk estimation with the required level of confidence. The conceptual model developed in previous stages should inform the design of the site investigation.
● Estimating risks using appropriate assessment criteria.

Risk estimation is usually supported by detailed site investigation data on the characteristics of the site and its setting.

Intrusive ground investigation

The overall aim of the site investigation at this stage is usually to obtain more representative data of site conditions than is likely to have been obtained during exploratory investigation.

The ground investigation should build on the exploratory work, collecting information on hazards (eg the type, concentration and distribution, laterally and vertically, of contaminants) as well as on risk factors. Such risk factors may relate to properties of the site and its setting (eg permeability of the ground, depth and direction of flow of groundwater) or to the characteristics of existing or proposed buildings, materials and services.

Assessors should follow good practice for site investigation as far as characterising hazards and site conditions are concerned. The intrusive ground investigation is an integral part of the site investigation and risk assessment process. Reliable estimation and evaluation of risks requires that data are obtained of the right type and of sufficient quality and quantity.

As soil is only one of the sources of contaminant exposure, and its effect, the cost of dealing with it needs to be kept in proportion with the total exposure to contaminants from all sources (CLEA Report CLR9). The intrusive ground investigation may be undertaken in a phased approach as follows:

Defra and Environment Agency. *Contaminants in soil: collation of toxicological data and intake values for humans.* CLEA Report CLR9. London, The Stationery Office, 2001

- limited (ad-hoc) sampling of soils, possibly during the site reconnaissance work,
- preliminary intrusive sampling involving targeted areas of the site,
- full intrusive investigation.

Soil contamination is usually heterogeneous and may be present in discontinuous lenses in various fill horizons across a site. Therefore, comprehensive fieldwork and laboratory analysis to allow fully validated and accountable risk analysis are equally important for the quality of the investigation.

The scope of the soil investigation required for brownfield land may include the following items (see Site Investigation Steering Group's series of books for more guidance).

- *Site investigation boreholes and/or trial pits* are almost always used in the intrusive investigation. Boreholes can be drilled by various means to significant depths into the ground, and enable sampling of ground and groundwater at different horizons. There is usually a range of borehole diameters available and the depths and methods of construction and choice of permanent lining and instrumentation materials may vary significantly from contractor to contractor. Trial pits are useful for depths of 4–5 m and can be extended into trenches, if necessary. The trial pit allows the make-up of the ground to be seen.
- *Monitoring and sampling boreholes* are used for on-going work involving assessment of the groundwater and soil.
- *Gas boreholes* are used to test for gas concentrations and flows in the ground and often involve long-term monitoring on a site.

Methods of ground investigation are set out in more detail in Tables 2.2–2.4.

It is essential to determine as early as possible, the specific requirements for each project (Site Investigation Steering Group 1993). Many sites could use trial pits rather than boreholes if the contamination is considered to be in the top 4 metres or so of the surface. However, these issues need to be specified based on the specific site needs.

The main purpose of the ground investigation is to reduce and control ground-related risks in the redevelopment of the site. Information from a properly conducted investigation will provide the basis for economic and safe design, and for meeting tender and construction requirements. It will also save money by avoiding the following problems (see BRE's Report *Brownfield sites: ground-related risks for buildings* for more guidance):

- redesign during construction on the site,
- over-conservative design of the land remediation processes and the protection necessary for buildings and other structures,
- delays and litigation costs due to unforeseen ground conditions,
- foundation problems post-construction over the lifetime of the building.

For geotechnical properties, the ground investigations classify the soils beneath the site into broad groups; each group contains soils with similar engineering behaviour. The simplest classification used by geotechnical engineers divides the ground into five categories:

- rock,
- granular soil (eg sands or gravels),
- cohesive soil (eg clays),
- organic soil (eg peats),
- fill or made ground.

The groundwater conditions should be assessed as they are significant in that they can affect construction in a number of ways as listed below.

- A high groundwater table can lead to extra costs and make construction more difficult, eg due to the need for increased support and dewatering of foundation excavations.
- The presence of chemicals can damage foundations (see BRE's Report *Performance of building materials in contaminated land* and the Environment Agency's Literature review 2000).

Site Investigation Steering Group (SISG). *Site investigation in construction (4 Volumes).* London, Thomas Telford, 1993

Charles JA, Chown RC, Watts KS & Fordyce G. *Brownfield sites: ground-related risks for buildings.* BR 447. Bracknell, IHS BRE Press, 2002

Table 2.2 Ground investigation techniques: Drilling techniques for visual inspection and description or soil sampling

Technique	Primary operation	Secondary operation	Strengths	Weaknesses	References
Hand augering	Obtain small undisturbed and disturbed samples	Provide access to soil at depth for simple in-situ tests	Light portable equipment Quick to perform	Limited diameter and depth Difficult to progress in stiff or gravelly soils or hard strata	BS 5930 Digest 411
Dynamic (window) sampling	Obtain disturbed samples		Lightweight equipment Rapid operation Relatively cheap	Limited depth (< ~10 m) Limited access in gravelly soils or hard strata	BS 5930
Dynamic sampling	Obtain disturbed samples		Lightweight equipment Rapid operation Good for limited access Continuous soil profiles Good for contaminated soils Cheap	Limited depth Limited access in hard strata	
Trial pits	Visually inspect near surface soil profile	Large disturbed and undisturbed samples Small in-situ tests	Full visual description of soil profile Assessment of 'diggability'	Limited depth (< ~4.5 m) Needs supports if >1.2 m depth for man entry Large volume and area disturbed needing backfilling Difficult in very soft clays and low density sands Unstable below water table	BS 5930 Digest 381
Cable percussion drilling	Obtain disturbed and undisturbed soil samples	Provides access for in-situ tests and instruments	Allows great depth (< ~75 m) Can cope with wide variety of ground conditions including chiselling through obstructions	Cannot drill through rocks Needs space for the plant to work	BS 5930 Digest 411
Mechanical augering	Obtain disturbed and undisturbed soil samples	Provides access for in-situ tests and instruments	Allows greater depths (< ~25 m)	Disturbed sample depth uncertain Difficult in hard strata	BS 5930 Digest 411
Rotary coring	Disturbed and undisturbed cores of soil or rock	Provides access for in-situ tests	High quality samples possible Deep holes possible	Relatively expensive Relatively large space requirements	BS 5930

British Standards Institution. BS 5930: 1999 *Code of practice for site investigations*

BRE. *Site investigation for low-rise building: direct investigations. Digest 411.* 1190

BRE. *Site investigation for low-rise building: trial pits. Digest 381.* 1989

Table 2.3 Ground investigation techniques: in-situ testing for soil profiling, variability and parameter assessment

Technique	Primary operation	Secondary operation	Strengths	Weaknesses	References
Small in-situ tests	Density Pocket penetrometer Hand vane		Useful guides		
SPT	Soil parameters correlated against results	Obtain disturbed samples	Large body of experience of results in UK	Operator dependency Variability of results Empirical correlations only Spacing of tests > 0.5m depth	
In-situ vane	Soil strength			Needs access for depth	
Cone penetration test (CPT)	Soil profiling and variability	Soil parameters correlated against results Direct design from results	Quick, Immediate results Near-continuous data Detects thin layers (< 100 mm) Measures strength and stiffness Special cones measure conductivity, temperature, seismic velocities Can detect contaminants	Sophisticated equipment High level technical support Results need validation for each site Penetration limited by hard layers, or skin Friction on rods	BS 1377-9 BS 5930
Dynamic probing (DP)	Soil profiling and variability	Soil parameters correlated Void detection	Quick Relatively cheap	Soil cannot be identified Cobbles and boulders give rise to unreliable results	BS 1377-9 BS 5930
Geophysics (many techniques available)	Geological assessment and soil parameters	Small strain	Generally non-intrusive Tests mass ground properties Small strain properties increasingly important in design	Choice of correct technique is essential Requires skilled interpretation	BS 5930 CIRIA 562
Pressure meter tests	Soil parameters	Direct design from results	Theoretically sound Full stress–strain curve for soil possible Tests larger volume of soil than other in-situ tests Can give high quality test data	Little experience in the UK Complicated procedures Requires high level of expertise Time-consuming Can be expensive Some devices fragile	BS 5930
Flat (Marchetti) dilatometer (DMT)	Soil parameters correlated against results	Direct design from results	Simple and robust Data every 200 mm	Sometimes needs pre-bored hole Correlations can be site-specific	
Loading tests	Soil parameters		Measures strength and stiffness Tests large volume of ground Plate load tests can give high quality data	Requires reaction to applied load Generally applicable only above water table Costly	BS 1377-9

British Standards Institution. BS 5930: 1999 *Code of practice for site investigations*. BS 1377-9: 1990 *Methods of test for soils for civil engineering purposes. In-situ tests*

McDowell PW et al. *Geophysics in engineering investigations.* CIRIA Report C562. London, CIRIA, 2002

Table 2.4 Ground investigation techniques: Common laboratory tests for physical properties

Technique	Primary operation	Secondary operation	Strengths	Weaknesses	References
Classification tests	Index tests Moisture content Density Grading Compaction	Soil description Simple tests, disturbed samples	Important guides to expected behaviour and appropriateness of laboratory tests and design method Compaction tests are also used as earthworks control.		BS 1377-2 BS 1377-4
Oedometer	Compressibility, collapse and swelling		Consolidation and other volume-change properties: used to predict settlements		BS 1377-5
	Permeability		Flow characteristics	Sample disturbance may be very significant	
Triaxial/shear box	Strength and stiffness		Undrained or drained strength parameters	May be complex and time-consuming but gives parameters for numerical analysis	BS 1377-7 and -8
	Permeability		Flow characteristics	Sample disturbance may be very significant	BS 1377-6
Laboratory vane	Shear strength		Undrained shear strength	Soft clay only	BS 1377-7

British Standards Institution. BS 1377: 1990 *Methods of test for soils for civil engineering purposes*
  Part 2: *Classification tests*
  Part 4: *Compaction-related tests*
  Part 5: *Compressibility, permeability and durability tests*
  Part 6: *Consolidation and permeability tests in hydraulic cells and with pore pressure measurement*
  Part 7: *Shear strength tests (total stress)*
  Part 8: *Shear strength tests (effective stress)*

- A high groundwater table implies that pore water pressure in the soil is high, and the soil will be correspondingly weaker; high pore water pressure can adversely affect the stability of slopes and pressures on retaining structures.

Where the potential for contamination requires an intrusive investigation, this should be carried out according to BS 10175. Where this is combined with a geotechnical investigation, the AGS *Guidelines* should be followed. On completion of the intrusive ground investigation, the assessor should be in a position to decide whether or not any of the pollutant linkages identified as potential hazards actually do exist and whether or not any could be described as 'significant'.

In relation to risks to human health or the environment, risk estimation often involves comparing the concentration of a hazardous substance in site soils or other media with numeric assessment criteria that are indicative of 'acceptable' concentrations of the substances in soils or media. Criteria may be:

- *generic*, ie derived by an authoritative body (such as a Government department) specifically for the purpose, or
- *site-specific*, ie derived by an assessor using an authoritative model and input values that reflect site-specific conditions.

Where observed or measured concentrations of the substances exceed the criteria, it can be assumed that there is sufficient evidence of the presence of

Paul V. *Performance of building materials in contaminated land.* BR 255. Bracknell, IHS BRE Press, 1994

Garvin S, Hartless R, Smith M, Manchester S & Tedd P. *Risks of contaminated land to buildings, building materials and services; a literature review.* Environment Agency 2000. Available as a pdf download from www.environment-agency.gov.uk

British Standards Institution. BS 10175: 2001 *Investigation of potentially contaminated sites. Code of practice*

Association of Geotechnical Specialists (AGS). *Guidelines for combined geoenvironmental and geotechnical investigation.* AGS, 2005

unacceptable risks to justify either remedial action or further work to refine risk estimates. In 2002 the CLEA soil values were published by the Environment Agency and Defra and this approach should be used where possible.

Risk evaluation

The purpose of risk evaluation is to consider all the risk-based information relating to the site to decide whether or not unacceptable risks exist and therefore should consideration be given to the use of remedial measures?

The relevant activities are as follows:

- collate and review all the available risk-based information,
- consider the sources and effects of uncertainty,
- decide, in the light of all the above, whether or not unacceptable risks exist,
- decide whether or not remedial action should be taken, given the broad costs and benefits of such action.

Where there is clear evidence that unacceptable risks exist, a decision to take remedial action is relatively straightforward. Where the situation is marginal or there is considerable uncertainty in risk assessment findings, it may be more difficult to make defensible decisions. In this case, it is possible to either carry out more detailed work in an attempt to reduce the level of uncertainty, or decide to take a precautionary approach by recommending that remedial works are carried out.

Risk management

Risk management is required only when it is demonstrated that unacceptable risks exist to defined receptors and action is needed to reduce or control those risks to acceptable levels. Usually, risk management does not begin until the full process of risk assessment has been completed. However, in short-term risk situations assessors may have to consider taking remedial action in advance of

Defra and Environment Agency. *Contaminants in soil: collation of toxicological data and intake values for humans.* CLEA Report CLR9. London, The Stationery Office, 2002

completion of the risk assessment. In some cases, for example where it is obvious even at the earliest stage of risk assessment that remedial action is required, it is possible to proceed directly to risk management rather than use resources to show comprehensively that unacceptable risks exist.

If unacceptable risks to the receptors exist, these need to be managed through appropriate remedial measures. The risk management objectives are defined by the need to break the pollutant linkages. Other objectives also need to be considered such as timescale, cost, remedial works, planning constraints and sustainability. There are three generic types of remedial measure that are briefly described below and discussed more fully later in this chapter in the section on *Remedial treatment for contaminated land*.

- *Removal.* This is often referred to as 'dig and dump' in which the defined area or volume of soil containing the hazardous contamination is removed from the site to a licensed landfill site. This method of remediating sites while still common is increasingly expensive as a result of the landfill directive and the increasing cost of remediating sites. The pollutant linkage is broken by removing the source of the contamination.
- *Containment.* This involves keeping contamination on the site, but containing it with cover systems or vertical in-ground barriers. The source of the contamination still exists, but the pollutant linkage is broken by placing a barrier in the pathway.
- *Treatment.* There is a variety of treatment options for a site: chemical, physical or biologically based treatments that can be used to destroy or change contaminants. Treatment breaks the pollutant linkage by changing or destroying the source of the contamination.

The selection of appropriate remedial measures is a specialist skill that requires the full knowledge of

the risk assessment and determination of the right balance between technical approaches and the costs and benefits of a form of remediation.

In reality, the remediation of a particular site may involve a mix of two or more types of site clean-up. The risks can also be managed by altering the site plan; for example, containing contaminated material in a landscaped area or underneath a car park, while placing the buildings on a cleared area of the site.

Generally, the risk assessment and risk management process will include recommendations on whether or not the proposed form of foundations and construction will be appropriate to the site conditions, and any precautions that should be taken. These recommendations should be reviewed against the proposed design and the implications reviewed with the client. Occasionally, specific problems will have been identified that may make the proposed development of the site non-viable or excessively expensive.

The site investigation report should contain information other than just contamination risk assessment; for example, the geotechnical properties of the soil. This information needs to be taken fully into account in designing the foundation system for the building. Potential risks from sulfate-bearing soils are widely recognised (see BRE's *Special Digest 1*) and specialist guidance should be consulted. The risks to buildings and building materials should be an integral part of the risk assessment and risk management process.

Site preparation

Ground treatment

In circumstances where the provision of deep foundations is prohibitively expensive (for example because poor ground extends to some depth) or where the risk of overall or differential settlement is unacceptably large, some form of ground treatment may be appropriate. Such ground improvements can be part of the risk management strategy.

BRE. *Concrete in aggressive ground.* Special Digest 1. 3rd edition. Bracknell, IHS BRE Press, 2005

Ground treatment techniques are intended to:
● improve the engineering properties of the ground,
● reduce the variability in engineering properties.

In some cases they also act as a form of site investigation to indicate variability of ground condition across the site.

On a potentially contaminated site, the use of ground treatment techniques that help mitigate the hazards posed by contaminants should be encouraged. Some forms of ground treatment have a generally beneficial effect with regard to several hazards, whereas other methods may help with one hazard but introduce other potential problems (see BRE's report *Brownfield sites: an integrated ground engineering strategy*).

Fill or made ground may include remnants from opencast mining, slags, ash, industrial and chemical wastes, and hydraulic fill, where slurry is pumped into cavities and the water is allowed to drain away, leaving solid matter as residue. It is relatively weak in terms of strength and stability, and may also include contaminants, so ground treatment may be needed to tackle several concurrent problems. BS 8004 states that:

'all made ground should be treated as suspect because of the likelihood of extreme variability'.

Natural ground which may require improvement includes the following:
● clay sites with slopes in excess of 1 in 10,
● soft soils or areas where long-term consolidation can be expected (eg peat soils),
● ground near ponds or watercourses (both water-logging and variability may be problems),
● areas around holes in the ground (eg swallow holes),
● areas liable to flooding,
● areas where sulfates or other natural contaminants are relatively concentrated (typically clay soils, but also soils overlaying certain rocks where radon may concentrate).

Skinner HD, Charles JA & Tedd P. *Brownfield sites: An integrated ground engineering strategy.* BR 485. Bracknell, IHS BRE Press, 2005
British Standards Institution. BS 8004: 1986 *Code of practice for foundations*

Figure 2.2 Dynamic compaction using vibro stone columns

Physical ground treatment

Treatment techniques can be distinguished by whether the properties of the soil are being modified (M), the treatment is reinforcing the soil (R) or a combination of both (MR). In each case, the requirements for performance of the treated ground must be considered before deciding on the appropriate treatment method for the soil types on site. Ground treatment may reduce subsequent movement, particularly differential movement; it cannot eliminate movement and, therefore, may not be an adequate solution for buildings that are particularly sensitive.

The information required to design an appropriate ground treatment will include the type of development being considered and its likely layout, as well as estimates of the following:

- required load-carrying capacity,
- limits on settlement and differential settlement,
- requirements for long-term performance.

Subsequent foundation design will need to be carried out using an understanding of the mechanics of the treatment process and the short- and long-term properties of the treated ground.

A brief review of the main ground treatment techniques and their applicability is found in Table 2.5. Specialist advice should be sought as to which technique is most appropriate in any given situation.

Before laying the foundations it is increasingly common to undertake ground improvement where site investigation has indicated the ground has poor load-carrying properties or where significant differential movements could occur over the area of a building. This is a specialist area, where a suitably experienced geotechnical engineer should design the works, but generally the methods listed in Table 2.5 are employed.

A significant proportion of ground treatment is carried out to improve the properties of made ground or fill. BRE's report *Building on fill: geotechnical aspects*, covers the application of these techniques to filled ground.

The effectiveness of each technique should be confirmed by an appropriate mix of the following:
- quality testing,
- testing of soil properties that have been modified by the process,
- performance (usually load) testing.

Basic specifications for ground treatment have been issued by the Institution of Civil Engineers Geotechnical Group, while the BRE reports *Specifying vibro stone columns* and *Specifying dynamic compaction* give more comprehensive specifications. See also chapter 4, *Foundations and basements*, for further guidance.

Remedial treatment for contaminated land

It is crucial to rectify a contaminated site with appropriate remedial action in accordance with site-specific target levels for contamination. Remediation options may include, but are not limited to, direct physical actions, such as treatment, removal, or destruction of contaminants, or other on-site risk management solutions, such as capping or containment of contaminants.

The remediation of contamination in the ground can be thought of in three contexts:
- the removal of contaminated soils to a licensed landfill site,
- containment of the contamination on site,
- processing of the contamination to destroy or render harmless the contamination.

Local planning authorities are responsible for ensuring that land contamination is dealt with through the planning system and that remediation takes place where it is required, while developers are responsible for carrying out the remediation and satisfying the local authority that necessary remediation has been carried out.

Remediation through landfill

The removal of contaminated soils to a licensed landfill site is commonly known as 'dig and dump'. This form of remediation has been and continues to be used more than other means of remediation in the UK for contaminated ground. It ensures that the problems of contamination and possibly poor engineering quality of the ground can be removed once and for all. The removed soils may need to be replaced by 'clean' soils depending on the land remodelling being carried out on the site during development.

BRE
Charles JA & Watts KS. *Building on fill: geotechnical aspects.* 2nd edition. BR 424. 2001
BRE. *Specifying vibro stone columns.* BR 391. 2000
BRE. *Specifying dynamic compaction.* BR 458. 2003

Institution of Civil Engineers Geotechnical Group. *Specification for ground treatment.* London, Thomas Telford, 1987

Table 2.5 Some commonly used remediation, treatment and containment techniques

Technique	Description	Objective	Application	Limitation
Preloading	Ground temporarily loaded by surcharge or fill	Consolidate and strengthen the ground	Loose fill; soft clay soils may need installation of vertical drains	Effective only to limited depth
Dynamic compaction	Heavy weight (typically 8 tonnes) repeatedly dropped from crane	Compact and strengthen the ground	Loose fill	Effective only to limited depth
Rapid impact compaction	Blows rapidly applied to articulating compacting foot using modified piling hammer	Compact and strengthen the ground	Loose fill	Effective only to limited depth
Vibrated stone columns	Depth vibrator forms cylindrical void and gravel or crushed rock is compacted into void	Reinforce ground with compacted columns of stone; may also densify ground	Loose fill; soft clay	Could facilitate entry of surface water into ground
Vibrated concrete columns	Concrete columns installed using similar equipment to that used for stone columns	Transfer foundation loads to stronger ground at depth	Most types of poor ground	Practicable where poor ground is of limited depth
Cover layers	Single or multiple layers of engineered fill and geosynthetic materials placed on ground surface	Reduce movement of contaminant	Forms element of a containment system	Surface clay layer may dry out, greatly increasing permeability
Vertical containment barriers	Formed by sheet pile wall, grouting, soil mixing or slurry trench wall	Reduce movement of contaminant by encapsulation	Forms an element of a containment system	Difficult to achieve very low permeability
Deep soil mixing	Stabilised soil columns formed using a mixing tool; binder may be dry or a slurry	Reinforce ground with columns of treated soil	Active stabilisers such as lime and cement produce a chemical reaction with the soil	Gain in strength of soil columns may be slow
Stabilisation/ solidification	Binding agent introduced into ground either by soil mixing augers or pressure injection	Immobilise contaminants by solidifying or binding into a matrix resistant to leaching	Heavy metals and some organic contaminants	Effectiveness highly dependent on physical soil properties
Grouting	Pumpable materials injected into ground (compensation, compaction, jet, infill and permeation grouting)	Increase strength, stiffness and density, reduce permeability, infill cavities	Very wide	Difficult to control where grout goes
Excavation and refilling	Ground is excavated and suitable material placed in thin layers with adequate compaction	Unsuitable material is replaced with engineered fill	Very wide	Difficulty and cost of removal of contaminated soil

Cont'd....

Table 2.5 Some commonly used remediation, treatment and containment techniques (cont'd)

Technique	Description	Objective	Application	Limitation
Leaching and washing	Contaminants removed in solution by leaching, and in suspension or an emulsion by washing	Remove contaminants that are carried or mobilised by liquids	Wide range of contaminants	Only applicable above water table
Bioremediation	Suitable bacteria or fungi introduced into ground for in-situ treatment; ex-situ methods employ treatment beds or bioreactor systems	Transform, destroy or fix organic contaminants by using natural processes of micro-organisms	Organic contaminants	Cannot degrade inorganic contaminants
Soil vapour extraction/air sparging	Gaseous contaminants removed in vapour phase; contaminated fluid then treated	Remove contaminants that can be carried by air	Volatile and semi-volatile contaminants	May be limited to sandy soils
Groundwater treatment	Ex-situ treatment involves extraction of groundwater and treatment above ground	Remove contaminants from groundwater	Contaminated groundwater	Dependent on hydrological and geological conditions
Thermal processes	Heat to 150 °C to volatise water and organic contaminants; vitrification needs higher temperatures	Remove or immobilise contaminants by heating	Vitrification has been used on VOCs, PCBs and heavy metals	Dewatering may be necessary
Chemical treatment	Chemicals introduced into ground via trenches, or pumped into wells for deeper treatment	Destroy/detoxify contaminants	Pesticides, solvents and fuels	May affect flora and fauna

Dig and dump can be a way of overcoming the risks from poor quality site investigations as it aims to remove the problem from the site. Inadequate site investigations may not have detected the full range of contamination on the site, but this may be rendered academic if a sufficient depth and area of contaminated soil is removed.

There are significant disincentives to using dig and dump as a means of remediation. The Landfill Directive and related legislation has resulted in increased costs for dig and dump as a lesser number of sites can now be used for landfill for contaminated soils.

The use of dig and dump should also not be an excuse for poor quality in either the site investigation and risk assessment stage, or during the remediation work. The volume of contaminated soil to be dug out and dumped needs to be carefully assessed during the risk management plan and careful control is required on site to ensure that the correct area and depth of soil is removed. Failure to achieve the proper preparation of the site or removal of soil will leave risks to users or uses of the site.

Containment

The term 'containment' is used for site remediation that does not involve the removal or change of the pollutant impacting on the site, but isolates the pollutant from the end-user, thus avoiding any potential exposure to the pollutant.

A variety of techniques is used to contain contamination in-situ on site. The intention is to restrict the potential for movement through the ground and thus protect potential users and uses of the site. Containment is the second most common form of site remediation for brownfield sites.

The most common form of containment is the use of cover systems. They are a relatively simple form of containment. The use of cover systems has developed over the years; they can range from simple systems to systems with multiple layers. Guidance documents and advice on cover system design has been available for a number of years, but recently BRE and others have produced state-of-the-art guidance on cover systems that includes design and construction guidance (eg BRE's booklet and CD ROM *Cover systems for land regeneration*).

This guidance should be used by those involved in designing, specifying and constructing cover systems and therefore the proper use of the guidance will be an indicator of quality in the remediation work.

The use of slurry cut-off walls as vertical containment from ground contamination is common, the technology was originally used for landfill sites to prevent leachate and gas migration. In the UK, a self-setting slurry of cement-bentonite is poured into a trench during continuous excavation. The depth of the slurry wall can be anything from 5–50 m. A high-density polyethylene (HDPE) sheet can be inserted into the trench before setting takes place where gas contamination of highly mobile groundwater is a problem.

Guidance on slurry cut-off walls has been available for a number of years (eg BRE's *Digest 395*). One of the essential issues concerned with slurry walls is to ensure that the material is consistent and reaches required values of permeability, strength and modulus of elasticity. Of these, the permeability is probably the most important aspect in restricting leachate migration.

In general, the design, construction and materials used for slurry walls are well known and understood and quality control measures have been well documented by the specialist contractors in this area. At the same time, there is still a need to ensure that the site investigation and risk assessment properly inform the selection of a slurry wall.

BRE
Cover systems for land regeneration: Thickness of cover systems for contaminated land. BR 465 (booklet and CD ROM). 2004
Slurry trench cut-off walls to contain contamination. Digest 395. 1994

Treatment in remediation

Process-based remediation has seen significant growth over the past 20 years as technologies have developed and contractors have become experienced in their use. The increased cost for landfill is pushing those involved in land remediation towards processes as they become more financially viable.

The major drawback of all processes is that they tend to deal with one type or form of contamination. Brownfield sites in the UK tend to contain heterogeneous contamination and therefore there is often a restriction on using some forms of process-based remediation. Examples such as bioremediation may only deal with certain types of organic contamination, but will not impact on inorganic contamination.

Process-based remediation possibly has the greatest risks of all the forms of remediation and all require significant quality measures. However, it is difficult to set quality measures that can be applied to all process-based remediation. Suppliers or contractors of technologies should therefore develop their own approach to quality.

Ground gases

Methane, carbon dioxide and other landfill gases

Most methane hazard comes from biodegradation of vegetable material within landfill, but gases may also be found in excavations in, for example, coal-mining areas. Methane is flammable, and carbon dioxide can asphyxiate at high concentrations in the air, so if either enters a building it can pose a danger to both health and safety. Landfill gas can enter buildings through gaps around service pipes, cracks in walls below ground and floor slabs, construction joints and cavities (Figure 2.3). It can also accumulate in voids and in confined spaces within buildings (eg cupboards and subfloor voids). Preventative measures include low-permeability membranes and high-permeability layers which allow gas to be extracted in a controlled manner.

In order to design counter-methane measures, it is necessary to know the concentrations of gases and have an estimate of the rate of emission from a site. This is usually done by monitoring specially constructed boreholes over a long period. Counter-methane measures may be required to prevent migration within or outside a site as well as into buildings. All counter-methane measures need to take account of the inevitable settlement of a site over time. Prevention of entry of gases to a building normally requires a combination of a gastight seal,

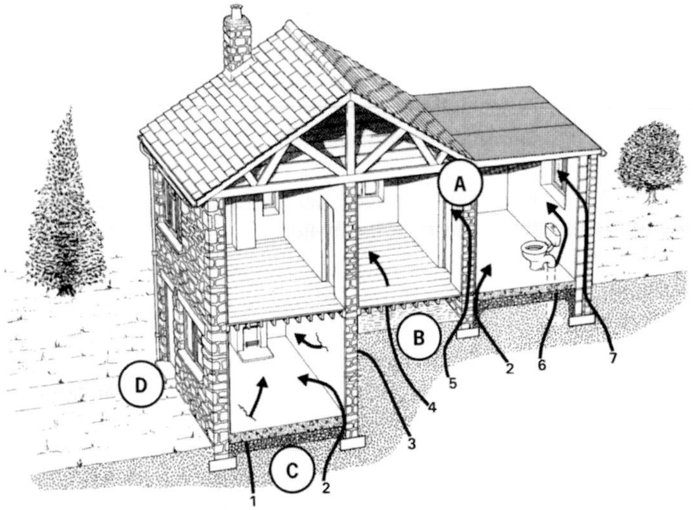

Key to ingress routes:
1 Through cracks in solid floors
2 Through construction joints
3 Through cracks in walls below ground level
4 Through gaps in suspended floors
5 Through cracks in walls
6 Through gaps around service pipes
7 Through cavities in walls

Possible locations for gas accumulation:
A Wall cavities and roof voids
B Beneath suspended floors
C Within voids caused by settlement or subsidence
D Drains and soakaways

Figure 2.3 Gas entry routes into buildings

under-building ventilation and attention to detailed design around services and foundations.

Although rare, there may be cases where the building being constructed is located on or adjacent to a landfill site or old coalfield affected by both radon and methane. In such cases, additional precautions may be needed to deal with methane. Appropriate measures for dealing with methane are described in the earlier section on *Remedial treatment for contaminated land*. The measures described exceed those required for radon, so where both methane and radon are present, methane-protective measures should be applied.

Radon

Radon is a colourless, odourless radioactive gas. It is formed where uranium and radium are present and can move through cracks and fissures in the subsoil, and so into the atmosphere or into spaces under and in buildings. Where it occurs in high concentrations it can pose a risk to health. While it is recognised that the air inside every building contains radon, some built in certain defined areas of the country might have unacceptably high concentrations unless precautions are taken.

To assist in identifying those areas of the country where radon is of concern, a series of risk potential maps have been prepared by the Health Protection Agency and the British Geological Survey.

To find out whether or not a site is within an area where radon-protective measures are required, the local building control authority can be contacted or a copy of the relevant BRE Report for the geographical area can be consulted (see chapter 13, *References and further reading*).

Requirements for England and Wales

The current requirement was introduced in 1999. Protection measures are described as being 'Basic' or 'Full'. Basic protection is applied in the lower risk areas and requires the provision of a radon barrier across the footprint of the building. Full protection, which is applied in higher risk areas, requires the provision of a radon barrier across the footprint of the building together with either a ventilated subfloor void or a radon sump.

The areas of England where measures are required include most of Cornwall, Devon and parts of Somerset, the Peak District in Derbyshire, an area running south from Lincoln through Lincolnshire, Rutland, Northamptonshire, North Oxfordshire and into Gloucestershire, as well as parts of the Yorkshire Dales, the Welsh Borders, the Lake District, and Northumberland, and a few scattered areas in South East England.

In Wales, the areas principally affected include most of Pembrokeshire, Ceredigion, Powys, Conwy, Denbighshire, Monmouthshire and the Vale of Glamorgan, as well as parts of Flintshire, Wrexham, Carmarthenshire, Anglesey and Gwynedd, and a few other isolated areas.

Requirements for Scotland

Radon-affected areas in Scotland have been designated by testing dwellings. Where tests on existing dwellings show that 1% of the dwellings in that area are likely to have a radon concentration above 200 Bq/m^3 (the action level), the area is designated as a radon-affected area. The areas affected are essentially the area around Deeside inland from Aberdeen in the Grampian Region and the area around Helmsdale in the Highland Region.

The Technical Handbooks to the Scottish Building Standards (Section 3) show maps of these radon-affected areas. Other localised maps may also show areas of radon risk.

Requirements for Northern Ireland

The current requirements for radon protection in Northern Ireland were introduced in 2001. The levels of protection are described as 'Zone 1' and 'Zone 2' protection. Zone 1 protection is applied in the lower-risk areas and requires the provision of a radon barrier across the footprint of the building. Zone 2 protection is applied in higher-risk areas, and requires the provision of a radon barrier across the footprint of the building, together with either a ventilated subfloor void or a radon sump.

The requirements apply to most of the area to the south west of a line between Londonderry in the west and Portaferry in the east, together with scattered areas around Ballycastle in the north west and Newtonards to the east.

Principles of protection against gas contamination

Gas enters a building primarily by airflow from the underlying ground. There are two main methods of achieving radon protection in new buildings:

- *passive system:* this involves providing a barrier to the gas. This can usually be achieved by increasing the general airtightness of the damp protection provided in floors and walls.
- *active system:* this involves providing natural or mechanical underfloor ventilation, or a powered gas extraction system, as an integral part of the services of the building. Either will incur running and maintenance costs for the life of the building.

Passive systems are to be preferred in new buildings, although they may need to be supplemented in some areas with provision for active protection.

In areas with a significant radon potential, sufficient protection will be provided by a passive system comprising a well-installed damp proof membrane (dpm) modified and extended to form a radon-proof barrier across the full footprint of the building.

New buildings in areas of higher radon potential should incorporate a passive system comprising a radon-proof barrier across the full footprint of the building, supplemented by provision for subfloor depressurisation or ventilation (either a radon sump or a ventilated subfloor void). The latter enables an active system to be introduced later should it prove necessary.

Gas-proof barriers

Generally, a membrane of 300 μm (1200 gauge) polyethylene sheet will be adequate. (It is acknowledged that some diffusion will occur through the sheet. However, as most gas entry is through cracks in the perimeter of the slab, this diffusion can be ignored.) Where there is a risk of puncturing the membrane, reinforced polyethylene sheet should be considered. The barrier can be constructed with other materials that match the airtightness and waterproofing properties offered by 300 μm (1200 gauge) polyethylene. A wide range of suitable materials is available.

With minor changes, standard dpms and damp proof courses (dpcs) can also provide radon protection. The additional detailing requires the dpc where it passes through a cavity wall to be in the form of a cavity tray to prevent gas entering the building through the cavity, and joints in the barrier and gaps around service penetrations to be sealed as well. In the case of stepped or basement construction, retaining walls will also need to be protected.

The gas-proof barrier should be continuous across the whole plan of the building, including taking it through or under internal walls and sealing around column bases. Semi-detached and terraced properties should be considered as single units with regard to the installation of the radon-proof barrier. The barrier will need to continue across party or separating walls where they occur.

If a basement or cellar is to be fully tanked using mastic asphalt or some similarly impervious barrier to prevent damp penetration, the tanking will also provide radon protection. There is no need to provide supplementary protection (eg a sump) in such cases.

Subfloor depressurisation and sumps

Where a ground-supported floor is to be constructed and full gas-protection measures are required, a sump should be provided (Figure 2.4). This would enable subfloor depressurisation to be introduced with relative ease if desired at a later date. (Subfloor depressurisation involves sucking gas-laden air from beneath a building and discharging it harmlessly into the atmosphere.)

Generally, the same rules apply to non-domestic buildings as to dwellings, one sump will have an influence over approximately 250 m^2 of floor area. There are, however, opportunities for economies to be made in terms of the number of sumps and the amount of pipework that needs to be provided. With larger floor areas several sumps can be manifolded together to reduce the amount of pipework and number of fans that might need to be fitted in the future. In cases where the fill beneath a floor is uninterrupted by foundation walls, a single sump is likely to be effective over far more than 250 m^2 of

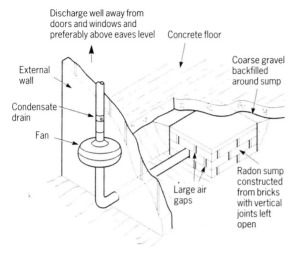

Discharge well away from doors and windows and preferably above eaves level

Concrete floor

External wall

Condensate drain

Fan

Coarse gravel backfilled around sump

Radon sump constructed from bricks with vertical joints left open

Large air gaps

Figure 2.4 An active sump system works by using a fan to reduce the flow of radon into a dwelling. During construction, Building Regulations (England & Wales) require only the sump and extract pipe to be installed. The fan and stack pipe are added later if the building has an elevated radon level

floor area. In such cases, it is likely that one sump will have an effect over an area of 500 m^2.

The end-use of the building will to some extent dictate where to locate the sumps. Areas of high ventilation and low occupancy such as warehousing will probably require less attention than warmer less ventilated areas such as offices that have higher occupancy levels. Warehousing or factory areas are also likely to have more robust floor construction (thicker and better jointed concrete), again reducing the risk in these areas.

In areas where it is known that the water table is particularly high, or the level fluctuates, there is a risk that sumps may become waterlogged and therefore ineffective. In such cases, tanking should be used to prevent water ingress and provide gas protection. There is then no need to provide a sump. It should be noted that, generally, water will act as a screen to radon. However, if the water level fluctuates, the ground pressure will also change, which in turn may drive more gas into the building.

Regulations and standards

Building Regulations

The loading imposed by a building must be transmitted to the ground safely and without resulting in excessive deflection. Ground movements caused by soil volume change, landslip or subsidence must not impair the stability of the building. This is covered in the Building Regulations for each part of the UK:

- **England & Wales:** AD A,
- **Scotland:** Section 1 of the Technical Handbooks, Domestic and Non-Domestic,
- **Northern Ireland:** Technical Booklet D.

These requirements mean that the site investigation must be sufficient to identify all the controlling soil parameters necessary for demonstrably safe design.

A number of more specific requirements are also articulated, relating to fill materials, variations in ground conditions, shrinkable clay soils, and ground instability which must be addressed in the site investigation.

Site preparation and resistance to contaminants, radon and moisture is covered in the Building Regulations for each part of the UK:

- **England & Wales:** AD C, covering small, domestic-scale buildings,
- **Scotland:** Section 3.1–3.4 of the Technical Handbooks, Domestic and Non-Domestic,
- **Northern Ireland:** Technical Booklet C.

In AD C guidance is given on the assessment of contamination that references the key Environment Agency documents.

Other relevant legislation

The Party Wall Act (1996) places responsibilities on a building owner carrying out works to inform adjacent building owners when carrying out excavations near to adjoining buildings and to provide temporary protection where necessary. This will require an understanding of the location, type,

HMSO. *Party Wall etc Act 1996*. Chapter 40. London, The Stationery Office

and depth of neighbouring foundations which should form part of the site investigation.

Construction must be carried out under the Health and Safety at Work Act (1974) and the Construction (Design and Management) Regulations (1994). These place an onus on designers to eliminate or reduce hazards where possible at design stage, and to pass on information on residual risks, such that they can be adequately managed during construction. The site investigation should take into account a need to identify potential hazards during construction and operation. This may include hazards to workers or occupants from soil contamination.

A significant amount of waste material is generated by demolition and groundworks processes, and the site investigation can form the first part of a strategy to minimise waste arisings through the design. This could be achieved by re-using existing foundations or by selection of alternative, spoil-free foundation systems.

The Environmental Protection Act 1990 requires an overall risk-based approach to dealing with contaminated sites. The requirements and associated waste management legislation govern the investigation, classification and remediation of contaminated land. Waste legislation governs movement and disposal of arisings. However, in 2002 the Environment Agency stated that geotechnical treatment of natural uncontaminated soils would not be considered a 'licensable activity' under the Waste Management Licensing Regulations (1994).

The Groundwater Regulations 1998 give the Environment Agency responsibility for enforcing controls over both direct and indirect discharges to groundwater.

Planning Policy

Policy Guidance Note 14 *Development on unstable land*, sets out the broad planning and technical issues relating to development on unstable land. The initial site investigation should identify whether or not hazards associated with unstable land will affect the development.

Policy Guidance Note 25 *Development and flood risk*, in which drainage is a material consideration for planning, may require sustainable drainage systems to be designed as part of the development. For developments on flood plains particularly, it will be important to ensure that the site investigation provides sufficient topographical and surface soils data for this to be carried out effectively.

Other regulations

Where housing developments are proposed, the NHBC or Zurich Building Guarantees requirements may also have to be met. These have specific requirements for the identification of ground-related hazards and site investigations.

The NHBC Standards Chapter 4.1 *Land quality: managing land conditions*, deals with the requirements for site investigation related to geotechnical and contamination issues. The extent of the site investigation is related to an understanding of the hazards present on site. Later sections provide descriptions of the key elements of the investigation and suitable persons for designing and carrying out investigations.

HMSO
Health and Safety at Work etc Act 1974. Chapter 37. London, The Stationery Office
The Construction (Design and Management) Regulations 1994. Statutory Instrument 1994 No 3140. London, The Stationery Office
Environmental Protection Act 1990. London, The Stationery Office
The Waste Management Licensing Regulations 1994. Statutory Instrument 1994 No 1056. London, The Stationery Office

HMSO. *The Groundwater Regulations 1998. Statutory Instrument 1998 No 2746.* London, The Stationery Office
DCLG. Planning Policy Guidance Notes. London, The Stationery Office
   14   *Development on unstable land*
   25   *Development and flood risk*
National House-Building Council. *NHBC Standards.* Amersham, Buckinghamshire, NHBC
Chapter 4.1 Land quality: managing land conditions
Chapter 4.6 Vibratory ground improvement

The NHBC Standards Chapter 4.6 *Vibratory ground improvement*, deals with the specific requirements for investigation, design and construction using vibratory ground improvement techniques. This includes ground considered unacceptable for treatment by these methods. Acceptable ground and foundations that can be accommodated are described.

Although vibratory ground improvement methods are the only prescribed method in the Standards, NHBC do allow the use of other methods. In these cases, acceptance by the NHBC is based on adherence to the Technical Requirements of their Standards, including the requirement that all structural design is carried out by a suitably qualified person in accordance with current British or European Standards. Additionally, there should be confirmation of the suitability of the method, validation testing carried out to confirm design and workmanship and appropriate foundation design.

Standards and codes of practice

Site investigations to determine geotechnical and geoenvironmental properties are covered by Codes of Practice BS 5930 and BS 10175, respectively. Associated testing standards can be found in BS 1377 Parts 1–9.

BS EN 1997-1 and DD ENV 1997-2, and a number of associated test and execution standards will eventually supersede parts of these British Standards, although not all. On publication of the relevant National Annexes, the removal of conflicting national standards will be mandatory after a minimum of a further three years.

Other relevant Standards and Codes of Practice are listed in chapter 13, *References and further reading*.

British Standards Institution

BS 5930: 1999 *Code of practice for site investigations*
BS 10175: 2001 *Investigation of potentially contaminated sites. Code of practice*
BS 1377: 1990 *Methods of test for soils for civil engineering purposes*
    Part 1: *General requirements and sample preparation*
    Part 2: *Classification tests*
    Part 3: *Chemical and electro-chemical tests*
    Part 4: *Compaction-related tests*
    Part 5: *Compressibility, permeability and durability tests*
    Part 6: *Consolidation and permeability tests in hydraulic cells and with pore pressure measurement*
    Part 7: *Shear strength tests (total stress)*
    Part 8: *Shear strength tests (effective stress)*
    Part 9: *In-situ tests*
BS EN 1997-1: 2004 *Eurocode 7. Geotechnical design. General rules*
DD ENV 1997-2: 2000 *Eurocode 7. Geotechnical design. Design assisted by laboratory testing*

3 Site environment and orientation

Building occupants and passers-by should not have to deal with wind-funnelling effects. The building envelope must be capable of withstanding the pressures of wind and snow loads, and prevent water penetration. The combination of the building fabric and the building services must ensure the comfort of occupants within the building. Whether or not natural lighting is available without glare, and whether or not natural ventilation is possible, will have a major impact on the comfort of occupants and on running costs. The site environment will also significantly affect the air quality within the building. It will also affect the rate of deterioration of materials, in particular those comprising the building envelope. The designer needs to consider the local site environment and the orientation of the building from the outset of the design.

Wind

Wind is the movement of air caused by thermal and mechanical stirring of the atmosphere. All buildings and structures are affected to a greater or lesser extent by the wind. Wind is often the dominant structural load on tall slender structures and for external cladding and roofing elements and their supporting members and fixings.

Wind speed varies depending on:
- the geographical location of the site,
- the site's altitude above sea level,
- the direction of the wind, and
- the season of the year.

More local effects such as height of the building above ground level, topography and terrain (Figure 3.1) also affect the speed and direction of the wind at a particular site. In the immediate vicinity of a building, or a group of buildings, the wind changes speed and direction rapidly, depending on the form and scale of the building(s).

The wind fluctuates randomly in both time and space due to turbulence. Turbulence increases with the roughness of the terrain because of frictional effects between the wind and features on the ground (such as buildings and vegetation) and at the same time these frictional effects reduce the mean wind velocity. Wind speeds in coastal regions are generally greater than in inland areas. For instance, the speeds near the coast of southern England are some 10% greater than they are at the same altitude in the centre of southern England. The highest wind speeds occur in the north of the British Isles; the highest effective gust wind speeds used in design are of the order of 56 m/s in the far north west of Scotland.

When the wind blows against a building it tends to be slowed down against the windward face

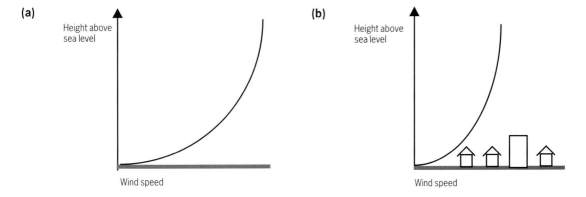

Figure 3.1 Wind speed varying with height: (a) smooth terrain, (b) rough terrain

generating a positive pressure. It will then accelerate up over the roof and around the sides of the building, usually producing negative pressures, ie suctions, especially just downwind of the eaves, ridge and corners. The greater the flow acceleration, the greater will be the suctions (Figure 3.2). In general, the more bluff (less streamlined) the building and the sharper the edges and corners, the higher the wind suctions will be.

On the windward slope of a roof, the pressure is dependent on the pitch. When the pitch angle is below about 30°, the flow separates and the windward slope can be subjected to severe suction. At steeper pitches, the roof presents a sufficient obstruction to the wind for the flow to remain attached and for a positive pressure to be developed on the windward slope. For the purposes of design, the varying pressures at different locations are normalised to give values which are independent of the site or building dimensions.

The highest extreme winds in the UK are expected in December and January. In the summer months of June and July, extreme wind speeds may be expected to be only about 65% of the winter extremes. Tables of seasonal factors are included in BS 6399-2; they can be helpful for planning construction works, particularly while the structure is not complete and therefore may have less resistance to wind gusting.

The flow and resulting pressure distributions around a building are strongly dependent on the

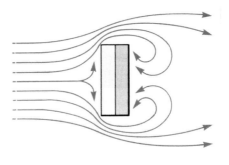

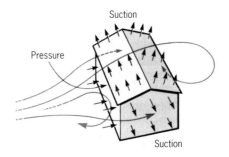

Figure 3.2 Wind flow around a typical building

British Standards Institution. BS 6399-2: 1997 *Loading for buildings. Code of practice for wind loads*

shape of the building. Small changes in the form of a building can have a large impact on the wind pressures. For example, changing the eaves detail of a flat roof from sharp to curved eaves can reduce the maximum suction by about 70%. Roof pitch and roof form can also make a big difference to the wind suctions.

The total wind force on the external envelope of a building depends on the difference in pressure between the outer and inner faces. Openings on the windward side of a building will increase the air pressure inside and this will tend to increase the loading on those parts of the building subjected to external suctions, such as the roof and corners (Figure 3.3). Conversely, openings in areas of the building that are subject to external suction will reduce the pressure inside the building, thus tending to increase the loading on those parts subjected to external (positive) pressure.

Calculating wind loads

Wind loads are calculated in accordance with BS 6399-2. The basic wind speed from Figure 3.4 will be adjusted in four main steps:
- determine the site wind speed,
- determine the site exposure,
- determine the effective (gust) wind speed,
- determine the pressure coefficients,
- determine the wind loads.

The detailed method is given in BS 6399-2.

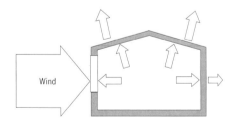

Figure 3.3 Internal pressures due to one dominant opening

British Standards Institution. BS 6399-2: 1997 *Loading for buildings. Code of practice for wind loads*

Other wind design considerations
Fixings

As the earlier diagrams indicate, both positive and negative pressures can result from wind loading, and fixings should be designed appropriately. This will particularly affect fixings of roof coverings and the connection of the roof structure to the walls. More details of specific requirements for resistance to wind loads are given in chapters 6 and 8 on walls and roofs, respectively.

Wind environment

> **Key Point**
> The designer of any new building should consider how wind speeds would affect the environment around the building.

An assessment of how wind speeds at ground level will affect pedestrians and users of the building will often be required, particularly when considering the effect of adjacent tall buildings. In general, the taller the building, the higher the wind speeds at its base. If the design allows these high wind speeds to reach down to ground level then they could create unpleasant or even unsafe conditions. Screens, fences, canopies, hard and soft landscaping, etc. can be used to reduce the impact of winds at low levels.

It is also important to ensure that a sufficient flow of air is maintained through courtyards and similar enclosed areas in order to stop pollutants being trapped and concentrated. Consideration should also be given to maintaining adequate dispersion of exhausts and fumes from plant rooms, kitchens, car parks, etc. and to ensure that these pollutants are not drawn back into the building through air-conditioning plant or air-intake openings or open windows.

Snow

Snowfall in the UK varies from year to year, and with geographical location and altitude. Snow loads tend to increase generally with altitude. Snow density can vary considerably (eg from 0.05 to 0.2 g/cm^3) and density can vary with compaction if snow lies for long periods. Compaction will not,

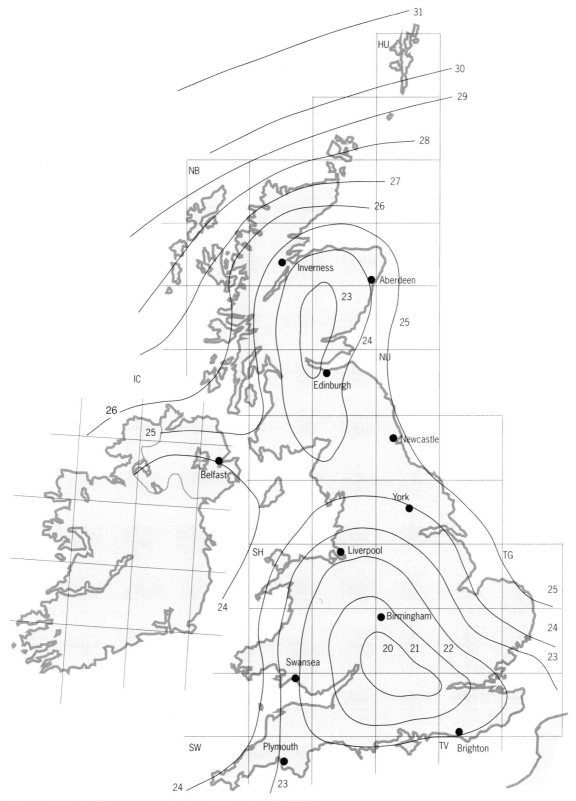

Figure 3.4 Basic UK wind speeds (V_b in m/s). Data from BS 6399-2

however, increase the actual loads on the roof unless rain saturates the unmelted snow.

Extremes of snow loads can arise in two ways: through heavy snowfall, and through drifting in windy conditions. Ground snow measurements exclude the effects of drifting. Snow in the UK rarely lies sufficiently long for successive falls to accumulate, and extremes are therefore most likely to ensue from a single snowfall. Sheet ice can also form in certain circumstances, giving maximum loads of 0.45 kN/m^2.

Snow tends to accumulate at obstructions to form drifts, and asymmetric snow loads can occur on some forms of pitched and curved roofs. On small roofs, where drifting is unlikely to occur, snow loading is taken to be a uniformly distributed load according to geographical location and specified minimum value. BS 6399-3 gives simplified loading conditions for design of small buildings where all the following criteria are complied with:
● no access except for cleaning and maintenance,
● total roof area no larger than 200 m^2, or if the roof is pitched and not wider than 10 m with no parapets,
● no other buildings within 1.5 m distance,
● no abrupt changes in height greater than 1 m (eg from dormers or chimneys) or, if there are such abrupt changes in height, the lower area is to be no greater than 35 m^2,
● no other areas on the roof which could be subject to asymmetric snow loading through, for example, manual or wind snow clearing.

Larger roofs, and those outside the criteria above, are governed by design procedures in BS 6399-3. BRE *Digest* 439 also applies. For the purposes of calculations, roofs up to 30° pitch are assigned the same maximum uniform snow loading as a flat roof. For steeper pitches, the asymmetric snow load case is different. It takes the uniform snow load, asymmetric snow load, local drifting and partial loading due to snow removal into account. In addition, other loads such as uniform distributed loads and access loadings must be considered. The loads are not additive, and local issues, such as parapets causing blocked drainage systems, will cause large localised loads. Snow boards may assist in keeping gutters clear.

Rainfall

> **Key Point**
> Walls and roofs must prevent snow and water reaching the interior of a building.

Within the national building regulations, control of moisture penetration is a functional requirement, and the building needs to be designed and built adequately to resist such penetration.

Two kinds of rain intensity need to be considered:
● vertical rainfall and
● wind-driven rainfall.

Both categories contribute to the total quantities of rainfall requiring disposal, but the second category particularly affects the weathertightness of lapped roof coverings, such as tiles, slates and larger sheets.

Vertical rain
Short spells of high intensity rain may be more significant than slight rainfall over longer periods. Generally, intense rainfalls are produced by convectional storms. These do not have the same 'exposed west, sheltered east' pattern of frontal-type rain. The highest incidence is in southeast England, and the lowest in northwest Scotland. Note, that climate change may affect the frequency and severity of rainstorms and it may therefore be appropriate to design cautiously for this aspect by, for example, upgrading the protection or probability of occurrence.

BS EN 12056-3 describes a method of calculating the adequacy of non-syphonic roof drainage along with performance requirements for syphonic

British Standards Institution. BS 6399-3: 1988 *Loading for buildings. Code of practice for imposed roof loads*
BRE. *Roof loads due to local drifting of snow.* Digest 439. Bracknell, IHS BRE Press, 1999

British Standards Institution. BS EN 12056-3: 2000 *Gravity drainage systems inside buildings. Roof drainage, layout and calculation*

systems. It introduces the concept of risk associated with the rainfall intensity rates measured in litres per second per square metre of roof. The rainfall intensity rate is multiplied by the risk factor to provide the rainfall intensity where statistical rainfall data do not exist. The supporting documents of the national building regulations provide design rainfall intensity figures:

- **England and Wales:** AD H,
- **Scotland:** Section 3 of the Technical Handbooks, Domestic and Non-Domestic,
- **Northern Ireland:** Technical Booklet N.

The risk factors are related to the consequence of an overflow and BS EN 12056-3 provides examples (Table 3.1).

The Meteorological Office can provide detailed information on rates and return periods if the advice in BS EN 12056-3 is insufficient. Figure 3.5 shows the period between rainstorms of 75 mm/hr for two minutes.

Table 3.1 Risk factor when considering rainfall intensity (Based on Table 2 from BS EN 12056-3)

Situation	Risk factor
Eaves gutters	1.0
Eaves gutters where water overflowing would cause particular inconvenience (eg over entrances to public buildings)	1.5
Non-eaves gutters and in all other circumstances where abnormally heavy rain or blockage in the roof drainage system could cause water to spill over into the building	2.0
For non-eaves gutters in buildings where an exceptional degree of protection is necessary, eg: • hospital operating theatres, • critical communications facilities, • storage for substances that give off toxic or flammable fumes when wet, • buildings housing outstanding works of art.	3.0

Driving rain

Design guidance on incidence of driving rain now takes account of the direction from which the wind is blowing. The results can be shown in map form (Figure 3.6).

Most locations in Britain experience one hour every 10 years, on average, in which the total driving rain is 30 mm. More tentatively, there is one hour in every 100 years when 80 mm falls.

For a given roof, local conditions of topography and terrain roughness (as described in the earlier section on *Wind*) will significantly affect the values given in maps. The following factors are included in calculations detailed in BS 8104:

- terrain roughness (from about 1.15 for open coastal areas to about 0.75 for built-up areas),
- topography (from about 1.2 for features likely to produce funnelling of the wind to about 0.8 for sheltered valleys),
- obstruction (to allow for localised affects of nearby obstructions, based on line of sight from the building).

Hailstones

The impact of small hailstones up to 10 mm in diameter does little or no damage to buildings. On rare occasions, hailstones of over 75 mm diameter can fall in severe local storms. These can weigh over 100 g and can smash roof lights and patent glazing, and can dent metal roof coverings. On average, these hailstorms occur somewhere in Britain about every 5 years, but they are very localised and designing for resistance to them would be excessive.

British Standards Institution
BS EN 12056-3: 2000 *Gravity drainage systems inside buildings. Roof drainage, layout and calculation*
BS 8104: 1992 *Code of practice for assessing exposure of walls to wind-driven rain*

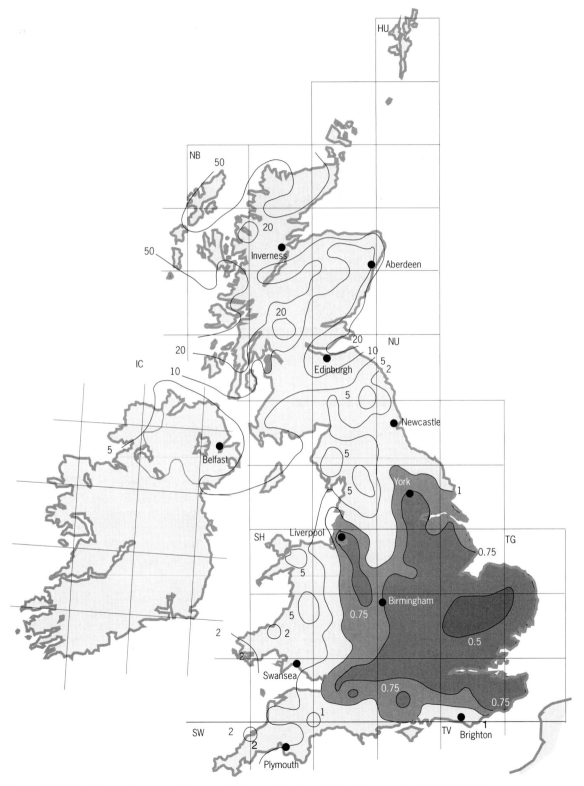

Figure 3.5 Period of years between rainstorms of 75 mm/h for two minutes (from meteorological data)

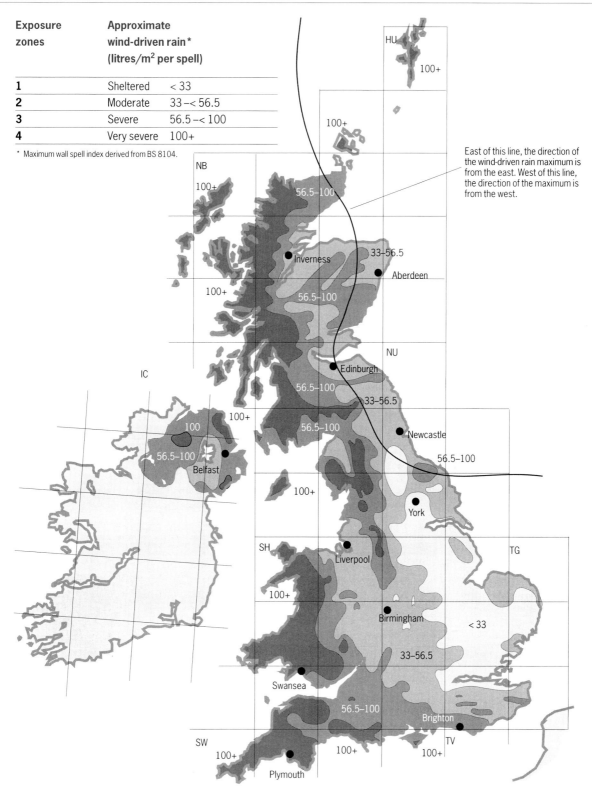

Exposure zones	Approximate wind-driven rain* (litres/m² per spell)	
1	Sheltered	< 33
2	Moderate	33 –< 56.5
3	Severe	56.5 –< 100
4	Very severe	100+

* Maximum wall spell index derived from BS 8104.

East of this line, the direction of the wind-driven rain maximum is from the east. West of this line, the direction of the maximum is from the west.

Figure 3.6 Categories of exposure to wind-driven rain and corresponding spell indices for worst direction at each location (from meteorological data)

Daylight and sunlight

In designing a new building or an extension, care should be taken not to obstruct existing buildings nearby. Both sunlight and sky light can be affected. The BRE report *Site layout planning for daylight and sunlight* and BRE *Information Paper* IP 5/92 give full details and are used by local authorities to help determine planning applications. The report also gives an introduction to Rights to Light; if these apply, the owner of the obstructed building may in some circumstances sue for damages or removal of the obstructing building.

BRE
Littlefair PJ. *Site layout planning for daylight and sunlight: a guide to good practice.* BR 209. 1998
Littlefair P. *Site layout planning for daylight.* Information Paper IP 5/92. 1992

Solar gain, temperature and glare

Typical design considerations might be as follows.
- **Review the need for sunlight and solar gain.**
  Will the occupants welcome or expect sunlight? Sunlight is seen as particularly important in dwellings and other residential buildings like hotels and hospitals, and some leisure buildings. Is the building likely to be heating-dominated or cooling-dominated? In buildings with high internal heat gains, solar gain is less welcome and more shading will be required.
- **Plan the layout of buildings and rooms** to maximise the benefits of sunlight and avoid problems. Where possible, have main façades of buildings facing north and south; this makes shading easier and allows maximum use of winter solar gain. Arrange groups of buildings to avoid mutual overshadowing (Figure 3.7); a large obstruction outside can block useful winter sun while letting through hot summer sun. Consider

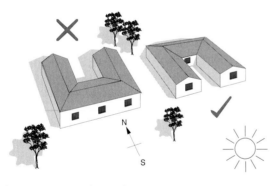

Open courtyard should face south

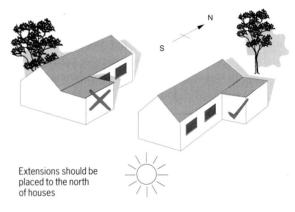

Extensions should be placed to the north of houses

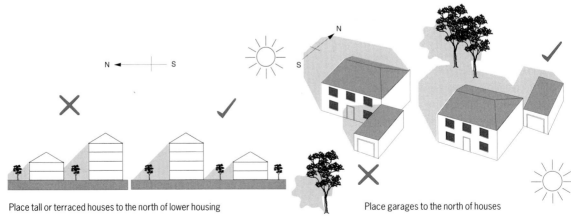

Place tall or terraced houses to the north of lower housing

Place garages to the north of houses

Figure 3.7 Techniques to ensure good sunlighting to buildings and open spaces around them

whether or not sunlight will reach gardens and other spaces around the buildings. Arrange the layout of rooms so that those that would benefit from sunlight face south, while those where sun would not be beneficial face north (Figure 3.8).

● **Choose appropriate window areas** to avoid overheating while allowing enough daylight to reach the interior (see section in chapter 5 on *Daylighting*). Avoid large areas of horizontal glazing, in particular, as this admits most solar gain in summer.

● **Choose appropriate shading devices** to control solar gain and glare. Shading can either be fixed or moveable. Fixed shading may be seasonally dependent in its effects; for example, an overhang may block high summer sun while allowing through low-angle winter sun. Other shading types like solar control glazing have a similar impact all year round, so may reduce useful winter solar gain. Occupants often prefer some moveable shading that they can control, particularly if glare is an issue.

Solar energy and temperatures
Solar radiation affects surface materials in two distinctive ways:
● ultraviolet radiation disrupting specific chemical bonds between materials and
● solar energy on surfaces causing their temperature to rise.

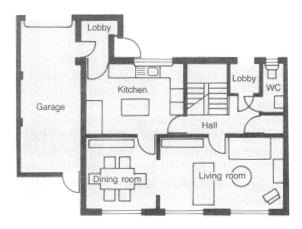

Figure 3.8 Living and dining rooms face south while other rooms, which do not need sunlight as much, face north. (The Linford low energy house, Milton Keynes)

Chemical changes are particularly important for organic materials such as plastics and coatings for sheet metals. Ultraviolet degradation may cause changes in the appearance or light transmission characteristics of organic materials, and ultimately lead to surface erosion and loss of strength. Chemical stabilisers are used to lessen effects such as fading.

Solar energy falling on surfaces causes their temperatures to rise considerably above air temperatures. Because glass is transparent to short-wave solar energy, there can be a build-up of internal temperatures as a consequence.

During design it is important to be able to predict the extreme temperatures which materials or components may reach, so that allowance can be made for the following effects:
● thermal expansion and contraction,
● softening or embrittlement,
● loss of plasticizers from plastics,
● photo-oxidation of plastics.

Light-coloured materials have high reflectance and therefore absorb less solar energy. But they also have good emittances for long waves. This gives the best combination for reducing solar gain and hence maintaining a cool surface. Bright sheet materials have high reflectance but their long-wave emittance is low so they can experience significant solar heating.

The obverse of solar overheating is cooling by night sky radiation. In these circumstances, surface temperatures as much as 8 °C below the outdoor air temperatures can commonly occur. Condensation will result, and can affect various types of roofs (see chapter 8, *Roofs*).

The areas with the highest maximum temperatures are in south east England, while the lowest minimum temperatures are in central Scotland, northern England, Wales and the Midlands. Temperatures in coastal areas are generally more equable. Long-term average temperatures are most useful for calculating seasonal heat losses, while shorter term averages are appropriate for sizing of heating, ventilating and air-conditioning plant. Threshold or extreme values

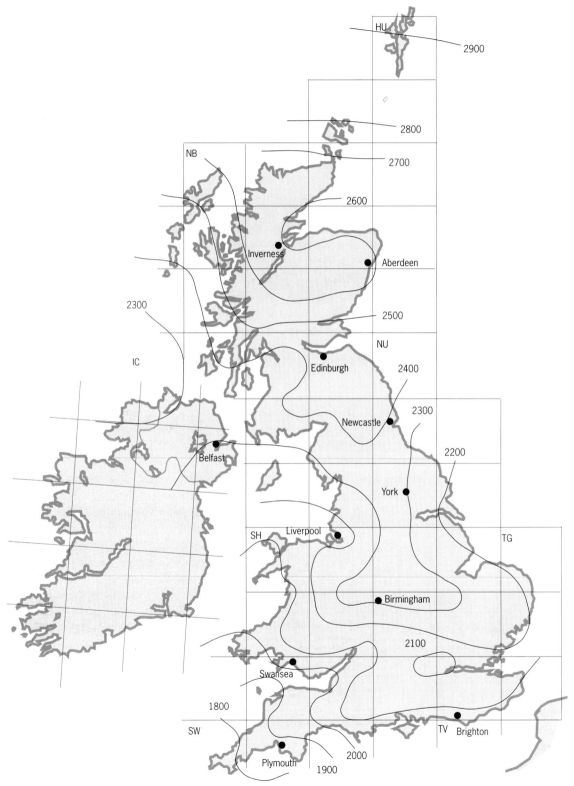

Figure 3.9 Accumulated temperature difference (degree day) totals, September–May average, 1957–76: data reduced to sea level. Design data from the CIBSE Guide A (1986) (out of print) reproduced by permission of CIBSE, London

may sometimes be important, for example, where material or mechanical failure may occur.

Seasonal accumulated temperature difference (or degree days) varies by a factor of about 1.6 over the UK between the south-west and extreme north when values are reduced to sea level. Figure 3.9 illustrates how geographical variation in temperature increases the importance of adequate thermal insulation in the colder northern latitudes of the UK. Totals are considerably increased at higher altitudes. However, these totals do not take much account of solar gains, and therefore careful design can reduce the impact of these variations. There is more information about degree days at http://www.vesma.com/ddd/.

Deterioration of building components frequently happens when several meteorological parameters act together. One example of this is frost damage to porous materials. This occurs when freezing conditions coincide with high moisture content, near saturation. The air temperature at which frost damages some materials usually needs to be well below freezing; the water in smaller pores will only freeze at several degrees below zero. Several factors affect frost damage, including:
- the precise porous characteristics of the material,
- the speed and duration of freezing,
- the number of faces frozen simultaneously,
- moisture content and its distribution,
- internal strength of the material or its parts.

Frost tests seek to model the complex interaction of factors, but no single test adequately models the risk of frost damage. Product manufacturers may provide results of a series of tests, including natural exposure tests, laboratory freeze/thaw cycling, crystallisation tests, measurements of absorption and strength characteristics, and characterisation of pore structure. Different tests are well-established for most traditional materials (eg roof slates, roof tiles, limestones and bricks).

Passive solar design
Passive solar buildings (Figure 3.10) are designed to make the most of ambient solar energy. These would normally have a main window-wall facing within 30° of due south and significantly larger windows

Figure 3.10 Passive solar housing on the Pennyland estate, Milton Keynes

on the south-facing wall compared with the north-facing one.

Trombé wall
A typical Trombé wall consists of a thick masonry wall coated with a dark, heat-absorbing material and faced with a single- or double-layer of glass. The glass is placed from about 30 mm to 150 mm away from the masonry wall to create a small air space. Heat from sunlight passing through the glass is absorbed by the dark surface, stored in the wall, and conducted slowly inwards through the masonry.

Thermosyphon-type water heater
A thermosyphon-type solar water heater has an insulated water storage tank mounted above flat-plate solar collectors. The collectors transfer heat from the sun to an antifreeze collector fluid. The hot collector fluid flows through a heat exchanger wrapped around the water storage tank, heating the household water inside the tank, and flowing back down into the collectors.

To encourage passive solar dwellings, the current Approved Document L1 of the Building Regulations (England & Wales) gives additional allowances for solar gain in its target U-value method, where the area of glazed openings on the south side of a dwelling exceeds that on the north side. This allows larger south-facing windows to be included in homes of this type. However, be aware that overheating can occur.

Site layout is important for managing solar gain. Orienting solar-collecting glazing to the south imposes constraints on building layout, and a low level of obstruction is also needed if sunlight is to be received for most of the winter.

Avoiding overheating

Solar overheating can be reduced or avoided by the techniques listed below. There is more information on the first two techniques in the section on *Windows* in chapter 5.

- **Limiting window area.**
- **Solar shading.**
- **Thermal mass.** An exposed heavyweight structure will tend to absorb coolness at night, resulting in lower peak temperatures on hot days.
- **Good ventilation** (Figure 3.11).
- **Reducing internal gains,** for example by specifying energy-efficient equipment or controls to switch off lighting and other equipment when it is not required.
- **Mechanical cooling or air-conditioning.**

Buildings other than dwellings

In England & Wales, the 2006 Building Regulations AD L include a requirement for 'limiting excessive solar gains'. For buildings that are not dwellings, guidance on this is given in AD L2A. It covers new non-domestic buildings, and large extensions (with floor area greater than 100 m^2 and greater than 25% of the floor area of the existing building).

The AD gives guidance on how to comply with this requirement. It explains that the requirement applies to naturally ventilated spaces as well as those that have mechanical ventilation or cooling. The idea behind this is to avoid the retrofit of cooling systems in naturally ventilated buildings that overheat.

The guidance in the AD applies only to occupied spaces. It mentions three possible design strategies:

- appropriate combination of window size and orientation,
- solar protection through shading and other solar control measures (Figure 3.12), or
- using thermal capacity with night ventilation.

For school buildings, the AD states that Building Bulletins 87 and 101 specify the overheating criterion and provide guidance on methods to achieve compliance.

Figure 3.12 Brise soleil used as a light shelf at the Scottish Executive, Victoria Key, Edinburgh

Figure 3.11 In the BRE Environmental Building air can circulate in the exposed concrete ceiling slab, reducing swings of temperature. Further control of overheating is given by external solar control louvres on the south side, ventilation in both daytime and at night, and by buried cooling coils circulating groundwater

Department of Education and Employment (Architects and Building Branch)
Guidelines for environmental design in schools. Building Bulletin 87. London, The Stationery Office, 1987
Ventilation of school buildings. Building Bulletin 101. London, The Stationery Office, 2006

For other building types, the AD gives two specific ways to comply with the requirement in a space:

- limit solar and internal casual gain,
- show the space will not overheat.

These are alternatives, so only one of them need be used to demonstrate compliance for a particular space.

Limiting gains

One way to comply is to show that the combined solar and internal casual gain on peak summer days would not be greater than 35 W/m^2 of floor area in each occupied space. AD L2A explains that the gains are averaged over a 0730–1730 BST time period. The solar gains are given as the entry for July in the table of design irradiances in Table 2.30 of the 2006 edition of *CIBSE Guide A*. The CIBSE publication *Designing for improved solar shading control* explains how to carry out the calculation of gains.

Overheating calculation

Compliance is also possible by showing that the operative temperature in the space does not exceed an agreed threshold for more than a reasonable number of occupied hours per annum. An exact definition of what constitutes overheating is not given in the Approved Document. The threshold temperature, and the maximum number of hours that it is to be exceeded, depend on the activities within the space. The 2006 *CIBSE Guide A* contains some guidance on this issue.

This is intended to provide a completely flexible method. It could be used, for example, in spaces with night cooling and thermal mass, or where innovative natural ventilation techniques, such as stack effects in tall spaces, are used. The AD does not specify a calculation procedure except to state that the building is tested against the CIBSE Design Summer Year appropriate to the building location. Any reputable calculation technique could be used.

Dwellings

The 2006 Approved Document L1A for dwellings also contains a requirement to limit 'excessive solar gains'. The requirement applies to all new dwellings, even those where full air-conditioning or comfort cooling is already planned. It does not apply to extensions or work in existing dwellings.

The AD explains that:

'Provision should be made to prevent high temperatures due to excessive solar gains. This can be done by an appropriate combination of window size and orientation, solar protection through shading and other solar control measures, ventilation (day and night) and high thermal capacity.'

SAP 2005 Appendix P12 contains a procedure that enables the designer to check on the likelihood of solar overheating. The method takes into account the following factors:

- heat gains through windows,
- internal gains from lighting, appliances, cooking and people,
- gains from hot water storage, distribution and consumption,
- heat loss through the fabric,
- natural ventilation,
- average external temperatures (depending on location within the UK),
- thermal capacity of the building.

Further information about techniques to avoid overheating in dwellings can be found in the Energy Saving Trust's *Reducing overheating: a designer's guide*.

CIBSE
CIBSE Guide A. Environmental design. 2006
Designing for improved solar shading control. TRM 37. 2006

BRE. The Government's Standard Assessment Procedure for energy rating of dwellings. 2005 edition. Available from http://projects.bre.co.uk/sap2005/

Energy Saving Trust. *Reducing overheating: a designer's guide.* Energy Efficiency Best Practice in Housing CE129. London, Energy Saving Trust, 2005

Regulations and Standards

Building Regulations

The Building Regulations require that combined dead, imposed and wind loads are sustained and transmitted to the ground safely. It is customary to establish limits for acceptable risks of failure in respect of the severity and frequency with which such loads can be anticipated. Although seismic loads occur in the UK, experience suggests their intensity is always less than wind forces.

The requirements are given in the supporting documents of the national building regulations:

- **England and Wales:** AD A,
- **Scotland:** Section 1 of the Technical Handbooks, Domestic and Non-Domestic,
- **Northern Ireland:** Technical Booklet D.

Control of moisture penetration is also a requirement, as is conservation of heat and power, leading to the need to consider temperatures and daylighting.

The requirements are given in the supporting documents of the national building regulations:

- **England and Wales:** AD C and L,
- **Scotland:** Section 3 and 6 of the Technical Handbooks, Domestic and Non-Domestic,
- **Northern Ireland:** Technical Booklet C and F.

Finally, there are requirements for disposal of rainwater (see chapter 8, *Roofs*).

Other relevant legislation

Other relevant legislation published by government departments (HSE and DCLG) is listed below.

The Stationery Office
HSE. *Workplace (Health, Safety and Welfare) Regulations 1992. Approved Code of Practice and Guidance.* L24
DCLG. *Housing Health and Safety Rating System: the Guidance (Version 1).* 2000

Standards and codes of practice

Buildings in the UK should be designed using BS 6399-2 to resist the highest wind loads likely to be experienced during their lifetime (generally taken to be once in 50 years).

The design procedures in BS 6399-2 apply only to buildings whose response to wind is static or mildly dynamic, ie structures that do not deflect significantly with the wind. The vast majority of common buildings including houses and low-rise masonry and concrete buildings are classed as static or mildly dynamic so BS 6399-2 will apply.

Other Standards to take note of are given in the footnote below.

BS 8206-2 recommends that interiors, where the occupants expect sunlight, should receive at least one-quarter of annual probable sunlight hours, including in the winter months (between 21 September and 21 March) at least 5% of annual probable sunlight hours. Here, 'probable sunlight hours' means the total number of hours in the year that the sun is expected to shine on unobstructed ground, allowing for average levels of cloudiness for the location in question.

British Standards Institution
BS 6399-2: 1997 *Loading for buildings. Code of practice for wind loads*
BS 8206-2: 1992 *Lighting for buildings. Code of Practice for daylighting*

4 Foundations and basements

This chapter deals primarily with parts of the building below ground level, ie foundations and basements. These underground elements will cause substantial disruption if any maintenance, repair or alteration are required during the life of the building. Investigation of defects will also tend to be disruptive to occupiers. The structural elements of the building (including the foundations) should be designed and supervised by competent professionals. Significant failures in major structural foundations are rare, but do occur, and are extremely expensive and potentially life-threatening when they occur.

BRE's book *Geotechnics for building professionals* provides an introduction and comprehensive overview of the key issues in soil mechanics of direct relevance to foundations. Other reference material has been published by BRE and this is listed in chapter 13, *References and further reading*.

Foundations

A building foundation provides a means by which the building loads can safely be transferred to the ground without excessive settlement. It also provides a means by which forces or movements within the ground can be resisted by the building.

Charles JA. *Geotechnics for building professionals*. BR 473. Bracknell, IHS BRE Press, 2005

In some cases, foundation elements can perform a number of functions: for example, a diaphragm wall forming part of a basement will usually be designed to carry loading from the superstructure.

> **Key Point**
> New construction must not put the stability of existing properties at risk.

If new foundations are placed close to those of an existing building, the loading on the ground will increase and movements to the existing building may occur. When an excavation is made, the stability of adjacent buildings may be threatened unless the excavation is adequately supported. This is particularly important with sands and gravels which derive their support from lateral restraint.

Both the type of construction and the subsequent use of the building are important determinants of foundation design. The choice of foundation type is determined by:
● the imposed loads or deformations,
● ground conditions,
● economics,
● buildability, and
● durability.

For example, strip footings are usually chosen for buildings in which relatively small loads are carried mainly on walls. Pad footings, piles or pile groups are more appropriate when the structural loads are carried by columns. If differential settlements must be tightly controlled, shallow strip or pad footings (except on rock or dense sand) will probably be inadequate so stiffer surface rafts or deeper foundations may have to be considered as alternatives.

Where a building is to be demolished and subsequent construction carried out, it may be economic to re-use existing foundations. This is particularly important on confined sites where removal of the existing substructure may be required if new foundations are to be installed.

According to a survey undertaken by HAPM in the early 1990s on new social housing, of over 2000 sites inspected, nearly three-quarters were of traditional concrete strip or trench fill, some 1 in 6 piled, 1 in 11 rafts, and 1 in 25 vibro-stone columns. There is no comparable information for other building types.

Shallow foundations

These are foundations constructed to a maximum depth of around 3 m. The applied loading is carried by the upper layers of soil, with most of the load carried by a bulb of soil of depth equivalent to twice the foundation width. Such foundations are suitable when these upper soil layers have sufficient strength ('bearing capacity') to carry the load with an acceptable margin of safety and tolerable settlement over the design life. On clay soils or granular soils with a high water table it is often the former requirement that determines the foundation dimensions. On granular soils with a low water table, a settlement limit will often form the key criterion.

BRE's *Good Building Guide 39* gives more guidance on simple foundations for low-rise housing.

HAPM. *Feedback from data 1991–1994.* TechNote 7. HAPM, 1997. Available from www.hapm.co.uk
BRE. *Simple foundations for low-rise housing.* Good Building Guide 39 (3 Parts). Bracknell, IHS BRE Press, 2000–2002

Strip and trench fill (Figure 4.1)

A traditional strip foundation consists of a minimum thickness of 150 mm of concrete placed in a trench, typically 0.8–1 m wide. Reinforcement can be added if a wider strip is required to bridge over soft spots at movement joints or changes in founding strata.

Trench fill foundations are narrower, generally the width of a JCB bucket, but significantly thicker (> 500 mm). These can take advantage of more competent bearing strata at greater depth. The foundation is excavated as a narrow trench and fully filled with in-situ concrete.

The width of a strip foundation will depend on the load-bearing capacity of the sub-soil and the loads from the building. However, the foundation width should also be sufficient to avoid oversail of any part of the supported wall and allow construction from it as required. The foundation depth will depend on:
- depth to a competent stratum allied to required capacity and width,
- minimum depth requirements for frost, clay shrinkage or heave.

To minimise the risk of excavation collapse, and hence improve construction safety, excavation

(a)

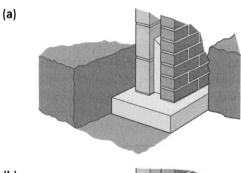

(b)

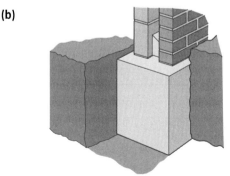

Figure 4.1 (a) Strip foundations, **(b)** trench-fill foundations

depths should be minimised when work is carried out in a trench which must be strutted. Limiting the extent of an excavation such that it remains above the water table also avoids the need for additional ground water control during construction. Minimising spoil reduces handling and disposal costs, but adequate investigation of the ground is imperative.

Note that where a change of level (step) occurs in strip foundations there must be an overlap of at least equivalent to twice the height of the step or 300 mm, whichever is the deeper (Figure 4.2a). In deep strip foundations, BS 8004 requires any step to be not greater than the concrete thickness and the lap to be at least 1 m or twice the step height, whichever is the greater (Figure 4.2b).

Prefabricated beam foundations are now available that obviate the need for on-site concreting and can be supplied ready to install either as shallow foundations or part of a pile and beam system. In many cases, these foundation elements are joined and post-tensioned in order to provide increased stiffness.

Pad

Pad foundations are discrete footings usually used to support a framed structure in which there are identifiable loading positions at the ground surface. The depth and area extent of the pad, as well as reinforcement details, will depend on the soil profile and loading conditions. A different footing would be required for a small vertical load on dense granular soil when compared with a higher load consisting of both vertical and horizontal components.

Pier and beam foundations are used when a number of column loads are applied in line sufficiently close that pad foundations for each would interact.

Raft

Raft foundations are generally used when the total area of pad foundations that would be required exceeds half the building area. Raft foundations can be stiffened with beams in order to help spread loading and stiff rafts are effective in reducing differential settlements. For housing, rafts are generally used when construction is over made

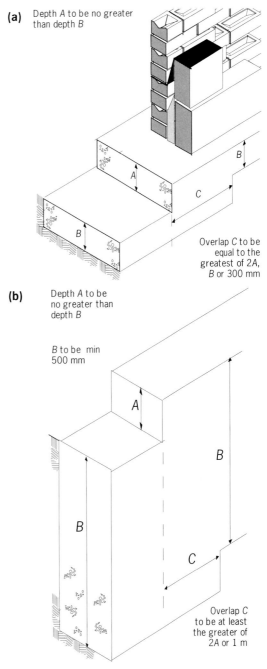

(a) Depth A to be no greater than depth B

B

A

C

Overlap C to be equal to the greatest of 2A, B or 300 mm

(b) Depth A to be no greater than depth B

B to be min 500 mm

A

B

B

C

Overlap C to be at least the greater of 2A or 1 m

Figure 4.2 Typical requirements for overlap at **(a)** step in concrete strip foundations, **(b)** step in deep concrete trench-fill foundations

British Standards Institution. BS 8004: 1986 *Code of practice for foundations*

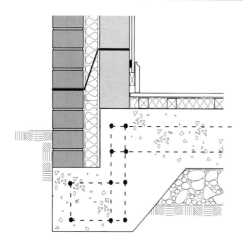

Figure 4.3 Typical edge detail of a reinforced concrete raft, commonly used where loads imposed on the soil must be limited

ground or fill and localised soft spots are anticipated. Raft foundations need careful structural design (Figure 4.3).

Deep foundations

Deep foundations are used when the soil at foundation level is inadequate to support the imposed loads with the required settlement criterion. Deep foundations act by transferring loads down to competent soil at depth and/or by carrying loading by frictional forces acting on the vertical face of the pile. Diaphragm walls, contiguous bored piles and secant piling methods are covered later in this chapter.

Pile

Piles are individual columns, generally constructed of concrete or steel, that support loading through a combination of friction on the pile shaft and end-bearing on the pile toe. The distribution of load carried by each mechanism is a function of soil type, pile type and settlement. They can also be used to resist imposed loading caused by the movement of the surrounding soil, such as vertical movements of shrinking and swelling soils.

Piles can be installed vertically or may be raked to support different loading configurations. Short-bored piles have been used on difficult ground for

low-rise construction for many years. They can be designed to carry loads with limited settlements, or to reduce total or differential settlements. They can have bases that are flat, pointed or bulbous, and shafts that are vertical or raked. In some circumstances, piles can be constructed of other materials, such as timber or plastics.

Piled walls or sheet piles are used to resist lateral movements, such as in forming a basement.

The piling technique used to install the piles will be determined by the ground conditions, loading requirements for the final pile as well as other factors such as access or proximity to other buildings and the need for noise reduction.

Pile types

There are two basic types of piles:
● cast-in-place (or replacement) piles and
● driven (or displacement) piles.

Box 4.1 describes the range of piling techniques.

Piled raft

Piled raft foundations, in which piles are tied into a stiff raft structure, can mobilise the bearing capacity of both the surface soils and those at depth. The use of piled rafts can lead to economic designs in which piles may be chosen to limit settlements of more heavily loaded areas of the raft.

Differential movement or settlement

If the building and its foundations are designed so that they all move together, to the same extent and at the same rate, then only connections of external services to the building are affected by differential movement. However, generally, provision for differential movement will be required in at least some locations. How much building distortion is tolerable depends on a wide range of factors, in particular the length-to-height ratio of the walls and the mode of deformation (ie hogging or sagging). The site investigation should reveal whether different soils or patches of fill are likely to cause the ground to behave variably under the building (Figure 4.5; see also chapter 2, *Site investigation and preparation*).

Box 4.1 Piling techniques

Cast-in-situ piles

Cast piles consist of concrete which is placed into a pre-formed hole before the insertion of reinforcement, or which is continuously pumped into the void created by a continuous flight auger (CFA; Figure 4.4), ie where the screw of the auger is as long as the pile. CFA piles are limited to around 30 m in depth by the auger length and torque required and diameters of between 300 mm and 1200 mm. Careful installation is required to ensure pile quality.

Bored piles

Bored piles are installed by drilling a hole (bore) in the ground with an auger, which may be open, cased or fluid-filled for support. The hole is filled with cast-in-situ concrete, and a steel reinforcement cage is often installed before concreting. The casing, if used, is withdrawn during the concreting. The diameter of bored piles can be increased near the base by 'under-reaming' to improve the end-bearing capacity. Grouting of pile bases has also been employed for this purpose.

Figure 4.4 CFA piling rig

Driven piles

Driven piles of pre-cast concrete or steel are hammered or vibrated into the ground, displacing the soil around them. The pile section is chosen on the basis of both the installation loading and design requirements. These provide for efficient load-carrying and produce no spoil. However, installation can cause nuisance due to excessive noise and vibration.

Rotary displacement piles

Rotary displacement piles are bored using an auger that displaces the ground. The pile is then constructed in the same way as a CFA pile. This method has advantages in that no spoil is produced, and can provide additional load-carrying capacity by improving granular soils.

In practice, engineers tend to rely on correlations between measurements of maximum total settlement and manifest damage to walls. Proposed design limits for total settlement are summarised in Table 4.1.

However, a widely held view is that the average masonry two-storey dwelling will not tolerate differential settlement exceeding about 25 mm and this figure is in general supported by BRE field observations.

The most common causes of subsidence damage are:
- clay shrinkage,
- erosion of soils and drain leakage,
- lowering groundwater levels,
- rising groundwater levels,
- compression of fill,
- collapse of mine workings and other cavities,
- construction processes,
- vibrations.

The effects of evaporation and moisture extraction by vegetation on clay soils are well-established. Clay shrinkage is by far the commonest cause of damage to low-rise buildings. However, the effects of evaporation are largely confined to the surface metre of soil. The types of building susceptible to

Table 4.1 Limits for maximum settlement

	Clay (mm)	Sand (mm)
Isolated foundations	65	40
Rafts	65–100	40–65

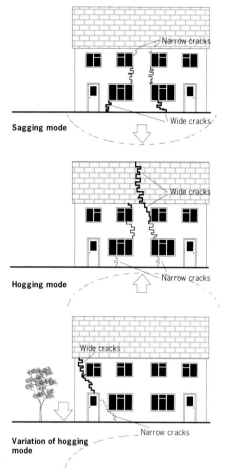

Figure 4.5 Crack patterns associated with different modes of distortion

damage as a result of clay shrinkage are largely restricted to:

● those that do not conform to BS 8103-1 requirement of a minimum foundation depth of 1 m in clay soils where there are no nearby trees or large vegetation, or
● those near trees or large vegetation that do not conform to current NHBC regulations. This can occur after construction, where owners plant trees that have the potential to grow too large too close to the structure, or
● extensions constructed on shallow foundations.

Percolation of water through granular soil can wash out fine particles or dissolve soluble minerals, creating voids or reducing the density of the remaining soil. Most of these problems are caused by fractured drains or water pipes.

Water abstraction can be a problem where certain soils (eg loose sands, silts and peat) reduce in volume and subside. This is a common problem in fen areas where buildings can be adversely affected by drainage schemes.

Rising groundwater levels can be a problem in certain cities (notably London, Birmingham and Liverpool), where water is rarely extracted now for industrial uses. Similar problems can arise in mining areas where pumping has stopped. In clay strata overlying aquifers (underground water reserves) the effect is to cause swelling which may result in vertical heave to buildings, particularly those founded partly on deep and partly on shallow footings (eg extensions and bays).

Shallow mine workings are prone to collapse as localised voids migrate upwards. A similar phenomenon can occur where there are limestone or chalk deposits containing natural cavities or solution features.

Another relatively common cause of differential movement is construction work, particularly excavation and tunnelling. A common design rule of thumb is that no movement occurs at distances greater than three times the depth of excavation. There are also rules governing new excavation adjacent to existing buildings (the Party Wall etc. Act 1996).

There is little evidence of ground-borne vibration causing damage to buildings in the UK except in extreme cases such as explosions, earthquakes, or driving piles close to buildings). It may, however, exacerbate existing damage.

Heave may be caused by the following four processes:

British Standards Institution. BS 8103-1: 1995 *Structural design of low-rise buildings. Code of practice for stability, site investigation, foundations and ground floor slabs for housing* **HMSO.** *Party Wall etc. Act 1996.* London, The Stationery Office, 1996

- removal of overburden,
- swelling of desiccated clays,
- frost heave,
- chemical reaction.

Unless large quantities of soil are being removed, heave due to removal of overburden is likely to be limited to a few millimetres. It may, however, be appropriate to design foundations to allow the underlying soil to swell without damage (eg by providing a void under the floor slab of a piled building).

Large trees on a site will remove water from the soil by a process known as desiccation. Reversal of desiccation can occur where large trees, or large numbers of trees are removed (eg during site clearance). The timescale for reversal of the process can extend to two decades.

Frost heave is unlikely to damage buildings founded below 0.5 m.

The most common chemical reactions result from sulfate attack on concrete supported on sulfate-rich fills (such as burnt colliery shale, brick rubble with adhering plaster and some industrial and mine wastes) (Figure 4.6).

In the UK it is rare for buildings to be located on slopes that are naturally unstable, though there are exceptions. If such a situation is identified in the site

investigation, steps should be taken to improve its stability (eg by excavating near the toe of the slope or loading the crest, or making local alterations to the slope or the groundwater regime).

Landslip occurs in one of two ways, either:
- a sudden bodily movement of a large mass of soil sliding on a shear plane (slope failure) or
- a gradual movement downhill of a surface layer (slope creep).

However, relatively shallow clay slopes (eg 7–12°) can migrate slowly downhill, which may be associated with cyclical clay shrinkage. Specialist geotechnical advice should be sought if this is thought likely to be a problem. There is more guidance in chapter 10, *External works*.

Issues requiring special consideration
Ground improvement techniques
Before commencing the process of laying the foundations, it is increasingly common to undertake ground improvement where site investigation has indicated the ground has poor load-carrying properties or where significant differential movements could occur over the area of a building. This is a specialist area, where a suitably experienced geotechnical engineer should design the works, but generally the methods described in Box 4.2 are employed. Further information can be found in chapter 2, *Site investigation and preparation*.

Durability of concrete subject to sulfate attack or frost
Care will be required in specifying mix materials and proportions, degree of compaction and general site workmanship. BS 8103-1 recommends foundations are taken to a depth of 1 m in clays subject to seasonal movements, and to a minimum of 450 mm in sands, chalk and other frost-susceptible soils. An increased depth is also recommended in upland areas and in areas known to experience long periods of frost. Rafts may crack as

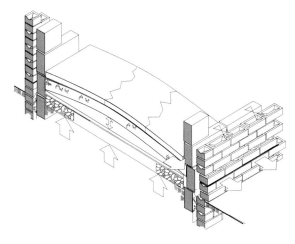

Figure 4.6 Sulfate attack from hardcore on site causes expansion in the concrete slab, leading, in turn, to the slab doming and displacing blocks and bricks in adjacent walls

British Standards Institution. BS 8103-1: 1995 Structural design of low-rise buildings. Code of practice for stability, site investigation, foundations and ground floor slabs for housing

Box 4.2 Ground improvement techniques

Dynamic compaction

Dynamic compaction is where repeated impacts of a heavy weight onto the ground surface compact the soil (Figure 4.7). It is rarely economic to treat areas of less than 5000 m², and inadvisable to use the method close to existing structures. Flying debris can be a hazard, and it will often be necessary to provide a granular blanket as a working platform for the crane, which may add substantially to the cost.

Figure 4.7
Dynamic compaction of fill

Rapid impact compaction

Rapid impact compaction is similar in principle to dynamic compaction with the weight dropping onto a 1.5 m diameter steel foot at a frequency of about 40 blows per minute (Figure 4.8). It can improve the engineering properties of a wide range of fill and natural sandy soils up to depths of about 4 m. Although there is a greatly reduced risk of debris or damage to other structures, vibration and noise may make this technique unsuitable for some urban sites.

Figure 4.8
Rapid impact compactor (Courtesy of Pennine Vibropiling)

Vibro stone columns (Figure 4.9)

Vibro stone columns are formed by using a large poker vibrator to penetrate the ground and form in its wake a continuous and dense column of suitable and durable broken stone from the required minimum depth of penetration at the competent stratum up to the ground surface. The stone is fed into the column either in a wet or dry state, depending on the circumstances and the particular design. In the UK, the dry method is normally used. Treatment typically comprises the placement of lines of stone columns beneath load-bearing walls or on a grid pattern beneath rafts. For small structures the treatment points along each line of footing are generally at about 2 m centres. The column must not be contaminated with spoil. Well over half the applications in the UK are in fill, with some in soft clay. Essentially, this is a hybrid between a foundation and a ground-improvement technique.

Figure 4.9
Vibro stone column ground improvement

Pre-loading

Pre-loading is where a temporary surcharge of soil improves the density of fill or soft clay through reduction in air voids and compression. Used on clay it will also require vertical drains to speed up the rate of consolidation and allow the water removed to drain away. This aspect requires specialist equipment, otherwise the process can be undertaken using normal earth-moving machines. It is essentially a whole site treatment technique.

Chemical stabilisation

Chemical stabilisation is where the behaviour of soft natural soils (or fill) is modified by additives which either modify the soil by physico-chemical processes or cement the soil together.

a result of incorrect positioning of reinforcement and in clay soils trench fill footings have been known to fail if compressible material is omitted from the trench sides.

Local building control departments should give advice on the presence of aggressive soil and groundwater levels in their areas. Information is also included in BRE *Special Digest 1*. Soluble sulfates are widely distributed in clays belonging to the London, Oxford, Kimmeridge, Lower Lias and Keuper Marl formations. These become dissolved in groundwater which can therefore be aggressive to concrete in foundations. Similar problems can occur on contaminated sites. Concrete subjected to water on one side (eg retaining walls) can be particularly liable to damage because fresh sulfates are carried by flowing water.

For less than 0.4% concentration of sulfates in the groundwater no special precautions are needed. Over this level, concrete mixes for use in the ground should be specified in accordance with BRE *Special Digest 1*. The recommendations provided in the 3rd edition take into account the thaumasite form of sulfate attack as well as the better known conventional form of sulfate attack, which produces the reaction products: ettringite and gypsum.

Industrial effluents and waste products may contain many chemicals that can attack concrete. More guidance can be found in BRE's Report *Performance of building materials in contaminated land* and other publications listed in chapter 13.

Performance testing of piles

Testing of piles covers both integrity testing and acceptability testing. Concrete pile integrity can be tested by sonic or vibration methods which reveal imperfections. A programme of testing at installation stage will determine their likely future performance. This should include:
- maintained load tests,
- constant rate of penetration tests,
- constant rate of uplift tests.

BRE

BRE. *Concrete in aggressive ground.* Special Digest 1. 3rd edition. 2005

Paul V. *Performance of building materials in contaminated land.* BR 255. 1994

Tests may reveal situations where piles continue to settle or remain settled on removal of load. Typical settlement limits used by the former Property Services Agency were 10 mm at design verification loading, and 8 mm residual settlement. All testing and monitoring of piles should be overseen by competent and experienced engineers.

Workmanship, mixes and piling

Pile heads of precast reinforced concrete piles should be protected during driving. Reinforcement of precast piles is necessary to resist tensile forces during handling, driving and in service. Unless otherwise specified in proprietary systems, concrete for driven, cast-in-situ piles and bored piles should be highly workable to avoid waisting as casings are withdrawn. Similarly, highly workable mixes are necessary for large bored piles without casings to allow the concrete to flow between the reinforcement. Where temporary casings have to be removed, vibrators must not be used because of the risk of lifting the compacted concrete and reinforcement with the casing.

Basements

Basements have long been a form of construction to provide additional space below the main rooms and, usually, within the same building footprint. Georgian and later housing in Bath, London and other major cities has many such properties. Another benefit is that sloping sites can be usefully developed with the retaining walls to provide the basement structure. In recent years, basements have been built in commercial developments, often for car parking or storage, but few in housing. The Basement Development Group (now The Basement Information Centre or TBIC) was formed in the early 1990s to encourage greater use of basements in housing. Details of their publications are available on their website.

Basements in smaller buildings can provide improved thermal performance and the resulting deeper foundations are advantageous on poor ground.

The Basement Information Centre. www.tbic.org.uk

Waterproofing obviously needs careful design and installation and to be an integral part of the structural design (see the next section on *Basement site preparation and resistance to moisture*).

Estimates made in 1991 by the Basement Development Group indicated that a typical three-bedroom house with a storey wholly below ground might cost 11.5% more than an equivalent house without a basement, while a house with a partial basement might cost only 4.1% more. However, the equivalent house site can be 3 m narrower, with a potential saving of 21% on land costs. Energy costs will also show about 5% savings. Given current pressures on land stocks and the drive to reduce CO_2 emissions we might therefore anticipate that basements will increasingly be considered in new buildings, particularly on smaller, or steeply sloping sites or where land values are high.

The information below relates mainly to newbuild construction for housing and other low-rise construction, but more details (eg remedial works for leaking basements) are provided in BRE's *Foundations, basements and external works*. It is critical to establish both present and future proposed uses for the basement with the client, to ensure the level of performance is clear. Table 4.2 shows the usages and levels of performance from BS 8102.

The TBIC's *Approved Document: Basements for dwellings* provides detailed guidance on the design and construction of basements for dwellings and ways of meeting the requirements of the Building Regulations including resistance to moisture, structure, fire, conservation of heat and power, ventilation, sound, stairs, glazing, drainage, heat-producing appliances, and access. Some key aspects are summarised in the sections that follow.

Harrison HW & Trotman PM. *BRE Building Elements. Foundations, basements and external works: Performance, diagnosis, maintenance, repair and the avoidance of defects.* BR 440. Bracknell, IHS BRE Press, 2002
British Standards Institution. BS 8102: 1990 *Code of practice for protection of structures against water from the ground*
The Basement Information Centre. *The Building Regulations 2000. Approved Document: Basements for dwellings.* Camberley, Surrey, TBIC, 2004

Table 4.2 Level of protection to suit basement use. Data from BS 8102				
Grade	Basement use	Performance level	Construction	Comment
1	Car parks, plant rooms, etc. (not electrical)	Some seepage and damp patches tolerable	Reinforced concrete to BS 8110	Groundwater check for chemicals
2	As for Grade 1 above but need for drier environment (eg retail storage)	No water penetration but vapour penetration tolerable	Tanked or as above or reinforced concrete to BS 8007	Careful supervision. Membranes well lapped
3	Housing, offices, restaurants, leisure centres, etc.	Dry environment required	Tanked or as above or drained cavity and dpms	As for grade 2 above
4	Archives and controlled environment areas	Totally dry environment	Tanked or as above plus vapour control or ventilated wall control and floor cavity and dpm	As for Grade 2 above. Check for chemicals in groundwater

Basement site preparation and resistance to moisture

Site preparation for basements broadly follows the guidance in chapter 2, including ascertaining the water table, ground conditions, presence of existing drains or other services and presence of contaminants. Note that the time of year is relevant to the presence of water in subsoil drains and identification of natural water tables and provision for natural drainage. The possible effects of contaminants on the materials used in basement construction must be considered (many membranes are vulnerable to degradation in the presence of certain contaminants, as noted in chapter 2). Where radon or other gases are present, the guidance in BRE's *Construction of new buildings on gas-contaminated land* should be followed.

Exclusion or control of moisture and in some circumstances, water vapour, are among the chief considerations for basements. Dampness in basements can be caused by condensation but design against condensation is covered more generally in chapter 5 on *Walls* and this guidance is also relevant to basement construction. In terms of specific design issues relevant to basements the focus is on dampness due to penetration of ground water and/or rising damp. The principle of tanking a basement requires both the floor and walls to be resistant to the passage of water. This may entail resistance to water under considerable pressure where water tables are high. Alternatively, on a relatively dry site, drained cavity construction can be considered where possible minor leakage will be managed.

The building should be orientated and designed to avoid the risk of increasing hydrostatic pressure on the basement (Figure 4.10). Where this is not practicable, the waterproofing system should be designed (where necessary) to withstand a full hydrostatic head, or provisions should be made for roddable subground drainage to control or maintain the external environment for which the waterproofing system was designed.

In terms of resistance to moisture, the requirement is that the environmental conditions

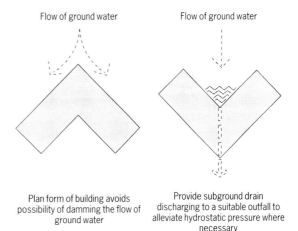

Flow of ground water   Flow of ground water

Plan form of building avoids possibility of damming the flow of ground water

Provide subground drain discharging to a suitable outfall to alleviate hydrostatic pressure where necessary

Figure 4.10 Effect of building orientation on flow of ground water. Reproduced from *Approved Document: Basements for dwellings* by permission of TBIC

should be appropriate to the proposed use, and that walls and floors below ground level should not be damaged by moisture from the ground.

BS 8102 specifies three types of construction:
- **Type A:** tanked protection,
- **Type B:** structurally integral protection, eg water-resistant concrete,
- **Type C:** drained cavity.

Construction method and waterproofing system need to be resolved and as can be seen from Table 4.3 that soil type will have an influence on the choices available.

The choice of construction waterproofing is basically between providing an external water-resistant wall, sheet or coating, or an internal one. The following steps should be taken to identify a suitable system.
- Determine the position of the water table relative to the underside of the lowest floor level (above, below or variable).
- Determine the drainage characteristics of the soil.
- Determine whether the waterproofing system needs to be continuous or can be discontinuous (discontinuity will require good drainage to below the lowest slab level, typically on a free-draining, sloping site).
- Select appropriate construction type for the water table and permeability of the soil (see guidance in

---

BRE. *Construction of new buildings on gas-contaminated land.* BR 212. Bracknell, IHS BRE Press, 1991

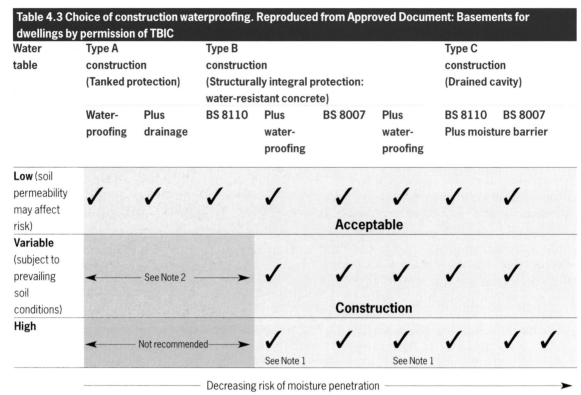

Table 4.3 Choice of construction waterproofing. Reproduced from Approved Document: Basements for dwellings by permission of TBIC

Water table	Type A construction (Tanked protection)		Type B construction (Structurally integral protection: water-resistant concrete)				Type C construction (Drained cavity)	
	Water-proofing	Plus drainage	BS 8110	Plus water-proofing	BS 8007	Plus water-proofing	BS 8110	BS 8007 Plus moisture barrier
Low (soil permeability may affect risk)	✓	✓	✓	✓	✓	✓	✓	✓
						Acceptable		
Variable (subject to prevailing soil conditions)	←— See Note 2 —→		✓	✓	✓	✓	✓	✓
						Construction		
High	←— Not recommended —→		✓ See Note 1	✓	✓ See Note 1	✓	✓	✓

———— Decreasing risk of moisture penetration ————→

Notes:

1 In high water table conditions, the effectiveness of the additional waterproofing system will depend on its application and bond characteristics when permanently under water. (Seek manufacturer's advice.)

2 Constructions may produce acceptable solutions when variability is due to surface water or other infrequent occurrence and not due to an actual rise in water table. (Seek manufacturer's advice.)

TBIC's *Approved Document: Basements for dwellings*).
- Select appropriate foundation type and assess its suitability for continuity of waterproofing.
- Select, confirm, install and protect the waterproofing system in accordance with the manufacturer's directions.

Special attention should be paid to waterproofing details at wall base, laps in membrane, changes in level of the slab, the head of the wall where it adjoins the superstructure, window openings which go below ground level and service entry points. The effectiveness of the waterproofing depends on the position of the system and the waterproofing protection system selected (tanking, structurally integral protection or internal drained cavity).

Waterproofing positions are:
- **External** (membranes 1, 3, 4, 5, 6 and 7 in Box 4.3),
- **Sandwich** (membranes 1, 3, 4, 5, 6 and 7 in Box 4.3),
- **Internal** (membranes 4, 6 and 7 in Box 4.3 and internal drained cavity),
- **Integral** (in-situ concrete with waterstops if necessary). See BRE's *Foundations, basements and external works*.

British Standards Institution
BS 8007: 1987 Code of practice for design of concrete structures for retaining aqueous liquids
BS 8110: Structural use of concrete
   Part 1: 1997 Code of practice for design and construction
   Part 2: 1985 Code of practice for special circumstances
   Part 3: 1985 Design charts for singly reinforced beams, doubly reinforced beams and rectangular columns

Harrison HW & Trotman PM. *BRE Building Elements. Foundations, basements and external works: Performance, diagnosis, maintenance, repair and the avoidance of defects.* BR 440. Bracknell, IHS BRE Press, 2002

Box 4.3 Categories of waterproofing protection systems

1 Bonded sheet membranes
2 Cavity drain membranes
3 Bentonite clay active membranes
4 Liquid-applied membranes
5 Mastic asphalt membranes
6 Cementitious crystallisation active membranes
7 Proprietary cementitious multi-coat renders, topping and
   coatings

Figure 4.11 indicates the three general types of
basement construction and their acceptability for
different water tables.

Structural requirements of basement walls

Approved Document (AD) A of the Building
Regulations (England & Wales) requires the

Key Point
Hydrostatic pressure can contribute uplift forces on the
structure.

building to resist loading, ground movement and
disproportionate collapse. The last of these is only
applicable to buildings having five or more storeys
(each basement level is counted as one storey).
Basements therefore require retaining walls of
masonry or in-situ reinforced concrete.

The maximum allowable storey height of the wall
containing the retaining wall should not exceed
2.7 m measured from the top of the basement floor to
the underside of the floor over the basement. With
the exception of single dwelling parking area or its
drive, the retaining walls should not be closer to a
road or other trafficked area than 1.5 times the depth
of the basement below ground level (Figure 4.12).

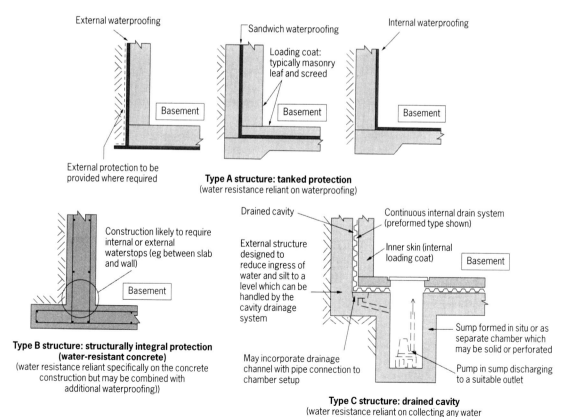

Figure 4.11 The three types of basement construction. Reproduced from *Approved Document: Basements for dwellings*
by permission of TBIC

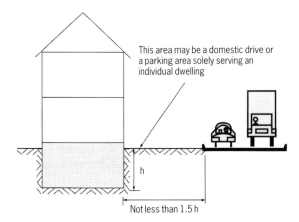

This area may be a domestic drive or a parking area solely serving an individual dwelling

h

Not less than 1.5 h

Figure 4.12 Distance of road from basement.
Reproduced from *Approved Document: Basements for dwellings* by permission of TBIC

Walls need to resist quite large lateral loads. Hydrostatic pressure can contribute uplift forces on the structure. Horizontal loads can be imposed on walls by nearby foundations. To assess theoretical load capabilities, actual wall, floor and foundation constructions will need to be identified. Site investigations are advisable to assess loadings from retained ground and groundwater.

The additional load on the ground caused by a building is reduced by providing a basement, the weight of which is less than the weight of the ground excavated. No additional load is applied to the ground until the total weight of the building and subsurface structure is greater than the total weight of ground excavated. Excavations for large basements may lead to clay heave in certain circumstances.

A basement constructed well before the superstructure may require continuous de-watering until the building is substantially complete. If tanked, upward displacement caused by flotation is a real possibility.

The Basement Information Centre. *The Building Regulations 2000. Approved Document: Basements for dwellings.* Camberley, Surrey, TBIC, 2004

Basement retaining walls

The TBIC *Approved Document* gives detailed requirements concerning retaining walls for basements.

● **Masonry retaining walls:** the requirements cover wall thickness, wall ties, reinforcement, bonding and mortar, compressive strength of bricks and blocks, concrete infill and cover to concrete.
● **In-situ concrete retaining walls:** the requirements cover wall thickness, reinforcement and concrete mixes.
● **Foundations:** the requirements cover design provisions of strip and raft foundations, reinforcement, concrete grade and area of steel and blinding.

Conventional methods for erecting temporary retaining walls to support site boundaries during excavation and construction are expensive, but it is now normal practice to install the permanent main wall structure before excavation starts. Temporary support will still be needed to support the walls before the permanent structure provides rigidity. All these methods will require design by an appropriately qualified engineer. There are three main methods in use:
● diaphragm walls,
● contiguous bored piles,
● secant piles.

Diaphragm walls

These are commonly used on clay and gravel sites. The resulting wall is substantially watertight. The method and sequence of construction are illustrated in Figure 4.13.

The advantages of this method are that:
● installation is free from vibration and excessive noise,
● walls are constructed with minimum disruption to adjacent areas,
● walls avoid the need for temporary sheeting to the excavation and become the final structural wall,
● walls are substantially watertight.

Note, however, that they will still require support, either from the permanent structure within the

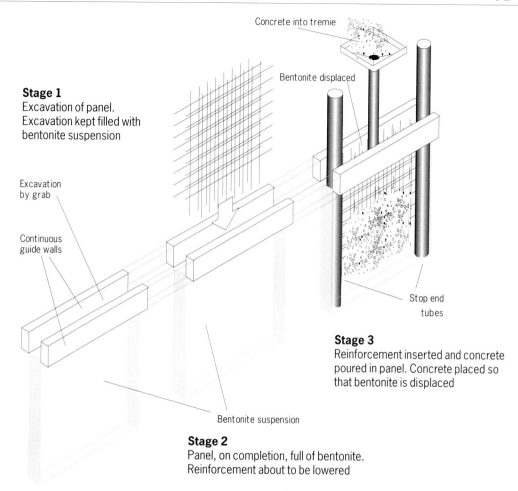

Stage 1
Excavation of panel.
Excavation kept filled with
bentonite suspension

Concrete into tremie

Bentonite displaced

Excavation
by grab

Continuous
guide walls

Stop end
tubes

Stage 3
Reinforcement inserted and concrete
poured in panel. Concrete placed so
that bentonite is displaced

Bentonite suspension

Stage 2
Panel, on completion, full of bentonite.
Reinforcement about to be lowered

Figure 4.13 Stages in the construction of a diaphragm basement wall using a bentonite slurry to maintain the walls of the excavation until the reinforcement and concrete are placed

basement or by ground anchors acting outside the walls.

Contiguous bored piles

These are commonly confined to ground conditions where naturally dry soil exists. The bored piles are installed as close together as possible to form the perimeter wall before any excavation takes place. The accuracy of the pile placement depends on the type of pile and the method of placing. Figure 4.14 shows cast-in-situ piles used in this way.

The advantages are similar to diaphragm walling, with the exception of water tightness. More efforts are required to provide an acceptable face finish to the walls. In cost terms, they are likely to be similar to diaphragm walls.

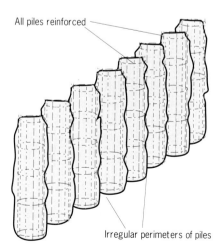

All piles reinforced

Irregular perimeters of piles

Figure 4.14 Cast-in-situ piles conform to the final shape of the excavation, leaving a rough surface when exposed

Secant piles

There are two different methods.

● *Libore secant piling* is a development of the bored pile principle, but it is normally only used for major engineering, although a lighter weight system involving less reinforcement is claimed to be competitive with contiguous bored piles.

● *Stent wall secant piling.*

In both methods, the adjacent piles cut into each other, forming a cut-out in the shape of a secant. The female piles are bored first, and the reinforced male piles follow. The two systems are illustrated in Figures 4.15 and 4.16.

Libore secant piling is more expensive than the other methods, but provides water tightness and strength in difficult boring conditions. Stent secant wall piling is claimed to provide a watertight wall no more expensive than the other systems.

Fire safety, insulation and services

Noise resistance

Basement walls, floors and stairs which separate dwellings are required to resist sound in accordance with relevant accompanying documents for the Building Regulations for each part of the UK:

● **England and Wales:** AD E (revised 2003),
● **Scotland:** Section 5 of the Technical Handbooks, Domestic and Non-Domestic,
● **Northern Ireland:** Technical Booklet G.

Care should be taken to detail the junctions of separating walls with other elements such as perimeter walls, floors, stairs and partitions. Where pre-completion testing is required it should be carried out in accordance with the requirements of AD E. Because undertaking remedial treatment in a basement is extremely difficult, acoustic advice should be sought at an early stage of design.

Thermal performance

Construction laid directly in or on the ground does not lose heat in the same way as walls and roofs. The ground acts to some degree as an accumulator of heat, and most heat loss occurs within a short distance of the edges adjacent to external walls.

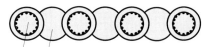

Light construction
Male piles reinforced with mild steel bars

Heavy construction
Both piles reinforced with rolled steel sections

Heavy construction
Male piles reinforced with helical binders

Figure 4.15 Light and heavy forms of Libore secant piling

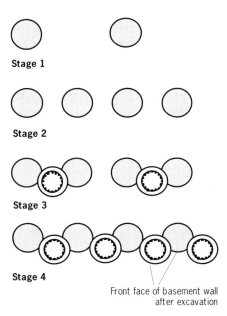

Figure 4.16 Stages in the formation of stent secant piling

Consequently, the average heat flow is dependent on the size and shape of the construction. For the purpose of determining standard U-values when using the Elemental method, basement retaining walls should be considered as exposed walls and basement floors should be considered as exposed floors. U-values can be calculated using the steady-state component averaged over the basement, which is adequate for most purposes. BRE's *Information Paper* IP 14/94 explains the procedure. Alternatively, Appendix 5A of TBIC's *Approved Document: Basements for dwellings* gives design U-values and base thickness of insulation layers for walls. Further tables are provided in the TBIC's *Approved Document* for solid floors of various target U-values, and there is also a calculation for the U-value of a basement as a whole.

Domestic basements are seldom entirely below ground, and temperature fluctuations adjacent to ground level will be similar to exposed walls. For this reason, wall insulation should extend well below ground level. As basements respond slowly to heating, insulation is necessary to provide a more rapid thermal response. Water saturation of insulation on external walls below ground level can occur when water-permeable insulation is used. Thermal bridging by other structural members, such as beams and columns, may occur. The insulation value of solid construction may be reduced if the structure is saturated.

Ventilation

> **Key Point**
> Basements tend to be difficult to ventilate properly and may often require special measures to avoid condensation.

Bathrooms and utility rooms often generate high moisture vapour levels. Cross-flow background ventilation can be provided, where possible, by approximately equal areas of adjustable background ventilation openings on opposite or adjoining sides of the basement. Ideally, there should also be a 10 mm gap under internal doors (or equivalent opening). Where ventilation openings are possible on one side only, supplementary ventilation by ventilation ducts, mechanical ventilation or passive stack ventilation should be considered. Remember also to think about the possible need for smoke control and sound insulation requirements.

Communal car parking areas in basements which do not have natural ventilation require mechanical ventilation in accordance with relevant accompanying documents of the Building Regulations for each part of the UK:

- **England and Wales:** AD F,
- **Scotland:** Section 3 of the Technical Handbooks, Domestic and Non-Domestic,
- **Northern Ireland:** Technical Booklet K.

Fire safety and means of escape

AD B1 of the Building Regulations (England & Wales) requires more onerous levels of fire protection for basements than superstructures, and also provides for venting of heat and smoke from all basements except the smallest and shallowest.

All construction separating the fire-fighting shaft from the remainder of the building needs two hours fire resistance, with one hour from the fire-fighting lobby. A floor over a basement of a dwelling in single occupancy with a basement floor area of less than 50 m^2 is required to have a full half-hour fire resistance, not a modified half-hour criterion. More exacting requirements apply to larger basements and those offering multiple dwelling accommodation. A compartment wall between basement flats should be of non-combustible material.

Means of escape from basements is covered in BS 5588-1. Where a flat is situated in a basement and does not have a separate entrance an alternative means of escape must be provided. Windows into light wells may not be large enough to act as an alternative means of escape. The requirement is for

Anderson B R. *U-values for basements.* Information Paper IP 14/94. Bracknell, IHS BRE Press, 1994

The Basement Information Centre. *The Building Regulations 2000. Approved Document: Basements for dwellings.* Camberley, Surrey, TBIC, 2004

British Standards Institution. BS 5588-1: 1990 Fire precautions in the design, construction and use of buildings. Code of practice for residential buildings

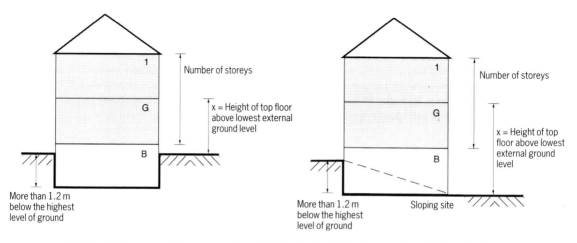

(a) When the basement floor is more than 1.2 m below the highest ground level adjacent to the outside walls, the basement storey is not included in the number of storeys

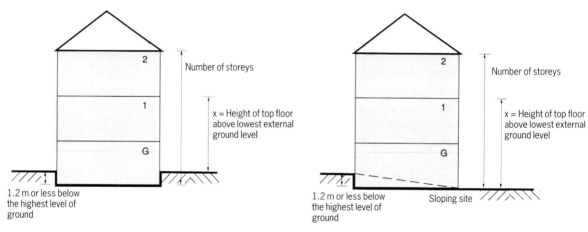

(b) When the basement floor is 1.2 m or less below the highest level of ground adjacent to the outside walls, the basement storey is included in the number of storeys

In both cases **(a)** and **(b)** the elements of structure should:
- when the number of storeys is 2 and where x is not more than 5 m, have 30 minutes fire resistance (modified 30 minutes for the upper floor except for floors over the garage),
- when the number of storeys is 3, have 30 minutes fire resistance,
- if x is not more than 4.5 m, the basement need not be separated from the rest of the building but escape windows or doors should be provided to all habitable rooms,
- if x is more than 4.5 m, the basement should be separated from the rest of the building and the provisions included in paragraphs 1.20, 1.21 and 1.22 of AD B should be followed.

Figure 4.17 Requirements for fire resistance and provisions for means of escape in dwellings containing a basement. Reproduced from *Approved Document: Basements for dwellings* by permission of TBIC

an opening of minimum dimensions of 850 mm ×
500 mm with a sill height of 600–1100 mm.

Figure 4.17 summarises the basic requirements
for fire resistance and provisions for means of
escape. More information (eg flats with one
common stair and flats with more than one common
stair) is given in TBIC's *Approved Document:
Basements for dwellings*.

Smoke alarms should be provided in the
basement area following guidance given in the
Building Regulations of each part of the UK:
- **England and Wales:** AD B,
- **Scotland:** Section 2 of the Technical Handbooks,
  Domestic and Non-Domestic,
- **Northern Ireland:** Technical Booklet E.

Daylighting
Previously, local authorities could order their own
requirements in respect of ventilation, lighting and
protection against dampness of habitable basement
rooms, but the Housing Act 1957 required an
average minimum floor height of 7 ft. Building
Regulations no longer control the height of rooms.
Windows in light wells will require guarding as
indicated in Figure 4.18.

Glazing within 800 mm of the external ground or
floor level not protected by barriers should be safety
glazing in accordance with BS 6262-4 and tested to
BS EN 12600.

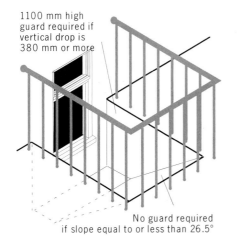

1100 mm high
guard required if
vertical drop is
380 mm or more

No guard required
if slope equal to or less than 26.5°

Figure 4.18 Provisions of guarding for light wells

Drainage
Pipework and drainage in a basement should comply
with the relevant accompanying documents for the
Building Regulations for each part of the UK:
- **England and Wales:** AD H,
- **Scotland:** Section 3 of the Technical Handbooks,
  Domestic and Non-Domestic,
- **Northern Ireland:** Technical Booklet N.

Where possible, if there are no drainage connections
at basement level, the soil stack should be
maintained on the external side of the basement
waterproofing system. The use of a macerator and
pump in basement toilets should allow connections
to be made at ground floor level.

Where surcharge can occur (ie if the top of the
external access point is above the upper surface of
the basement floor) drainage connections which
connect at basement levels should be through a
flood-valve that complies with BS EN 13564-1.
Connections through a basement structure should
make allowance for differential movement of the
pipe and structure.

Access areas at basement level and light wells
should be provided with surface water drainage.
Gullies should be designed to be resistant to debris.

Heat-producing appliances
Heat-producing appliances should be installed in
accordance with the relevant accompanying
documents for the Building Regulations for each
part of the UK:
- **England and Wales:** AD J,
- **Scotland:** Section 3 of the Technical Handbooks,
  Domestic and Non-Domestic,
- **Northern Ireland:** Technical Booklet L.

LPG storage containers should not be located in
basements. When LPG installations are fitted in

British Standards Institution
BS 6262-4: 2005 *Glazing for buildings. Code of practice for
safety related to human impact*
BS EN 12600: 2002 *Glass in building. Pendulum test. Impact
test method and classification for flat glass*
BS EN 13564-1: 2002 *Anti-flooding devices for buildings.
Requirements*

buildings with a basement, an LPG detector is required not more than 200 mm above the floor of the basement, unless low-level direct ventilation is possible.

Air supply to appliances must be provided as required by the relevant national building regulations or the appliance should be room-sealed. When room-sealed appliances are provided, flue terminals must be at least 300 mm above external ground level. Balanced flue terminals should not be located in light wells or similar areas where the combustion products could become trapped as a result of wind pressure on the building.

Vehicle access

Vehicle access ramps to single-family dwellings should not be steeper than 1 in 6. Ramps to basement car parking should not exceed 1 in 10, or 1 in 7 for short lengths with a transition length at the top and bottom of the ramp. The transition points should be eased to prevent vehicles grounding. Ramps should have a textured or ribbed surface. Ramps falling towards the basement should incorporate drainage channels.

Regulations and standards

Building Regulations

The Building Regulations (and AD A) in England & Wales require that the loading imposed is transmitted to the ground safely and without resulting in excessive deflection. Ground movements caused by soil volume change, landslip or subsidence must not impair the stability of the building. Clearly, foundation design must ensure that these criteria are met.

Basement design must comply with the various parts of the Building Regulations which govern the performance requirements of buildings. The TBIC's *Approved Document: Basements for dwellings* details requirements from the building regulations and gives guidance on methods to meet those requirements.

The Basement Information Centre. *Approved Document: Basements for dwellings.* Camberley, Surrey, The Concrete Centre, 2004

Relevant accompanying documents for the Building Regulations for each part of the UK include:

- **England and Wales:** Part C (moisture), Part A (structure), Part B (fire), Part L (conservation of heat and power), Part F (ventilation), Part K (stairs), Part N (drainage), Part J (heat-producing appliances), Part M (access),
- **Scotland:** Section 3 (moisture), Section 1 (structure), Section 2 (fire), Section 6 (energy), Section 3 (ventilation), Section 4 (stairs), Section 3 (drainage), Sections 3 and 4 (heat-producing appliances), Section 4 (access) of the Technical Handbooks, Domestic and Non-Domestic,
- **Northern Ireland:** Technical Booklet C (moisture), Booklet D (structure), Booklet E (fire), Booklet F (conservation of fuel and power), Booklet K (ventilation), Booklet H (stairs), Booklet N (drainage), Booklet L (heat-producing appliances), Booklet R (access).

Other relevant legislation

Construction must be carried out under the Health and Safety at Work Act (1974), the Construction (Design and Management) Regulations (1994) and the Party Wall etc. Act 1996. These place an onus on designers to eliminate or reduce hazards where possible at design stage, and to pass on information on residual risks, such that they can be adequately managed during construction. A suitable foundation design should enable construction to be carried out safely.

Water is supplied in the UK by statutory water undertakers under the control of The Water Supply

The Stationery Office
HSE. The Health and Safety at Work etc. Act 1974
DCLG. The Construction (Design and Management) Regulations 1994. Statutory Instrument No. 3140
HMSO. Party Wall etc. Act 1996
Water Supply (Water Fittings) Regulations 1999. Statutory Instrument 1999 No 1148. 1999
The Water Supply (Water Quality) Regulations 2001
The Water Supply (Water Quality) (Scotland) Regulations 2000
The Water Supply (Water Quality) (Amendment) Regulations (Northern Ireland) 2003
Private Water Supply Regulations 2000
Water Authorities Enforcement Regulations

(Water Quality) Regulations (see references in footnote). Installations in buildings are required to comply with the Water Supply (Water Fittings) Regulations 1999.

Other regulations

Where housing developments are proposed, the NHBC or Zurich Building Guarantees requirements may also have to be met. These have specific requirements for the design and construction of foundations.

The NHBC specifies ground and foundation requirements over and above those stipulated in the Building Regulations, as a condition of compliance with their warranty schemes. NHBC Standards Chapter 4.2 deals with housing construction near trees and in particular the required depth of foundations in shrinkable clay soils. Chapters 4.4 and 4.5 address the requirements for strip and trench fill, raft and pile foundations, respectively. Chapter 4.6 deals with vibratory ground improvement techniques.

Standards and codes of practice

BS 8004 covers foundation design for shallow and deep foundations. BS EN 1997-1 and DD ENV 1997-2 will eventually supersede this British Standard. Other relevant Standards are included in chapter 13 *References and further reading*.

NHBC. *NHBC Standards*. Chapter 4.2 Building near trees. Chapter 4.4 Strip and trench fill foundations. Chapter 4.5 Raft, pile, pier and beam foundations. Chapter 4.6 Vibratory ground movement. Amersham, Buckinghamshire, NHBC

British Standards Institution
BS 8004: 1986 *Code of practice for foundations*
BS EN 1997-1: 2004 *Eurocode 7. Geotechnical design. General rules*
DD ENV 1997-2: 2000 *Eurocode 7. Geotechnical design. Design assisted by laboratory testing*

5 External walls, windows and doors

This chapter deals with external walls, openings through them and applied finishes. It deals in outline with framed structures but the main focus is on masonry walls. The first part of the section on *External walls* forms a brief introduction to the key performance requirements of walls, and is followed by more details in subsequent sections.

External walls

Table 5.1 describes critical design issues for various wall construction types.

Performance requirements

Structure
The predominant factor in the design of a wall is whether it is load-bearing or serves only as an enclosing feature. All walls need to be sufficiently strong to carry the self-weight of the structure, together with imposed loads such as wind.
The minimum requirement for loads to be carried by walls is set out in the Building Regulations for each part of the UK:
- **England & Wales:** AD A,
- **Scotland:** Section 1 of the Technical Handbooks, Domestic and Non-Domestic,
- **Northern Ireland:** Technical Booklet D.

The following loads are relevant.

Dead loads. These must be carried to the ground via a structural frame or a load-bearing wall.

Applied loads. Loads due to furniture, fixtures, fittings, live occupants, snow and wind acting on all surfaces must be carried to the ground via the frame or a load-bearing wall.

Seismic loads. These generate similar forces to wind loads but are not explicitly designed for in the UK, as experience suggests that their intensity in the UK is always less than wind load forces.

> **Key Point**
> The key issues for the designer are the connections between walls and the rest of the structure and the compressive strength of the materials used.

Key issues for walls within this category are connections between walls and the rest of the structure (eg wall ties and strapping) and the compressive strength of the materials used, as well as their ongoing resistance to the loads. This will also be the issue where the connections between the walls and the foundations (and ultimately the ground below) are considered, bringing in issues dealt with extensively in other parts of this book.

Table 5.1 Wall construction types and design issues

Construction	Typical construction	Design issues
Cavity masonry	100 mm brickwork, stone or rendered block outer leaf + cavity (with full fill insulation or 50 mm clear cavity with partial fill insulation) + 100 mm or more block. Various grades of block can be used according to end-use and thermal requirements. Thermal insulation is normally installed within the cavity, but either external or internal insulation can be used. Block outer leafs can also be finished in a variety of ways (eg render, tile-hung, timber-clad).	Whether full fill or partial fill insulation is used will depend on insulation material and exposure. Typically, with dense blockwork 90 mm of insulation will be required to achieve a U-value of 0.35 in England & Wales and 0.3 in Scotland. The inner leaf will normally carry the floor and roof loads.
Solid masonry	Nowadays, this is usually lightweight blockwork, sometimes with additional thermal insulation either externally or internally. Render or tile hanging on the outside provide the primary weather-resisting layer.	Detailing and workmanship around openings are critical to prevent moisture penetration.
Timber frame	Load-bearing timber stud inner leaf with insulation between studs + 50 mm clear cavity + 100 mm brick (or rendered block) outer leaf. Alternative claddings are common, such as tile hanging or timber cladding.	Construction is able to achieve good U-values from relatively slender wall, particularly with claddings other than brick.
Light steel frame	Load-bearing, cold-rolled galvanised steel stock used to manufacture panels or volumetric units. A variety of external claddings is possible: conventional 100 mm brick outer leaf, rendered, tile-hung, timber-clad, external insulation and polymer render. Volumetric units are frequently finished with factory-applied brick slip systems.	This type of construction benefits from the use of jigs and templates in a factory environment. Volumetric units can be fully finished internally. Normally, a large number of similar or identical units are needed for cost-effective manufacture, so traditional markets are hotels, flats, student accommodation, prisons, etc. Ensure frame is 'warm' by positioning the insulation on the outer face of the steel frame.
Concrete frame	Reinforced concrete either poured in-situ or pre-cast; variety of infill or cladding systems.	Fire and moisture protection to the reinforcement is usually provided by the concrete cover.
Steel frame	Hot-rolled steel sections with a variety of infill or cladding systems which usually provide corrosion protection.	Protection of steel against corrosion and fire/fixings and effectiveness of vapour control layers (vcls).
Rammed earth	Load-bearing using moist soil placed in layers and compacted. Walls are typically 300–450 mm thick but greater thicknesses can be used according to design requirements.	Soil needs to be of a specific composition. Special formwork required. External or internal insulation needed to meet current U-values.
Earth building	Mix of soil, straw and water built up without formwork to wall thickness of approximately 600 mm. Alternative methods use blocks normally air-dried and laid using an earth mortar.	Soil needs to be of a specific composition. External or internal insulation needed to meet current U-values unless thicker wall is designed.
Hemp	Mix of hemp, lime and water poured between temporary plywood sheets fixed to a timber frame.	External cladding required to resist weather.

Fire safety

Fire safety is covered by the Building Regulations for each part of the UK as follows:
- **England & Wales:** AD B,
- **Scotland:** Section 2 of the Technical Handbooks, Domestic and Non-Domestic,
- **Northern Ireland:** Technical Booklet E.

There is more information about the national building regulations in chapter 1.

The performance of a wall is judged by its ability to satisfy specified given criteria for a given test duration. Different criteria apply to different building types, heights of walls and the relative position of the wall within the building and to the boundary of the site. Both the resistance to the spread of fire within the building and to adjacent buildings, and the ability of the wall to withstand fire are covered in the supporting documents of the national building regulations. All external walls must be fire-resisting to some degree, which may limit options particularly in respect of openings and their construction, but small residential buildings are allowed different limits from other building groups, and it is these that are covered in this book.

Resistance to weather and to ground moisture

External walls are required to be resistant to weather and ground moisture. The minimum requirements are covered in the Building Regulations for each part of the UK:
- **England & Wales:** AD C,
- **Scotland:** Section 3 of the Technical Handbooks, Domestic and Non-Domestic,
- **Northern Ireland:** Technical Booklet C.

These requirements include issues of dimensional stability (eg movement joints) as well as exclusion and disposal of rain and snow. Although dimensional stability affects structural strength and stability, it is easier to relate it to the exposure of the walls to environmental agents (as described in chapter 3). There are basically three ways of designing a wall to resist driving rain:
- as a continuous impermeable water barrier,

- as a sufficiently thick structure of absorbent material that allows drying out between rainstorms,
- as a continuous cavity that permits drainage of the water within the thickness of the wall(s).

It is difficult to achieve the first way as movements will tend to cause joints or gaps to open. Solid masonry walls (with or without renders) can successfully achieve the second way, but economy of materials often means that the third way, ie cavity walling, is adopted with the associated structural requirements such as wall ties and lintels, as well as dpcs and cavity trays. Each of these approaches brings its own challenges, in particular in respect of detailing of windows and doors (and any other openings).

The requirement to resist ground moisture leads to the provision of continuous damp proof courses (dpcs) between masonry (or other permeable materials or materials prone to damage from moisture) in contact with the ground and the structure above. These must in turn be linked to the equivalent damp proof membranes (dpms) in ground floors, together with radon measures where necessary.

The acoustic environment

The acoustic environment in buildings is affected by the:
- noise climate outside,
- façade sound insulation,
- internal activities,
- size and shape of rooms,
- furnishings and finishes,
- noise sources such as air-conditioning and appliances,
- sound insulation between different internal areas,
- sound insulation between buildings with separating walls.

Because of the range of uses for which buildings are designed, it is not useful to try to identify a single rule for optimising the internal acoustic environment in all buildings. Desirable acoustic conditions in restaurants are unlikely to be considered appropriate in bedrooms. Therefore, to create a quality building, it is essential that designers identify the internal acoustic conditions that are most appropriate for the intended use for each space. They should be able to find relevant guidance on criteria to be met and on forms of construction that ensure that those criteria are achieved in practice. The advice of an acoustics specialist is recommended.

The *minimum* requirements for sound insulation are set out in the Building Regulations for each part of the UK:

● **England & Wales:** AD E,
● **Scotland:** Section 5 of the Technical Handbooks, Domestic and Non-Domestic,
● **Northern Ireland:** Technical Booklet G.

Thermal properties

> **Key Point**
> Heat loss from buildings through materials, gaps and ventilation are key issues in respect of walls.

Fabric transmission losses depend on the total U-values of the various parts of the building envelope, and the Building Regulations lay down minimum performance requirements, which can be expected to become increasingly challenging. For domestic buildings, these are described by target U-values or by energy rating, and in non-domestic buildings by element, by calculation or by energy use. Although trading-off is permissible, the sheer scale of walls makes it important to achieve effective resistance to the passage of heat, to permit adequate openings, ventilation, etc. The insulation value of the wall can be reduced by wetting the insulation materials, by thermal bridges and by air movement.

The minimum requirement for energy conservation is set out in the Building Regulations for each part of the UK:

● **England & Wales:** AD L1 and L2,
● **Scotland:** Section 6 of the Technical Handbooks, Domestic and Non-Domestic,
● **Northern Ireland:** Technical Booklet F.

British Standards Institution. BS 5250: 2002 *Code of practice for control of condensation in buildings*

Condensation on walls

Condensation is usually seen as a completely unacceptable phenomenon in modern construction, particularly when it is accompanied by mould growth; the risks of its occurrence merit the closest attention on the part of designers.

At any temperature, air is capable of containing a limited amount of moisture as an invisible vapour; the warmer the air, the more water vapour it can contain before it becomes saturated and liquid water appears.

Two other important parameters of water vapour in the air are also important and can be illustrated on a psychrometric chart. This graphically represents the interrelation of air temperature and moisture content and is a basic design tool for building engineers and designers (BS 5250).

Relative humidity. This is simply the ratio, normally expressed as a percentage, of the actual amount of water vapour present, to the amount that would be present if the air was saturated at the same temperature. Relative humidity is especially important as it is the parameter that determines the absorption of water by porous materials, governs the growth of moulds and other fungi and is the basis of many systems for measuring the water vapour content of air. Relative humidity depends on both the temperature and the vapour pressure of the air and may therefore be modified by changing either or both of these. A relative humidity of 100% is necessary for condensation, with deposition of liquid water, to occur, for example, on the inside of single-glazed window panes on winter mornings. However, moulds can grow on an internal wall surface when the relative humidity at the wall surface is only 80%.

Dewpoint temperature of the air. This is the temperature at which condensation of liquid water starts when air is cooled, at constant vapour pressure. It is important to realise that, although the dewpoint is expressed as a temperature, it depends only on the vapour pressure; warming or cooling the air makes no difference to the dewpoint.

The occurrence of condensation is a function of four factors:

● heating,
● ventilation,
● insulation, and
● occupant activity.

All four factors need to be in balance or under control to eliminate condensation (and, therefore, mould growth), ie a reasonable level of heating and ventilation should be provided, thermal bridging in the building's materials and components should be absent, and the occupants should avoid producing excessive moisture. It is quite possible that correction or enhancement of any one of these factors subsequent to construction by placing too much reliance on the servicing subsystems, will not necessarily reduce the problem but will encourage a wasteful use of energy (this is clearly unacceptable in the present climate of energy-saving measures). Such a case might be a poorly heated dwelling in which washing is routinely dried indoors.

A comprehensive examination of condensation and its prevention is included in BRE's book *Understanding dampness*.

Surface condensation

Surface condensation and mould growth produce immediately visible problems on the inside surface of buildings. This occurs when the relative humidity of the air in contact with the wall reaches 100% (the dewpoint). The most common situations in which condensation occurs on the external envelope are:

● single-glazed windows (rare in new construction),
● walls with high thermal capacity where the temperature is unable to follow rapid changes to the air temperature, and falls below the dewpoint,
● window and door reveals,
● inside fitted wardrobes adjacent to external walls.

The design issues associated with these requirements include:

Trotman P, Sanders C & Harrison H. *Understanding dampness.* BR 466. Bracknell, IHS BRE Press, 2004

● thermal insulation to reduce heat losses,
● vapour control layers (vcls),
● cross-ventilation of cavities (other ventilation issues are dealt with in chapter 9 on *Building services*),
● thermal efficiency of glazing and doors.

Condensation can be a problem at window and door reveals where there may be locations with reduced wall thickness and thermal bridging. Use of robust details should obviate this risk. Impervious external wall coatings may prevent or restrict drying out of water vapour from the wall and lead to dampness and an increased risk of condensation. Changes to heating, ventilation and occupancy patterns that were not foreseen at the design stage can often lead to condensation.

Interstitial condensation

Interstitial condensation is more insidious; before it becomes visible, it can already have caused severe structural damage.

Water vapour generated in a building creates a vapour pressure difference which drives the vapour through the material of the walls and roof. This causes a gradient of vapour pressure, and therefore dewpoint, through the structure, which in turn depends on the relative vapour resistances of all the materials in the wall. At the same time, there is a gradient of temperature through the wall depending on the distribution of thermal resistances. If the main thermal resistance is on the warm side of the main vapour resistance, the temperature falls faster than the dewpoint, until a point is reached where they are equal: condensation then occurs. The dewpoint becomes fixed equal to the temperature and condensation continues.

Effects of interstitial condensation

In some circumstances, interstitial condensation is unimportant, for example:

● condensation commonly occurs in the outer leaf of masonry cavity walls, but in small amounts compared with the effect of wetting by rain;
● condensation can occur overnight as a fine mist on the underside of the outer sheet of metal roofs; this usually clears rapidly; problems arise only

when so much condensate accumulates that it starts to drip or run into areas where it can cause damage.

In many circumstances, severe damage can result from sustained condensation. Persistent timber moisture contents in excess of 20% (by mass) can lead to decay. Over a winter season, absorbent and hygroscopic materials are likely to accumulate moisture; during the summer, this moisture tends to evaporate. It is difficult to calculate the rate of this evaporation but it should be borne in mind when assessing whether or not condensation is harmful.

Accumulation of condensate within thermal insulation will significantly increase the thermal conductivity of the insulation. Dimensional changes, migration of salts and liberation of chemicals can also result. Although interstitial condensation usually occurs when water vapour is diffusing out from the interior of a building, there are circumstances in which the interior is cooler and drier than outside. Water vapour will then enter the structure from outside. An example is an air-conditioned building in warm, humid weather. There have been many cases of severe damage in air-conditioned buildings in the Middle East or the southern states of the USA, which have impermeable linings such as vinyl wallpaper.

Controlling interstitial condensation
To minimise interstitial condensation, do one or more of the following.
- Reduce the vapour pressure within the building by ventilation and/or reduced moisture input; this is the driving force for interstitial condensation. An additional benefit is a reduced risk of surface condensation or mould growth within the building.
- Use materials of high vapour resistance near to the warmer side of the construction, for example the polyethylene vcl within the timber-framed wall.
- Use material of low vapour resistance near the colder side of the construction; replacing the high-resistance plywood sheathing with a low-resistance fibreboard sheathing could eliminate the need for a vcl in a timber-framed wall.

- Include a ventilated cavity on the warm side of the main vapour resistance; for example, if an impermeable sarking felt, such as a 1F felt is used in a pitched roof, the loft space must be ventilated to prevent severe condensation on the felt.
- Use materials of high thermal resistance near to the colder side of the construction; external insulation to walls keeps the whole structure above the dewpoint and eliminates the risk of condensation.

Thermal bridges
Thermal bridges are areas of the building fabric where, because of the presence of high conductivity materials or the geometry of the detail, there is significantly higher heat loss than through surrounding areas. Besides leading to increased energy use, they lower the internal surface temperature and are therefore sites for condensation and mould growth.

Air permeability

> **Key Point**
> Particular attention should be paid to sealing gaps in masonry, particularly around floor joist bearings.

Air permeability is defined and discussed in general terms in chapter 1. So far as airtightness of external walls is concerned, the designer must identify a continuous plane, both within the flat plane of the wall and within its returns, including any changes of materials or components and its intersections with roofs and floors, where the chosen seal can be made effective, and can be linked to adjacent seals. In addition, designers should define exactly whose responsibility it will be to make it so. Particular attention should be paid to sealing gaps in masonry, particularly around floor joist bearings. Some masonry is air-permeable if not sealed. Joints between sheet cladding materials, though they may be fewer both in number and total length, also demand the closest attention. And whatever the material of the external wall, external and re-entrant corners, window-to-wall joints, eaves and verges should all be given high priority.

Other performance requirements

Issues such as drainage pipes or heating appliance flues passing through the wall relate to the primary performance requirements of the wall, but there are some specific design issues involved.

Structure

External masonry walls

The required thickness of masonry walls in coursed brickwork and/or blockwork for houses up to three storeys is given in Table 5.2.

The leaves of a cavity wall must be tied together with wall ties in accordance with national building regulations. For example, in England & Wales, Table 5 in AD A and BS 5628-3 provide guidance on the permissible types of wall tie dependent on the width of the cavity. Spacing of wall ties should be in accordance with BS 5628-3 and should normally be spaced vertically at 450 mm centres and horizontally either at 750 mm or 900 mm centres depending on the thickness of the thinner leaf. In addition, ties placed not more than 300 mm apart vertically should be provided within 225 mm[†] of the sides of all openings with unbonded jambs and either side of movement joints.

[†] NHBC require such ties to be within 150 mm of the opening.

British Standards Institution. BS 5628-3: 2001 *Code of practice for use of masonry. Materials and components, design and workmanship*

Mortars

Fine (soft) sands, preferred for mortar preparation in the UK because they impart good plasticity and workability, are largely supplied from quarries unwashed. In this state they contain some silt and clay particles with the only requirement being a limit of 8% passing through a 75 mm sieve for the general purpose (G) grade and a 5% limit for the (S) grade. Sands with excessive amounts of clay fines increase the amount of water required in the mix which can lead to lower strengths and higher shrinkage.

> **Key Point**
>
> It is a common misconception that mortars always need to be as strong and as rigid as the bricks they join.

Provided mortars are sufficiently strong to resist weathering, even relatively weak mixes will normally be strong enough for 2-storey work. It is only when large loads have to be carried, as, for example, in frameless load-bearing masonry multi-storey flats which are outside the scope of this publication, that the question of matching mortar strengths to units arises. On the other hand, if there is insufficient lime or cement to bind the sand, durability problems will usually ensue.

Retarded ready-to-use mortars perhaps offer the least room for error. The complete mixing process is carried out in the factory and mortar can be used as received. Since a Portland Cement (PC) binder is

Table 5.2 Minimum thickness of external walls, compartment walls and separating walls		
Height of wall (mm)	Length of wall (mm)	Minimum thickness* of wall (mm)
≤ 3500	≤12 000	190
> 3500 ≤ 9000	≤ 9000	190
	> 9000 ≤ 12 000	290 from base for one storey, 190 for remainder of height
> 9000 ≤ 12 000	≤ 9000	290 from base for one storey, 190 for remainder of height
	> 9000 ≤ 12 000	290 from base for two storeys, 190 for remainder of height

- Minimum thickness* of wall is 1/16th storey height or as given in Table 5.2 (left), whichever is greater
- Minimum thickness of either leaf of a cavity wall is 90 mm
- Minimum cavity width is 50 mm

* For external walls, compartment walls and separating walls of cavity construction, thickness is calculated as the combined thickness of both leaves + 10 mm.
For solid walls constructed of uncoursed stone, flints, clunches, etc. thickness should be 1.33 x the figures given in Table 5.2 (left).

used, a chemical retarder has to be added to delay the setting process. This is usually varied to control the set to a period of 2–4 days. This system obviously gives the highest level of quality control, but problems are still possible if further additions are made on site.

Pointing technique, in addition to altering the character of a brick or block wall, can also play havoc with its weather resistance. Recessed joints, for example, could well lead to reduced weathertightness, and should be avoided in all but the most sheltered locations (ie Zone 1, Figure 5.3).

Strapping

The current Building Regulations and Codes require that building components providing lateral restraint (eg floors) must be tied (or strapped) to walls to give resistance to wind suction forces and to outward bowing due to cyclic movement.

Fire safety

The requirements for the provision of cavity barriers are set out in the Building Regulations for each part of the UK:
- **England & Wales:** AD B,
- **Scotland:** Section 2 of the Technical Handbooks, Domestic and Non-Domestic,
- **Northern Ireland:** Technical Booklet E.

Table 13 of AD B of the Building Regulations (England & Wales) summarises the requirements for the provision of cavity barriers in masonry construction. Some cavity walls are excluded from the provisions for cavity barriers (Figure 5.1).

In situations where services pass through a cavity barrier, measures may need to be taken to prevent the spread of fire. Guidance can be found in Section 11 of AD B (or the equivalents in Scotland and Northern Ireland).

Resistance to weather and ground moisture

External cavity walls will meet the requirements for resistance to moisture if the outer leaf is separated from the inner leaf by a drained air space. Where the

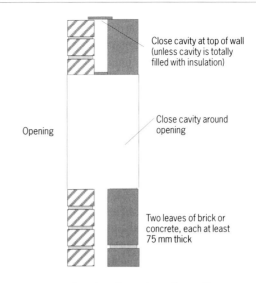

Section through cavity wall

Close cavity at top of wall (unless cavity is totally filled with insulation)

Close cavity around opening

Opening

Two leaves of brick or concrete, each at least 75 mm thick

Note: Combustible material should not be placed in or exposed to the cavity, except for:
(a) timber lintels, window or door frames, or the end of timber joists,
(b) pipe, conduit or cable,
(c) dpc, flashing, cavity closer or wall tie,
(d) thermal insulating material,
(e) a domestic meter cupboard, provided that:
    – there are no more than two cupboards per dwelling,
    – the opening in the outer wall leaf is not more than 900 × 500 mm for each cupboard, and
    – the inner leaf is not penetrated except by a sleeve not more than 80 × 80 mm, which is fire-stopped.

Figure 5.1 Cavity walls excluded from provisions for cavity barriers

cavity is partially filled, a cavity of at least 50 mm is needed. Alternative construction methods that prevent precipitation from reaching the inner leaf are acceptable provided they follow the recommendations of BS 5628-3, LPS 1132-4.1 and Figure 5.2.

Alternative insulation materials can be used provided they are the subject of a current third-party certificate. The work should be carried out in accordance with the terms of that document and by operatives who are either directly employed by the

British Standards Institution. BS 5628-3: 2001 *Code of practice for use of masonry. Materials and components, design and workmanship*
Loss Prevention Certification Board (LPCB).
Requirements and tests for LPCB approval of wall and floor penetration and linear gap seals. LPS 1132-4.1. Available from www.RedBookLive.com

Solid walls

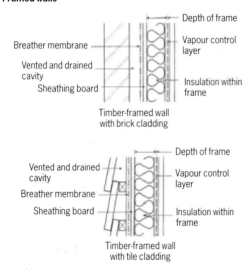

External
protective
system

Insulation

Cavity

Insulation

Internal
protection

External insulation   Internal insulation

Cavity walls

50 mm
residual
cavity

Insulation

Insulation

Partial fill insulation   Full fill insulation

Framed walls

Depth of frame

Breather membrane

Vented and drained
cavity

Sheathing board

Vapour control
layer

Insulation within
frame

Timber-framed wall
with brick cladding

Depth of frame

Vented and drained
cavity

Breather membrane

Sheathing board

Vapour control
layer

Insulation within
frame

Timber-framed wall
with tile cladding

Figure 5.2 Examples of insulated external walls

holder of the document, or employed by an installer approved to operate under the document.

Solid wall masonry may meet the requirements in some circumstances, depending on exposure and construction.

Table 5.3 gives the recommended exposure zones for insulated masonry walls. Figure 5.3 gives the category of exposure to wind-driven rain and corresponding spell indices for worst direction in a given geographical area. The designer should also consider local conditions that could increase or reduce the exposure. For example, on open hillsides

and valleys where the wind may funnel, add 1 to the exposure zone, whereas if the wall is facing away from the prevailing wind, reduce the exposure zone value by 1.

Renders

With appropriate choice of mixes and finishes, external renderings should provide good service for many years, but their longevity is dependent on the quality of the background as well as the quality of both materials and workmanship.

> **Key Point**
> If the render is too strong, it may crack and lead to rain penetration so weaker renders often perform better.

Masonry backgrounds to be rendered direct with a cement-based render should be stronger than the rendering. Each subsequent coat should not be stronger than that preceding.

Renders undoubtedly have a significant effect on reducing rain penetration into walls. The following points observed during experimental work by BRE will have a bearing on its effectiveness.

● 1:1:6 and $1:1/2:41/2$ renderings were effective in reducing the passage of water into brick backgrounds, and the addition of a dry-dash finish further improved the performance. Cracking, or loss of dash by erosion, might significantly alter this protective property.

● Rendering did not reduce the passage of water into aerated concrete backgrounds to the same extent as observed with clay brick backgrounds. Rendering does, however, help to prevent rain penetration through the joints.

● Although evaporation rates were generally much lower than rates of absorption, there was no significant build-up of water within the materials.

● Rainwater absorbed intermittently did not penetrate deeply before it was lost by evaporation.

Renderings that are protected by features of the wall's surface, such as string courses, hood moulds and bell-mouths over window and other openings, will tend to be more durable than those on elevations

Table 5.3 Maximum recommended exposure zones for insulated masonry walls

Wall construction — Maximum recommended exposure zone for each construction

Insulation method	Min. width of filled cavity or clear cavity (mm)	Impervious cladding		Rendered finish		Facing masonry		
		Full height of wall	Above facing masonry	Full height of wall	Above facing masonry	Tooled flush joints	Recessed mortar joints	Flush sills and copings
Built-in full fill	50	4	3	3	3	2	1	1
	75	4	3	4	3	3	1	1
	100	4	4	4	3	3	1	2
	125	4	4	4	3	3	1	2
	150	4	4	4	4	4	1	2
Injected fill not UF foam	50	4	2	3	2	2	1	1
	75	4	3	4	3	3	1	1
	100	4	3	4	3	3	1	1
	125	4	4	4	3	3	1	2
	150	4	4	4	4	4	1	2
Injected fill UF foam	50	4	2	3	2	1	1	1
	75	4	2	3	2	2	1	1
	100	4	2	3	2	2	1	1
Partial fill								
Residual 50 mm cavity	50	4	4	4	4	3	1	1
Internal insulation								
Clear cavity 50 mm	50	4	3	4	3	3	1	1
Clear cavity 100 mm	100	4	4	4	4	4	2	2
Fully filled cavity 50 mm	50	4	3	3	3	2	1	1
Fully filled cavity 100 mm	100	4	4	4	3	3	1	2

Exposure zones	Approximate wind-driven rain* (litres/m² per spell)	
1	Sheltered	< 33
2	Moderate	33 –< 56.5
3	Severe	56.5 –< 100
4	Very severe	100+

* Maximum wall spell index derived from BS 8104.

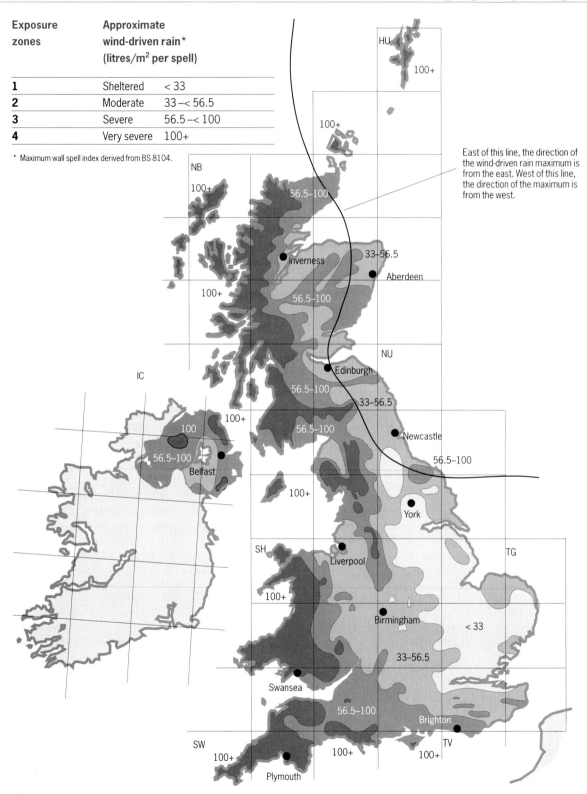

East of this line, the direction of the wind-driven rain maximum is from the east. West of this line, the direction of the maximum is from the west.

Figure 5.3 Categories of exposure to wind-driven rain and corresponding spell indices for worst direction at each location (from meteorological data)

without such surface modelling. Plain finishes should be wood floated (steel-floated finishes should be avoided). A key aspect of durability is adequate control of suction. Some rendered surfaces will tend to dirty unevenly, and may be prone to algal growth. Further information is available in BRE's *Walls, windows and doors*.

Workmanship

The relevant features of good practice are well known: the protection by a flashing or throating of even the smallest projection (Figure 5.4), the throwing of the render onto the wall surface, ensuring a rough-textured finished surface, the use of cement:lime:sand (not cement:sand) mixes, the careful selection of well-graded gritty sands, and the use of gauge boxes rather than the inaccurate and inconsistent batching 'by the shovel'.

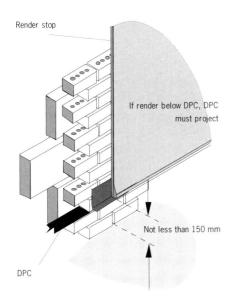

Render stop

If render below DPC, DPC must project

Not less than 150 mm

DPC

Figure 5.4 Criteria for rendering at the base of a wall

Harrison HW & de Vekey RC. *BRE Building Elements. Walls, windows and doors. Performance, diagnosis, maintenance, repair and the avoidance of defects.* BR 352. Bracknell, IHS BRE Press, 1998
British Standards Institution. BS 8000-10: 1995 Workmanship on building sites. Code of practice for plastering and rendering

Control of suction

Open textured surfaces may have such high suction in hot weather that the final water:cement ratio will be affected, and the coat weakened. Other points to remember are:

- the wall surface should be wetted but not soaked in dry weather to reduce suction,
- walls to be rendered must not have been subject to prolonged rain in the previous 48 hours.

Mechanical keys

- All joints in masonry should be raked out to a depth of 10–12 mm unless a bonding agent is specified.
- Expanded metal lathing fixed in accordance with BS 8000-10 can be used over most surfaces.
- For exposed conditions, austenitic stainless steel to BS 1449 Grade 304 will be suitable.
- To promote adhesion, apply SBR latices, acrylic polymers, or EVA emulsions, mixed with cement or cement and sand and well scrubbed into the surface.
- Always render directly over the wet surface.

Insulating renders

Insulating renders can combine a measure of insulation with an acceptable rendered finish. Such renders usually incorporate lightweight aggregates such as polystyrene or expanded minerals (perlite) in a cement-based mix in total thicknesses of as great as 100 mm and exist either with or without reinforcement.

Precautions to be taken include avoiding water vapour transfer from inside the building that condenses on the external insulation system by carefully selecting permeabilities or providing vcls as appropriate.

Plastics materials used for fixing external rendered thermal insulation systems to walls are likely to melt or burn in a fire and this can lead to deformation of the cladding and extension of damage. Loaded fixings fail at much lower temperatures than unloaded fixings.

Cavity trays and dpcs

Horizontal dpcs should be installed at least 150 mm above any surface likely to have water falling onto it

(Figures 5.5–5.8; insulation not shown for clarity). Such dpcs should be continuous with other dpcs and dpms so as to form a continuous barrier to rising or penetrating dampness.

Care should be taken at joints and laps to ensure that water is directed towards the outside of the building (see Table 5.4 for methods of joining dpcs).

Weep holes should be provided in the perpends of the first course of masonry immediately above a cavity tray at 1000 mm maximum centres, with a minimum of two above an opening (BS 8215).

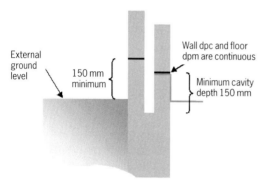

Figure 5.5 dpc requirements at ground level for cavity walls

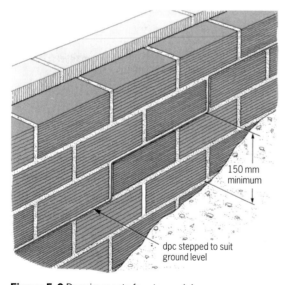

Figure 5.6 Requirements for stepped dpcs

British Standards Institution. BS 8215: 1991 *Code of practice for design and installation of damp-proof courses in masonry construction*

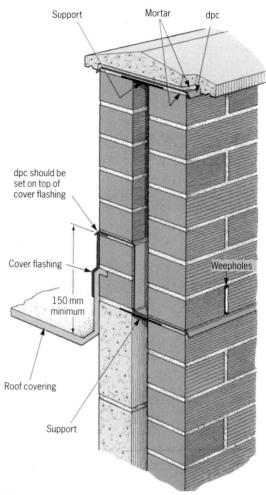

Figure 5.7 Requirements for cavity wall dpcs and flashings at junction with roof

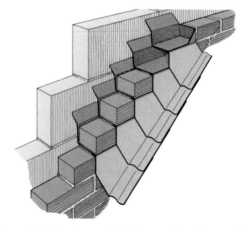

Figure 5.8 Requirements for dpcs and flashings at sloped abutment

Table 5.4 Jointing dpcs	
Material	Jointing method
Bitumen	Lap 100 mm minimum and seal with cold-applied roofing felt adhesive
Pitch and bitumen polymer	Lap 100 mm minimum and seal with adhesive according to manufacturer's instructions
Lead and copper:	
Solid wall	Form welts
Cavity wall	Form single-welted upstands in the perpends at 1.5 m maximum centres
Polyethylene	Lap 100 mm minimum and seal according to manufacturer's instructions

However, some proprietary products for forming weepholes may require closer spacing than 1000 mm so the manufacturer's details should be checked. NHBC and Zurich may also require closer spacing of weepholes in some circumstances.

Cavity trays should be continuous or be provided with stop ends fully sealed to the ends of the tray (Figure 5.9). Cavity trays above lintels should extend at least to the ends of the lintel, ie beyond the opening width. The rise requirements are illustrated

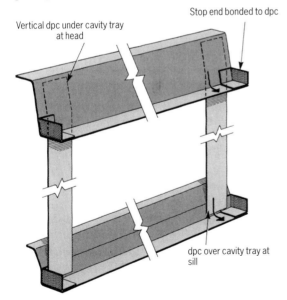

Vertical dpc under cavity tray at head

Stop end bonded to dpc

dpc over cavity tray at sill

Figure 5.9 Requirement for stop ends to cavity trays at window openings

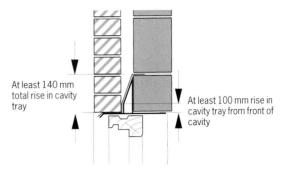

At least 140 mm total rise in cavity tray

At least 100 mm rise in cavity tray from front of cavity

Figure 5.10 Requirement for slope of cavity tray from inner to outer leaf of cavity

in Figure 5.10. For complex details (eg internal and external corners, steps and staggers) preformed trays are preferred. Where a cavity is bridged (eg by services, airbricks, etc.) a cavity tray should be provided above, and should extend 150 mm either side of, the bridge.

Internal noise levels in buildings

The publications listed in Box 5.1 give guidance on appropriate noise environments in buildings along with guidance on noise control, appropriate constructions and standards for sound insulation performance. Other sources of guidance are referenced where appropriate.

BS 8233 contains guidance on noise control for buildings and on indoor ambient noise levels in different types of buildings. Indoor ambient noise levels in buildings are defined as those noise levels in buildings due to external noise and noise from services such as air conditioning but not due to activities related to occupation. Some of the guidance is reproduced in Table 5.5.

BS 8233 also gives guidance on appropriate indoor ambient levels in different types of (unoccupied) spaces where privacy is important such as open-plan offices, reception rooms and restaurants (Table 5.6). However, noise control must be considered before discussing concepts such as privacy, and sound insulation is important in controlling noise.

British Standards Institution. BS 8233: 1999 *Sound insulation and noise reduction for buildings. Code of practice*

Box 5.1 Published guidance on sound insulation

- **British Standards Institution.** BS 8233: 1999 *Sound insulation and noise reduction for buildings. Code of practice*
- National Building Regulations :
  - ❑ **England & Wales:** AD E
  - ❑ **Scotland:** Section 5 of the Technical Handbooks, Domestic and Non-Domestic
  - ❑ **Northern Ireland:** Technical Booklet G
- **Robust Details.** *Robust Details Handbook.* 2nd edition. Part E, Resistance to the passage of sound. Milton Keynes, Robust Details Ltd. Updated 2005

- **DfES.** *Acoustic design of schools.* Building Bulletin 93 (BB 93). Revised 2003. Available as a pdf from www.teachernet.gov.uk or as a publication from The Stationery Office, London
- **BRE and CIRIA.** *Sound control for homes.* BR 238. Bracknell, IHS BRE Press, 1993
- **BRE.** *Quiet homes: a guide to good practice and reducing the risk of poor sound insulation between dwellings.* BR 358. Bracknell, IHS BRE Press, 1998
- **BRE.** *Specifying dwellings with enhanced sound insulation: a guide.* BR 406. Bracknell, IHS BRE Press, 2000

Table 5.5 BS 8233 guidance for indoor ambient noise levels in spaces when they are unoccupied

Criterion	Typical situations	Design range $L_{Aeq,T}$ (dB)*	
		Good	Reasonable
Reasonable speech or telephone communications	Department store	50	55
	Cafeteria	50	55
	Corridor	45	55
Reasonable conditions for study and work requiring concentration	Library	40	50
	Cellular office	40	50
	Staff room	35	45
Reasonable listening conditions	Classroom	35	40
	Lecture theatre	30	35
	Concert hall	25	30
Reasonable resting/sleeping conditions	Living rooms	30	40
	Bedrooms	30	35

Table 5.6 BS 8233 guidance for indoor ambient noise levels in spaces when they are unoccupied and where privacy is also important

Criterion	Typical situations	Design range $L_{Aeq,T}$ (dB)*
Reasonable acoustic privacy in shared spaces	Restaurant	40–55
	Open-plan office	45–50
	Night club, public house	40–45
	Reception room	35–40

* $L_{Aeq,T}$ is the value of the constant A-weighted sound pressure level in decibels that contains the same acoustic energy as the sound under consideration that varies with time. A-weighted sound pressure level measurements are made with a sound level meter that has an A-weighting filter, and roughly correlate with people's subjective assessments of loudness.

Sound insulation

The sound insulation of the building envelope and of walls and floors within buildings is important in controlling the passage of sound. Examples of airborne sound are voices, or sound from a television. The airborne sound insulation of façades can be measured. Airborne sound insulation can also be measured between a pair of rooms having a common wall or floor. The principle of the measurement method is to create a noise in one room (or outside a façade) and to measure this noise and the resulting noise in the receiving room. The difference between the two measurements gives the airborne sound insulation. Because the sound pressure level in the receiving room is affected by the amount of acoustic absorption present (due to furnishings, for example) this is normally measured and a correction applied to the sound level difference. To do this, reverberation times are measured in receiving rooms. The reverberation time is the time taken for sound to decay by 60 dB.

The airborne sound insulation is the difference between the source noise and the received noise. Therefore, the greater the numerical value the better the airborne sound insulation; that is, 55 dB $D_{nT,w}$ describes higher sound insulation than 50 dB $D_{nT,w}$. The impact sound insulation is the noise received from the tapping machine. Therefore, the greater the numerical value, the worse the impact sound insulation; that is, 60 dB $L'_{nT,w}$ describes lower sound insulation than 55 dB $L'_{nT,w}$.

Building Regulations

This section is based on AD E of the England & Wales Building Regulations. The requirements for Scotland and Northern Ireland building regulations are found in:

- **Scotland:** Section 5 of the Technical Handbooks, Domestic and Non-Domestic,
- **Northern Ireland:** Technical Booklet G.

AD E contains the performance standards for airborne and impact sound insulation that must be achieved for compliance with the Building Regulations for dwellings and rooms for residential purposes such as hotel rooms. AD E also contains guidance on controlling reverberation in corridors and stairwells. AD E gives:

- minimum values of airborne sound insulation for walls, floors and stairs that separate dwellings $(D_{nT,w} + C_{tr})$.
- a minimum value of airborne sound insulation for internal walls and floors within dwellings (R_w).
- maximum values of impact sound pressure levels $(L'_{nT,w})$ for floors and stairs that separate dwellings.

It is important to note that the performance standards for sound insulation in AD E are the minimum values that must be achieved for compliance with the Building Regulations. These minimum standards *do not* ensure that occupants of rooms cannot hear noise generated by normal activities in adjacent rooms or dwellings. Because of this, it cannot be assumed that the minimum standards of sound insulation are appropriate for all buildings covered by the Building Regulations. This is particularly important to remember when the design brief is to produce a quality building.

Controlling external noise

Noise transfer into buildings from outside can be reduced by:

- quietening or removing the external noise source,
- installing barriers and/or acoustic absorption between the noise source and the building,
- improving the sound insulation of the building envelope, particularly the sound insulation of windows and ventilators.

Producing buildings with good acoustic environments

> **Key Point**
> It is advisable that any design team responsible for creating a good acoustic environment includes a specialist who is qualified to advise on appropriate acoustic conditions in rooms and suitable constructions and layouts to achieve them.

The acoustic environment in a building can be considered to be part of its ambience. It can affect people's response to a building as can lighting,

Box 5.2 Guidance for enhanced acoustic conditions in buildings

BS 8233 contains guidance on most aspects of noise control in buildings and gives recommendations for appropriate acoustic conditions in buildings. However, other documents contain guidance for specific types of buildings and some of this guidance is discussed below.

Dwellings

There is no regulatory guidance in AD E on the sound insulation of external walls of domestic accommodation or of the elements in them such as windows, doors and ventilators. To specify appropriate building envelope sound insulation, BS 8233 recommends that an assessment of the noise outside the building should be made and desirable indoor ambient noise levels identified. Calculations can then be conducted to determine the building envelope sound insulation required and to specify the sound insulation of elements such as windows and ventilators.

For bedrooms, an indoor ambient noise level of 30 dB $L_{Aeq,T}$ is considered to give good conditions for sleeping in terms of internal noise levels. Therefore, the guidance in DCLG's Planning Policy Guidance 24 at the time of writing is that where night time façade noise levels from road traffic are greater than 66 dB $L_{Aeq,T}$, planning permission (for dwellings) should normally be refused.

Healthcare premises

At the time of writing, guidance for acoustic conditions in hospitals is contained in a series of Health Technical Memoranda (HTM) produced by NHS Estates. For example, HTM 2045 contains guidance on sound insulation and acoustic conditions in healthcare premises, and HTM 56 on sound insulation of partitions.

To assist in specifying appropriate airborne sound insulation between rooms in hospitals or health centres, the concept of privacy is used. Privacy between adjacent rooms is quantified by considering the airborne sound insulation between them and the noise level in the (unoccupied) receiving room. A Privacy Factor (PF) defined by:

$$PF = R'_w + B$$

where R'_w = weighted apparent sound reduction index and B = total NR noise (mechanical services noise + intrusive noise, eg from traffic).

NR is a value derived from a set of curves (NR curves) used by mechanical services engineers to quantify noise from mechanical ventilation systems. BS 8233 states that NR ≈ dB (A) – 6 where dB (A) is the value of $L_{Aeq,T}$ in decibels.

HTM 2045 also contains guidance on reverberation times (the time taken for sound to decay to inaudibility in a room) and noise from mechanical services (in terms of NR values).

Offices

Guidance for offices can be found in BS 8233 and the Commercial Offices Handbook (RIBA Publishing). At the time of writing, both documents use the concept of privacy between offices to aid the specification of suitable separating elements (such as partitions) between them. The approach adopted in the two documents is similar: both documents use the weighted sound level difference between rooms (D_w) to define privacy. D_w is measured in a similar manner to $D_{nT,w}$ but no correction is made for the acoustic absorption in the receiving room.

Schools

Regulation E4 in AD E deals with acoustic conditions in schools in England and Wales. It states:

'each room or other space in a school building shall be designed and constructed in such a way that it has the acoustic conditions and the insulation against noise appropriate to its intended use.'

There is more guidance for schools in Building Bulletin 93.

In schools, speech intelligibility is of primary importance. In cellular classrooms and laboratories, compliance with the performance parameters for indoor ambient noise level, reverberation time and sound insulation should provide suitable conditions for speech communication. However, these parameters alone cannot ensure good speech communication in open-plan teaching areas where a minimum value for the speech transmission index has to be achieved for compliance with the England & Wales Building Regulations.

The acoustic conditions necessary in rooms used for teaching in schools depend on the use to which the rooms are put. The requirements for music rooms, laboratories, classrooms and for rooms used for pupils with hearing impairment are different. Compliance with minimum standards for (say) acoustic conditions in classrooms in schools does not necessarily guarantee a 'quality' acoustic environment. It is now accepted that most pupils suffer temporary hearing difficulties due to illness for periods during their time at school. Also, the need to integrate pupils with permanently impaired hearing into mainstream school activities can mean that minimum standards for acoustic conditions may not be appropriate for all.

NHS Estates. *Acoustics: design considerations.* Health Technical Memorandum (HTM) 2045. 1996

NHS Estates. *Building components: partitions.* Health Technical Memorandum (HTM) 56. 1997

Battle T. The commercial offices handbook. London, RIBA Publishing, 2003

DfES. *Acoustic design of schools.* Building Bulletin 93 (BB 93). Revised 2003. London, The Stationery Office

DCLG. *Planning and noise.* Planning Policy Guidance 24. 1994. Available at www.odpm.gov.uk

thermal comfort and decor. After a period of time, those using a building may reach a consensus on how successful it is. However, for practical and contractual reasons it is advantageous to measure objective criteria when a building is complete so that its acoustic performance can be compared with design specifications. Appendix B describes an approach adopted in BREEAM to the assessment of acoustic criteria in buildings.

Achieving good acoustic performance is complex so the designer should refer to published guidance (see Box 5.1) and seek advice from an acoustics specialist.

Masonry walls

Some areas relating to masonry cavity walls that can affect acoustic performance are:
- wall ties in particular (eg in separating and external walls and between separating walls),
- mass of the inner leaf of the external flanking wall,
- plasterboard linings on external walls,
- external wall junctions with separating walls and floors.

Wet plaster

Wet plaster on masonry walls generally gives better sound insulation than plasterboard on dabs. Wet plaster finishes can also help with reducing air leakage.

Plasterboard linings on external walls

In the case of lightweight masonry ($< 120 \, \text{kg/m}^2$), two layers of plasterboard at a lining to the inner face will help to reduce flanking transmission.

External wall junctions with separating walls and floors

> **Key Point**
>
> With masonry construction, resistance to airborne sound depends mainly on the density of the structure (usually specified as the mass per unit area of wall).

> **Box 5.3 Key principles for achieving improved sound resistance in masonry constructions**
>
> - Fully fill all mortar joints and beds to eliminate air pockets and paths.
> - For solid separating walls the masonry units must be laid flat to the full thickness of the wall.
> - Masonry or concrete separating walls and floors should be bonded into the inner leaf of an external cavity wall, rather than abutting it. If a tied junction is preferred, the inner leaf should abut both sides of the separating wall. Joist hangers are preferred to building in joist ends.
> - Wet plaster finishes are preferred to dry finishes.
> - Cavities in separating walls should always be kept clear of mortar and rubble.
> - Use shallow sockets and chases for electrical services.
> - Use appropriate wall ties and flexible closers to achieve maximum acoustic separation of elements.

Construction detailing and quality of build also have a significant impact on the final performance of the structure.

The main principles for achieving improved resistance to the passage of sound in masonry constructions are listed in Box 5.3. There is more about achieving good acoustic performance in separating walls in chapter 7 *Separating and compartment walls and partitions*.

The satisfactory performance of a separating wall can be compromised by inappropriate specification of flanking constructions such as external walls.

Other performance requirements

Drains

The main issue as far as external cavity wall construction is concerned for drainage relates to the situation where drains run through a wall. The two situations given below are permitted.
- The drain passes through an arch formed in the wall, or through a lintelled opening such that a 50 mm gap is left all around the pipe. In this case, the gap should be masked on both sides with rigid sheet material to prevent entry of fill or vermin.
- A length of pipe (as short as possible) is built into the wall with its joints either side of the wall as close as possible (at most within 150 mm). This

short pipe is connected on both sides to rocker pipes of maximum length 600 mm, with flexible joints.

Fireplace

When constructing fireplace recesses in a cavity wall both leaves must be at least 100 mm thickness.

External claddings

This section deals with a number of external claddings that can be applied to a range of substrates including masonry, concrete, timber frame and steel frame. Types of cladding covered in this section are:
- half-brick,
- vertical tile hanging (vertical being defined as pitched 75 or more degrees from the horizontal),
- slate hanging,
- timber boarding, and
- render.

Half-brick cladding

Brick cladding is common in the UK, if only because it is used extensively as the external leaf of a cavity wall. Movement joints in clay brickwork are to accommodate expansion, so must be able to close by prescribed amounts using compressible packing and non-setting mastics.

Cladding to timber-framed buildings

Cladding to timber frames presents two key aspects of design if a durable construction is to be produced. The first is the treatment of cavity trays where it is important that trays are tucked under the breather paper so that any moisture in contact with the inner leaf is directed away from the timber frame (Figure 5.11).

Another crucial element of design is to allow for differential movement between the timber frame and the brickwork outer leaf. Considerable shrinkage of horizontal members of the timber structure (floor joists, plates, etc.) can occur after construction as the structure dries out. The differential movement needs to be accommodated at a number of points, principally at the eaves and where components rigidly fixed to the timber frame (eg windows, door frames, service pipes) pass

Figure 5.11 The cavity tray is not tucked under the breather paper so any water running down it will be fed to the sole plate. There are other defects too: the barriers are too short and too small to fill the cavity

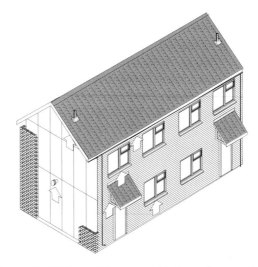

Figure 5.12 Differential movement between frame and cladding needs to be allowed for at a number of points

---

Key Point
It is important that cavity trays are tucked under the breather paper so that any moisture in contact with the inner leaf is directed away from the timber frame.

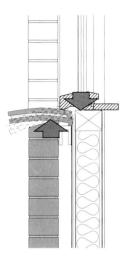

Figure 5.13 Rotation of a sill caused by the shrinkage of a timber frame

through the brick cladding (Figure 5.12). Allowance should also be made for movement to occur at flashings, cavity trays and wall ties, and any mastics or sealants used around door and window frames should be capable of accommodating the shear movement that will occur at the sides of those openings.

The magnitude of the differential movement will depend on construction and the initial moisture content of the frame (Figure 5.13). The relative movement will also increase with height. As a rule of thumb, allowance should be made for 6 mm of frame shrinkage per storey, plus an additional 3 mm if the timber frame is supporting a suspended timber floor. Thus, at the eaves of a two-storey dwelling with a suspended timber floor, a total movement of 15 mm should be allowed for.

Vertical tile hanging

'Traditional' tile hanging uses similar techniques and materials to those used in tile roofing (ie tiles are nailed to treated timber battens), although there are some systems that are fixed to a metal frame. These systems are described later in this chapter in the section on *Rainscreen claddings*.

The basic construction would be treated timber battens fixed to the substrate, either directly or on counter battens. A vcl and/or thermal insulation boards may be incorporated between the substrate

Table 5.7 Tile-hanging methods

Substrates suitable for direct fixing of battens

New common bricks	Corrosion-resistant fixings at 450 mm centres
New lightweight concrete blocks	100 mm aluminium or mild steel cut nails at maximum 450 mm centres
Timber-framed walls (studs and plywood sheathing)	Round wire nails, galvanised or better fixed to each vertical stud, **not** to sheathing
Light gauge steel-framed structures	Corrosion-resistant self-tapping screws into steel studs

Substrates suitable for fixing of battens to counter battens

Dense concrete blocks	Counter battens at
Pre-cast concrete panels	maximum spacing of
Bricks	600 mm fixed with
Old lightweight concrete blocks	corrosion-resistant fixings
Stone	

The number and size of fixings will depend on predicted wind pull-out loads (see BS 5534 and BS 6399-2 for calculation methods; nail pull-out resistance figures can be found in BS 5268-2).

British Standards Institution
BS 5534: 2003 *Code of practice for slating and tiling (including shingles)*
BS 6399-2: 1997 *Loading for buildings. Code of practice for wind loads*
BS 5268-2: 2002 *Structural use of timber. Code of practice for permissible stress design, materials and workmanship*

and battens if required. In some cases, battens may be fixed directly to the substrate if it is sound and true. In other cases, battens need to be fixed to counter battens (see Table 5.7).

Internal and external corners are best constructed with special corner tiles, but mitred joints are possible with metal soakers at every course. The bottom edge of the soakers should be flush with the bottom of the tile course above, and the top edge of the soaker folded over the top edge of the tile. If mitred corners were employed then more robust detail to prevent rain penetration in areas with a high probability of wind-driven rain would be to construct a hidden gutter, protected by the overlapping tiles.

Durability

Tile or slate hanging can be a durable weather-resistant finish for low-rise buildings in relatively sheltered locations, but not at ground-floor or access levels because of poor impact resistance. It would not be advisable to specify tile hanging on buildings over three storeys high because of possible dislodgement by wind conditions, unless special arrangements were to be made for fixing and clipping, especially at the corners. Mathematical tiles are more robust, and they may be available with integral shaped thermal insulation boards, but evidence would be needed of the security and longevity of attachment arrangements.

Specification of materials

Battens should be:
● produced from type A or type B (BS 5534) timber of cross-section 38 mm × 25 mm for the battens or 38 mm × 38 mm for counter battens (can be thinner for counter battens on timber-frame construction),
● treated, if required, according to BS 5268-5,
● no shorter than 1200 mm in length,
● fixed at a maximum 114 mm gauge and minimum 88 mm gauge, either directly to the wall where

appropriate, or to counter battens fixed at maximum 600 mm centres.

Table 5.8 provides guidance on fixings for various applications.

Leadwork should have:
● soakers that are at least code 3, and soakers extended to the bottom edge of the tile course above, and
● flashings that are minimum code 4.

All exposed leadwork should be coated with patination oil.

Underlays should be water vapour permeable if used over boarding or plywood, and be of types 1F or 5U (BS 747) or of other suitable materials above brick or blockwork.

British Standards Institution
BS 5534: 2003 *Code of practice for slating and tiling (including shingles)*
BS 5268-5: 1989 *Structural use of timber. Code of practice for the preservative treatment of structural timber*
BS 747: 2000 *Reinforced bitumen sheets for roofing. Specification*
Note: This Standard will be replaced by BS EN 13707

Table 5.8 Fixings for tile hanging		
Application	**Type of fixing**	**Specification**
Fixing tiles	Clout nails	Stainless steel nails 2.65 mm diameter complying with BS 1202-1; Copper nails 2.65 mm diameter complying with BS 1202-2; Aluminium 3.35 mm diameter complying with BS 1202-3. Minimum length 38 mm. Specify two nails per tile.
Fixing battens to counter battens	Round wire nails	Complying with BS 1202-1 and using hot dipped galvanised or stainless steel.
Fixing battens or counter battens to wall	Cut clasp nails (for aerated concrete blocks)	Complying with BS 1202-1.
	Wall anchors (for dense concrete, brickwork, stone and precast concrete panels)	According to fixing manufacturer's recommendations.
	Self-tapping screws (for light gauge steel structures)	Corrosion-resistant according to fixing manufacturer's recommendations.
	Round wires nails (for fixing to timber-frame structures)	Galvanised or better to BS 1202-1.

British Standards Institution
BS 1202: *Specification for nails*: Part 1: 2002 *Steel nails*; Part 2: 1974 *Copper nails*; Part 3: 1974 *Aluminium nails*

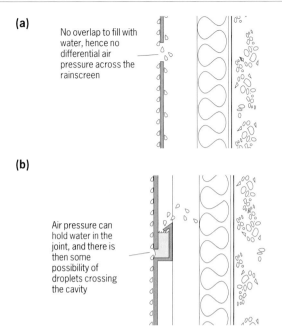

(a)

No overlap to fill with water, hence no differential air pressure across the rainscreen

(b)

Air pressure can hold water in the joint, and there is then some possibility of droplets crossing the cavity

Figure 5.14 The rainscreen principle: **(a)** drained and back-ventilated system, **(b)** equalised air pressure system

Rainscreen claddings

Essentially, a rainscreen is a relatively thin open-jointed screen spaced away from an inner wall. The rainscreen principle is illustrated in Figure 5.14.

Drained and back-ventilated system

In its simplest form, the rainscreen can be sheet materials spaced apart that allow rainwater to drain down the back face of the sheets, with run-off dripping from one edge to another over the horizontal joint. The cavity is fully ventilated and not limited in size to enable rapid drying of any water crossing the cavity.

Equalised air pressure system

A more sophisticated version of the rainscreen is to attempt to equalise air pressures both within and outside the cavity by carefully controlling both the sizes of the cavity and the sizes of the open joints. There must also be a complete air seal at the back of this cavity. The rainscreen skin catches most of the droplets, and for those few drops that get past the screen, because the cavity is open to the external air (though limited in size) the pressure inside and outside is practically equal and there is, therefore,

neither energy nor air stream available to drive the droplets across it to the inner face.

The width of joints must be accurately controlled, especially where catchment trays at the rear of both vertical and horizontal joints are dispensed with. The widths of these trays are directly related to the width of the joints. It may be possible to combine the tray with the vertical members of the support system. Unlapped horizontal joints are unlikely to fill with water, and air pressure inside and outside the cavity is therefore more or less the same. Lapped horizontal joints will need to be provided with sufficient upstand such that they are not likely to fill with water. It is also important that vertical joints do not fill with water since there is a risk that water will overflow inwards instead of outwards, as a result of blocking the ventilation slots.

It is also important to appreciate that water can and will run down the back of some designs of rainscreen panels and any stiffening or damping applied to the back of the screen will get wet. This will also affect fixings on the back of the screen, and, depending on the design, the underlying insulation, which could be protected by a breather membrane.

There is further information concerning rainscreen cladding systems in LPS 1181-4.

Cladding to light gauge steel-framed buildings

Cladding to light steel frames follows the same principles as those to timber frames. The frame needs to be protected from the elements to ensure its durability and the use of a cavity is recommended. Insulation should be applied to the outer face of the steel frame to minimise thermal bridging and create a 'warm' frame. Additional insulation may be provided between the studs, where the risk of condensation is low. The location of the main insulation on the outer face of the steel framing will mean that wall ties and other cladding fixings will need to be fixed through the insulation to the

Loss Prevention Certification Board (LPCB).
Requirements and tests for external thermal insulated cladding systems with rendered finishes (ETICS) or rainscreen cladding systems (RSC) applied to the face of a building.
LPS 1181-4. Available from www.RedBookLive.com

structure, or preferably on counter battens in steel or timber which have been positioned on the face of, and fixed through, the insulation.

Cladding systems used on timber framing can be just as easily applied to light steel frames, ie masonry, tile hanging, timber or PVC boarding, render, etc. Vertical settlement of steel framing is less than than that which needs to be allowed for in timber framing.

A good easy-to-use source detailing light steel framing is Gorgolewski et al's *Building design using cold formed steel sections*.

Applied external finishes

Many external walls are self-finished, or at any rate finished as part of the process of construction of the wall, eg the pointing of brickwork or stonework to form a fair face. Renders and tile hanging have already been dealt with earlier in this chapter. This section, however, deals with other finishes which are applied after the construction of the wall, on the external surfaces, such as adhered tiling or painting, whether on walls, windows or doors, and on the internal surfaces of walls, partitions and trim, such as plastering and painting.

External tiling

Thin ceramic or metal tiling covering large areas of the façades of buildings has largely been a phenomenon of the 20th century, and the finish has proved extraordinarily popular with designers, though with mixed results. Ceramic tiles, butt-jointed and laid and pointed in cement or proprietary mortars form the most usual construction. The mortars are often compounded with rubber and plastics additives to reduce shear modulus and improve resistance to differential movement.

Moisture-induced irreversible initial expansion, coupled with temperature and moisture-induced changes in the wall to which the tiles are applied, can impose compressive loads in the plane of the

---

Gorgolewski MT, Grubb PJ & Lawson RM. *Building design using cold formed steel sections: Light steel framing in residential construction.* Ascot, Berkshire, Steel Construction Institute, 2001

> **Key Point**
> Major movement joints in a wall should not be bridged by tiling.

---

tiling, leading to cracks and displacement. Movement joints not less than 6 mm wide are recommended to be incorporated at not more than 4.5 m centres both horizontally and vertically, and at vertical external and re-entrant corners in large areas; major movement joints in the wall itself should not be bridged by the tiling. It should be possible to match the size and colour of the ordinary joint to make the appearance of the movement joints acceptable.

Timber and plastics cladding
Timber
Timber battens and sidings or ship lap will most likely be of treated softwood sections, laid horizontally, of either redwood or whitewood, finished externally with a stain or paint system. Boards may be either feather-edged or plain, and fastened so that the upper board covers the fixings. Timber feather-edged shiplap has a comparatively good record for weathertightness provided it does not rot, warp or split. Boards mounted vertically will tend to show less run-off staining than boards mounted horizontally, and may therefore be more durable.

Sidings and ship lap cannot always be used on the ground storey as they will not in general meet the vandal-resistance criteria for impact, or other abuse (eg prising off or even unscrewing accessible parts of the fixing system).

Timber boarding may be treatable to obtain Class 0 surface spread of flame without significantly affecting appearance. Without such treatment, it is likely not to better Class 3.

Plastics
Plastics sidings will commonly be of PVC-U, although sidings could be of extruded aluminium or

---

BRE. *Thermal insulation: avoiding risks.* 2002 edition. Bracknell, IHS BRE Press, 2002

sheet steel with a variety of surface finishes. Since the materials are impermeable, installations need to be provided with ventilated cavities to prevent condensation. Further information is available from BRE's report *Thermal Insulation: avoiding risks.* PVC-U sidings are unlikely to suffer too much from defects in the absence of deliberate vandalism, but there can be surface deterioration.

PVC-U in fires behaves differently from either wood or metal, and any necessary assurances must be sought from manufacturers.

Paints and other liquid or plastic finishes
Paints for masonry
Paint systems for use on the external surfaces of masonry are normally one of three kinds:
- cement-based mixtures, which might also contain other additives such as lime, pigments or accelerators in a water base,
- oil- or resin-bound emulsions also in a water base, some of which contain mineral fillers,
- conventional oil or acrylic paints in a solvent base.

Cement-based paints will need to be thoroughly cleaned of algae and redecorated every 5–7 years, depending on conditions and a satisfactory substrate. Chlorinated rubbers and exterior emulsions should last about the same, whereas sprayed textured coatings should last 10 years or even more.

Water repellents for masonry
Colourless, or nearly colourless, liquids designed to increase the ability of masonry to resist driving rain have been available for many years. They must, however, be chosen with care.

Water repellents are distinguishably different from waterproofing treatments to walls; the latter provide a thin impervious film to block the pores of masonry and prevent all entry of water, making the surface impermeable also to water vapour. Such treatments may consist of paints or bitumen solutions, which completely alter the surface appearance.

On the other hand, water repellents brushed or sprayed onto the surface have the effect of reducing the amount of water entering the pores of masonry, but leaving it still permeable to water vapour. The two most common types are metallic stearates and silicones or silanes. BRE tests have shown that silicone-based treatments were effective on both calcium silicate and clay masonry, but aluminium stearates were slightly less effective.

Paints and stains for wood
Some paints, both solvent- and water-borne, are available which are described as microporous or breathing paints. There is no evidence that the long-term moisture contents of wood under these paints differs significantly from those under conventional systems, though there may be a short-term benefit. Durability can be expected to be similar to that obtained from more conventional systems and may need less surface preparation for re-coating.

Water-borne paints are more permeable to moisture than conventional paints and have high levels of film extensibility which is retained during weathering, and hence tend to be more durable than solvent-borne alkyd paints. They are slow-drying in adverse weather conditions.

Exterior wood stains consisting of lightly pigmented dispersions of resins and fungicides are available in zero-build, low-build (10–15 µm) and high-build (30–40 µm) versions. The zero-build and low-build versions are moisture permeable and tend to erode rapidly. High-build stains are more resistant to moisture transmission, and the substrate is therefore less affected by moisture-induced movements.

Varnishes of unpigmented resin solutions give coats of up to 80 µm thicknesses. These have, in the past, tended to be of relatively short durability, although improved durability formulations have been developed. They do not, however, prevent long-term colour changes to the underlying substrate.

Paints for metals
Paint systems primers include:
- *Two-pack pretreatments,* as washes or etching primers, are typically polyvinyl butyral resin solution with phosphoric acid (as a separate component) and with a zinc tetroxychromate

pigment, and are usually yellow in colour. One pack are typically polyvinyl butyral/phenolic resin solutions with tinting pigments and are usually blue in colour. The main function of these primers is to improve the adhesion of paint systems to non-ferrous metals. They can provide temporary protection to blast-cleaned steel and sprayed-metal coatings. The two-pack types generally give superior performance but may be less convenient to use. They do require, however, the application of a normal type of primer on top of the treatment. Most pretreatment primers may be used in conjunction with conventional and specialist coating systems.

- *Red oxide primer.* These are typically drying-oil or resin-type binders with red oxide pigmentation and are usually red-brown in colour. These primers for iron and steel are quicker in drying and hardening than red lead primers to BS 2523 Types A and B are therefore more suitable for use in maintenance work or when early handling or recoating is necessary. They should not be left exposed for too long without top coats.

Nothing is for ever, and paint systems are no exception. Preparation of the surface should be the watchword. Exterior-quality solvent-borne paints will require maintenance every 4–6 years, and exterior-quality water-borne paints slightly longer at 5–8 year intervals.

Applied internal finishes

Ceramic tiles
Ceramic tiles have been made for many years from a mixture of clays and other materials, for example crushed flints, which are then ground and pressed in a semi-dry state into the shape and size required. After the first firing, the colouring and glaze is then added to the biscuit, to produce the finished products.

The following points should be observed:
- evenly coating the tile to substrate joint with adhesive is a prime requirement to prevent loss of adhesion,

British Standards Institution. BS 2523: 1966 *Specification for lead-based priming paints*

- tiling should not proceed on new render for at least 14 days, and on new plaster for at least 28 days,
- thin bed adhesives should not be used when the substrate surface varies by more than 3 mm under a 2 m straight edge. Problems will also occur for thick bed adhesives if the deviation is more than 6 mm,
- tiles without spacer lugs should not be butt jointed,
- grouting should proceed between 24 and 72 hours after fixing.

Plastering
Plastering is one of the relatively few manual processes remaining in construction which are not amenable to completion or partial completion off-site, though a degree of in-situ mechanisation may be available for projection plasters. In spite of the growth of the use of filled joint board finishes, wet plastering remains a popular technique, as long as relevant skills are available. There is some evidence that selection of appropriate materials and techniques is not always thorough, particularly with respect to shrinkage of backing coats and resultant cracking and detachment of surfaces. Rigorous design procedures and quality control on site are therefore of major importance in reducing, if not eliminating, defects.

Most wall plasters are retarded beta hemi-hydrates of gypsum. Some may be suitable for gauging with sand, some are available premixed with barytes aggregate for use in X-ray equipment areas, or some premixed with vermiculite or perlite for additional thermal insulation properties. The other important class is those based on cement:sand mixtures.

Autoclaved aerated concrete blocks need careful control of suction before rendering. For heavier

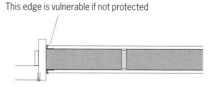

This edge is vulnerable if not protected

Figure 5.15 Plaster abutting doorways ought to be protected

plaster systems, a cement-based undercoat can be used or for lighter weight systems gypsum browning can be used with success, and the finish in either case can be gypsum or lightweight gypsum. A plastered wall surface, whether painted or not, will meet the Class 0 surface spread of flame criterion.

High suction backgrounds should be dampened with water before proceeding with plastering undercoats. Mechanical keys will be needed on backgrounds with low suction. These can best be provided by, for example, raking out joints or by applying cement spatterdash or PVA bonding agents. On the other hand, an important factor in achieving a good bond between the finishing coat and the undercoat is that the undercoat has developed sufficient suction. It is important that both the background and undercoat should have dried sufficiently to allow this suction to develop. (see also BRE's *Good Building Guide 65*).

Sheet and board finishes

Plasterboard drylining may be either mounted on timber or metal battens (straps) or on plaster dabs. Normal practice is to use tapered edge boards able to take scrim and filling to a surface flush with the remainder of the board.

There is a considerable variety of plasterboards available which are designed to give different strength characteristics and improved performance in respect of damp conditions. Laminated plasterboard and urethane boards are also available which give improved thermal insulation. Boards incorporating a vapour control layer are also available. Dry lining of external walls will need to meet increased standards of airtightness, and, in consequence, continuous strips of fixing plaster will be needed at head, foot and re-entrant corners to seal the cavity.

Where plasterboard is used as the wall lining, the frequency of fixing battens or plaster dabs affects impact resistance. All joints between boards should ideally be nogged, and the boards supported at intermediate points at least every 400 mm for 9.5 mm board and 600 mm for the thicker 12.5 mm board. In principle, similar criteria apply to all board materials — the closer the fixings the better the resistance to impacts or bowing.

Sheet surface finishes are frequently chosen for their resistance to high humidity environments, the occasional risk of water spillage, and the need to be able to clean the surface from time to time. The most waterproof are the welded plastics finishes, but the melamines and the vinyls also perform well provided the adhesive is sound, and provided there is no movement large enough to destroy the welds.

Paints and other liquid or plastic finishes

Most of the paints in common use internally are applied in multiple coats, though some, for example some of the textured paints, are applied in single coats. Formulations include, for example, emulsions, vinyls, urethanes, elastomerics and gloss paints applied by spray. Some of the textured spray finishes are available with multicoloured flecks, which go some way towards discouraging graffiti in vulnerable interiors.

Gypsum plasters need to be quite dry before painting, but a wide variety of paints is then suitable. Some lathing plasters contain additional lime and are slightly alkaline. Over-trowelled surfaces, particularly of anhydrous types, may lead to poor adhesion of paints. Lightweight plasters take longer to dry out. Sealing the surface may be needed with uneven trowelling. Although water-based emulsion paints generally are permeable to water vapour, the gloss varieties are less so and should not be used on plasters that are not completely dry.

Anti-fungal paints need special precautions in application.

Gloss paints tend to show surface imperfections in plastered walls and ceilings, especially when seen against the light, more so than matt paints.

Harrison H. *Plastering and internal rendering.* Good Building Guide 65 (2 Parts). Bracknell, IHS BRE Press, 2005

Windows

Windows have two prime purposes:
- the admission of daylight and sunlight, and
- the admission and emission of air needed for ventilation purposes.

Windows (and doors) so far as possible, should not compromise the main performance requirements selected for the walls in which they are to be incorporated (see the earlier part of this chapter and, for example, the DTLR and Defra report *Limiting thermal bridging and air leakage*). Nevertheless, any puncturing of the external wall carries with it a risk of reduced overall performance with respect to strength and stability, weather exclusion and thermal and sound insulation.

In this section, we are concerned mainly with daylighting and aspects of performance including, for example, materials, durability (including vulnerability to unauthorised entry) and means of escape.

Daylighting

> **Key Point**
>
> Surveys have shown that in nearly all building types, people prefer to work by daylight.

Daylight is also highly valued in dwellings. In addition, people generally like sunlight. The sun provides light and warmth, and is seen as having a health-giving effect. Sunlight can also be a source of energy, providing solar gain in the winter which can reduce the need for heating.

Daylighting design can be viewed as a four-stage process:
- **Choose an appropriate building form.** The centre of a deep-plan building will be poorly daylit unless it can be lit by rooflights. Where possible, window walls should be sited away from external obstructions (Figure 5.16).

Figure 5.16 Good daylighting may be difficult to achieve if there are large external obstructions nearby

- **Size and position windows** to provide good daylighting and view out.
- **Choose appropriate glazing and shading** (see later section on *Solar gain, temperature and glare*) to provide adequate control of heat gain and glare while maintaining daylighting.
- **Select appropriate electric lighting controls** to ensure that lighting is switched off when daylight is sufficient.

Designing to maximise daylighting

The overall amount of daylight in a space is quantified by the average daylight factor (DF). BS 8206-2 recommends a DF of 5% for a well daylit space and 2% for a partly daylit space. In housing, BS 8206-2 also gives minimum values of 2% for kitchens, 1.5% for living rooms and 1% for bedrooms.

DF is given by:

$$DF = \frac{W\,T\,\theta\ (\%)}{A\,(1{-}R^2)}$$

The key factors in the equation are:
W = the glazing area. The equation can be inverted to give the required window area for a particular average daylight factor.

DTLR and Defra. *Limiting thermal bridging and air leakage: robust construction details for dwellings and similar buildings.* London, The Stationery Office, 2001

British Standards Institution. BS 8206-2: 1992 *Lighting for buildings. Code of practice for daylighting*

T = the glass transmittance. For clear double or low-emissivity glazing, T is around 0.65.

Tinted glazings often have much lower transmittances and give poorer daylight levels. USA studies have shown that when T falls below 30–35% people find the view out looks dull and gloomy.

θ = the angle (in degrees) of sky visible from the centre of the window measured in a vertical section through the window (Figure 5.17). With no obstruction, θ is 90° for a vertical window. Large obstructions outside can lower θ significantly.

A = the area of all room surfaces (ceiling, floor, walls and windows).

R = their average reflectance. In a light-coloured room, R is about 0.5.

These key factors are illustrated in Figures 5.17 and 5.18. The effect of outside obstructions and of the

use of sloping glazing to increase the visible sky angle can be clearly seen. Beyond the no sky line, points at desk or table-top height do not have a direct view of the sky. These areas of the room will tend to look gloomy.

Atria can give an attractive daylit space in the centre of a deep building. However, often daylight

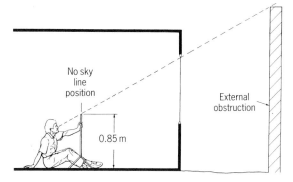

Figure 5.18 The no sky line

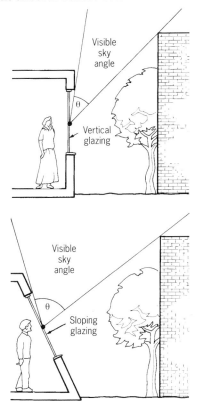

Figure 5.17 θ is the visible sky angle

Figure 5.19 Light-coloured atrium walls and fabric diffusers aid daylight penetration to surrounding spaces (Anglia Polytechnic University building)

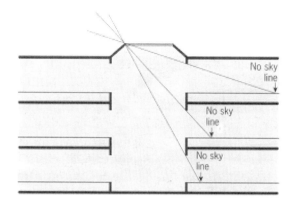

No sky line

No sky line

No sky line

Figure 5.20 On lower floors daylight will not penetrate far from the atrium

from the atrium does not penetrate far into adjoining spaces (Figures 5.19 and 5.20). This can be improved by:
- light-coloured atrium walls (Figure 5.19),
- limiting glazed openings to rooms at the top of the atrium with larger openings on lower floors,
- raising the opening head height on each floor,
- limiting the amount of planting.

Effective control of electric lighting is essential to make the most of daylight and there is a wide choice of systems available. Further details are given in chapter 9 on *Building services*.

Solar gain, temperature and glare

> **Key Point**
> Excessive solar gain can lead to overheating in summer or high air-conditioning loads.

The sun can have negative impacts too: excessive solar gain can lead to overheating in summer or high air-conditioning loads. Glare can also cause problems, particularly in interiors with computer screens. In most buildings, careful design is required so the benefits of sunlight can be enjoyed while any problems are effectively controlled.

HSE. *Workplace (Health, Safety and Welfare) Regulations 1992. Approved Code of Practice and Guidance.* L24

Avoiding overheating

In most buildings, measures should be taken to prevent overheating. This is a requirement of the Workplace (Health, Safety and Welfare) Regulations 1992, which state:

'During working hours, the temperature in all workplaces inside buildings shall be reasonable'.

However, the Regulations do not specify a particular maximum temperature.

Solar overheating can be reduced or avoided by the following techniques.

- **Limiting window area.** Solar heat gain is roughly proportional to the window area. However, reducing window area can also limit daylight and view out (see previous section on *Daylighting*). For offices and similar side-lit buildings, BRE's Report *Environmental design guide for naturally ventilated and daylit offices* gives tables that show the impact of window area (and other design parameters) on peak temperatures, and other useful design guidance.
- **Solar shading.** BRE's Report *Solar shading of buildings* gives detailed advice. The British Blind and Shutter Association also publish a guide detailing the different shading types and giving a list of suppliers. The ability of a shading device or window system to reduce incoming solar gains is given by its total solar transmittance (g value). A low g value means the device is good at blocking solar gain. Generally, external shading devices and reflecting glazing are more effective at blocking solar gain than internal shading devices. Mid-pane shading has an intermediate level of performance. The BRE Trust Report *Summertime solar performance of windows with shading devices* gives guidance.
- Thermal mass.
- Good ventilation.
- Reducing internal gains.
- Mechanical cooling or air-conditioning.

BRE
Rennie D & Parand F. *Environmental design guide for naturally ventilated and daylit offices.* BR 345. 1998
Littlefair P J. *Solar shading of buildings.* BR 364. 1999
Littlefair P. *Summertime solar performance of windows with shading devices.* FB 9. 2005

There is more guidance in chapter 3 in the section on *Avoiding overheating*.

Controlling glare

The Health and Safety Regulations for display-screen equipment apply here. They require that:

'windows shall be fitted with a suitable system of adjustable covering to attenuate the daylight that falls on the workstation'.

Usually, this will mean adjustable blinds that the occupant can control. People often resent shading systems that operate automatically. Further guidance on the design of daylit spaces for display-screen equipment is given in BRE *Information Paper* IP 10/95.

Glare from windows can happen in various ways. The most common source of glare is the sun itself. Less often, glare may come from a bright patch of sky, or by reflection from a building opposite, or a bright patch of sunlight inside a room (Figure 5.21). People can experience glare either directly from the source, or when it is reflected from a surface indoors such as a computer screen.

If direct sunlight causes glare, transparent shading devices like tinted glazing usually do not help much because the sun is so bright. Opaque shading, for example venetian blinds, is best. Translucent shading, such as a thin light-coloured fabric blind, gives some protection but may itself become uncomfortably bright under sunlight. The BRE Report *Solar shading of buildings* gives further guidance.

Health & Safety Executive (HSE). *Health and safety (Display Screen Equipment) Regulations 1992*
BRE
Littlefair P J. *Daylighting design for display-screen equipment.* Information Paper IP 10/95. 1995
Littlefair P J. *Solar shading of buildings.* BR 364. 1999

Figure 5.21 Reflection of a bright window in a computer screen. Reflected sunlight in light-coloured surfaces in the room could also cause glare here

Glare or dazzle can also result outside the building, when sunlight is reflected from a façade or roof, either from glazing or shiny cladding. This can cause problems for motorists or for people in adjoining buildings. The extent of any problem should be assessed at the design stage. Glare can also be experienced from nearby buildings. Solar dazzle can be reduced by:

- reducing areas of glazing or shiny cladding,
- altering the type of cladding,
- substituting clear or absorbing glass for reflective glass,
- reorienting the building,
- using opaque screening, although this usually needs to be quite large.

Performance requirements

Materials and durability
Wood

Hardwoods such as oak from managed forests are being specified in addition to softwoods, which are comparatively easy to treat with preservatives. Finishes may be selected from a wide variety of formulations.

Most paints for external use on timber windows are now based on solvent-borne alkyd resin systems, and a three-coat system (primer, undercoat and top coat) is still the most popular finish. Some paints, both solvent- and water-borne, are available which are described as microporous or breathing paints. There is no evidence that the long-term moisture contents of wood under these paints differs significantly from those under conventional systems, though there may be a short-term benefit. Durability will be expected to be similar to that obtained from more conventional systems.

Wood stains provide suitable finishes for windows. whether zero-, low- or high-build, and these were described in an earlier section, *Applied external finishes*.

Metal

Although, hot-rolled galvanised steel sections in the past have not been suitable (in domestic sizes) for double-glazing units, more recent designs are suitable, such as W20 steel windows which can take a unit up to 14 mm thick and W40 steel windows which can take a 24–28 mm unit. Polyester powder factory-applied coatings that overcome the problems of site painting of galvanised steel and regular repainting are the normal finish.

Aluminium is a good conductor of heat (ie a poor insulator) and thus, aluminium frames need to incorporate a 'thermal break' plastics insert to reduce the amount of heat that is transmitted to the outside of the building (Figure 5.22). This helps prevent condensation on the frames in cold weather.

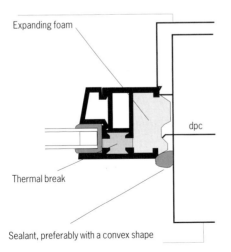

Expanding foam

dpc

Thermal break

Sealant, preferably with a convex shape

Figure 5.22 A thermal break window

Since the late 1980s, aluminium frames fitted directly into openings have largely superseded the types which required timber subframes. The normal finish for aluminium frames is a white polyester powder coating which should be durable and need no maintenance other than washing. Other colours may be available.

Plastics

PVC-U windows give good overall performance, with hardly any maintenance of the frame required. PVC-U compound is extruded into lengths of 'profile', which are then cut to length and assembled into window frames. Corner joints are heat welded, while other joints may also be welded, or may use screw fixings. Most types of window can be obtained in PVC-U; inward or outward opening, sliding sashes, bay windows, door frames and conservatories. They do not usually require sub-frames, but suitable load-bearing elements will be needed for bay windows.

Basic formulations of PVC-U are rather brittle, particularly when cold. PVC-U for windows is specially formulated for improved toughness. Other additives are used to limit the deterioration caused by ultraviolet light from the sun. Different manufacturers use different formulations, and so the quality of the PVC-U is important. PVC-U is not as stiff as wood or metal, so profiles are made with a fairly large cross-section. Metal reinforcement should be placed within the profile to stiffen it. The corner welds are a potential weakness.

Many materials can have their colour changed by exposure to sunlight. It is a potential problem for white PVC-U and imitation wood grain, just as for paints or polymeric finishes on window frames of other materials. In all cases, the manufacturers need to test that their material will maintain its colour well. Dark colours need to be selected with due regard to their potential for thermal distortion under prolonged exposure to sunlight.

Ultraviolet light from the sun will tend to break down the surface layer of white PVC-U. At first, it will become rather more brittle followed by 'chalking' away of a thin layer. The window will lose its shine and fresh white pigment will be revealed. If suitable stabilisers have been added to the PVC-U, the loss of thickness will be too slow to matter.

There have been concerns that, being plastics, PVC-U windows would be particularly dangerous in the event of fire. In fact, tests carried out on behalf of the British Plastics Federation have shown that PVC-U windows do not create new hazards in fires.

Glazing

Annealed glass is the commonest glazing medium, in spite of its vulnerability to breakage. However, safety glasses are increasingly being specified in situations where the risk of breakage is greater.

Laminated glass. Ordinary laminated glass is, thickness for thickness, no stronger than annealed glass; its virtue is that broken shards tend to be held by the interlayer and, if recommended glazing techniques have been followed, the glass should be retained until safe removal is possible. Wired glass is weaker than unwired annealed glass of corresponding thickness but, as with laminated, the shards tend to be held. The best of both worlds can be obtained with sheets of laminated toughened glass.

Laminated glass for domestic work is commonly made with two sheets of glass laminated together under heat and pressure with a polyvinylbutyral (PVB) interlayer. Laminates with acrylic interlayers may also be used.

Thermally toughened glass is stronger than annealed, both in terms of impact resistance and of thermal stress. When broken, the glass will not hold together though the pieces are relatively small.

The Workplace (Health, Safety and Welfare) Regulations 1992 are retrospective, and require all glazing in relevant non-domestic buildings that is at a vulnerable height either to be protected or to be of a safety glazing material. Annealed glass may be suitable provided it is thick enough.

Sheet glass is available in many varieties which it is not practical to list in detail here, though they range from methods of strengthening (toughening, laminating and wiring) to coatings to alter the light and heat transmission characteristics (eg low E), and surface patterning and modelling.

Plastics glazing materials are less prone to breakage than glass, but most are softer and more liable to surface damage such as scratching. They may also have quite large thermal movements. Those most commonly used are polycarbonate and acrylic.

Polycarbonate is not prone to discoloration so much as surface erosion due to weathering. This tends to lead to loss of surface gloss and reduction of clarity of transmitted images. Polycarbonate has relatively good retention of impact resistance and dimensional stability on weathering. Surface-coated varieties are available with significantly enhanced weatherability.

Acrylic sheet (polymethyl methacrylate) has very good retention of optical clarity on exposure to weathering but can be prone to crazing and embrittlement in the long term, especially when subjected to undue stress.

There is more information on types of glass and their breakage characteristics in BRE's report *Sloping glazing: understanding the risks*.

Insulating glass units

Insulating glass or double-glazing units have a lower heat transmittance (U-value) than single glazing, and can improve the comfort of building occupants in areas of rooms near to windows. Little difference is made to noise insulation.

A double-glazing unit normally comprises two panes of glass held at a fixed distance apart by a

HSE. *Workplace (Health, Safety and Welfare) Regulations 1992 Approved Code of Practice and Guidance.* L24

BRE
Kelly D, Garvin S & Murray I. *Sloping glazing: understanding the risks.* BR 471. 2004
Garvin SL. *Insulating glazing units.* Digest 453. 2000
Garvin SL & Blois-Brooke TRE. *Double-glazing units: a BRE guide to improved durability.* BR 280. 1995

Figure 5.23 Sealed double-glazed units mounted in timber frames

continuous spacer bar located around the perimeter of the glass which is then sealed (Figure 5.23; see also BRE's *Digest 453* and BRE's guide *Double-glazing units* for further information). Most types of glass can be used. The spacer is normally manufactured from mill-finished aluminium, although other materials such as galvanised steel and plastics are used. If the unit is to perform satisfactorily over a long period, the sealant or combination of sealants must be appropriate and the unit must be properly glazed in a suitable frame. If only one sealant is used, the unit is termed a 'single-seal' double-glazing unit, but if there is an additional seal compressed between the spacer bar and the glass, the unit is termed a 'dual-seal' double-glazing unit.

In the UK and Europe, types of sealant that are used in considerable quantities include polysulfide, polyurethane, silicone, hot melt butyl and polymercaptan.

In all double-glazing units a desiccant is held within the hollow spacer bar. This desiccant absorbs water vapour sealed in the double-glazing unit at the time of manufacture and also absorbs moisture that permeates through the edge seal. The desiccant is intended to prevent misting within the air space during the service life of the units. Typical desiccants used in the UK are molecular sieves, silica gel or blends of both types.

The design life of a sealed unit is intended to be 30 years, but the service life can vary between 5 and 35 years. Third-party certification is recommended.

Glazing techniques

The two principal glazing systems for both single- and double-glazing are known as:
- drained and ventilated systems (Figure 5.24), and
- fully bedded systems (Figure 5.25).

The drained and ventilated is the preferred method, particularly for installing double-glazing units of all types into frames and uses either a drained or a drained and ventilated glazing system. In these systems, weather-proofing is provided by gaskets or foam strips with a sealant capping in intimate contact with the glass. Any moisture which breaches these seals is not allowed to pond in the rebate but is drained away via a sloping rebate and drainage holes. Holes may also be provided in the upper section of the frames to encourage air to flow around the perimeter of the unit and to equalise air pressures in the glazing rebate. Particular attention should be

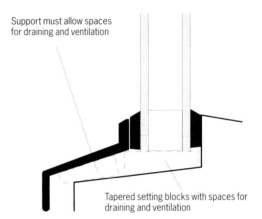

Figure 5.24 The drained and ventilated glazing method

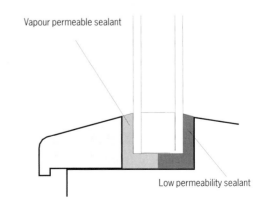

Figure 5.25 The fully bedded glazing method

paid to the type and position of setting blocks, location blocks and distance pieces to ensure that they do not inhibit either drainage or ventilation (see Figure 5.24).

Drained and ventilated glazing systems can be designed for all types of window frame, including those made of aluminium, steel, PVC-U and timber.

The alternative method, the 'fully-bedded' method means just that. In order to function correctly, and keep water from lying in contact with the edge seal, the bedding must be perfect. Fully bedded glazing systems should only be used where drained and ventilated systems are impractical, such as some timber and W20 steel windows, or are unavailable.

Particular attention should be paid to the type and position of setting blocks, location blocks and distance pieces, especially in the case of drained glazing to ensure that they do not inhibit either drainage or ventilation. Similar conditions apply also to the installation of laminated glazing. On no account should linseed oil putties be used.

BRE's *Digest 453* gives more guidance on insulating glazing units.

Thermal insulation

Typical double-glazing units have a thermal transmittance value (U-value) of around 3.0 or 4.2 W/m²K, depending on frame materials and air gap sizes, compared with a typical single-glazed figure of 5.45 W/m²K. To achieve even lower transmittance figures low-emissivity glazing can be used, and the space between the panes filled by inert gases such as argon or krypton. The low-emissivity coatings allow light to pass through the glazing in one direction, but reflect the majority of long-wave radiation or heat, re-emitted from inside the building. This increases the 'greenhouse effect' of glazing, allowing solar radiation into the building but trapping heat inside. Filling the cavity between panes with special gases reduces the heat loss through this gap. Coatings and inert gases reduce the overall U-value of the glazing part of the window to 2.0 W/m²K or even as low as 1.4 W/m²K.

Sound insulation

BS 8233 gives the following sound insulation values for typical windows measured in the field:
- any type of window in a façade when partially open: 10–15 dB R_w,
- single-glazed windows (4 mm glass): 22–30 dB R_w,
- thermal insulating units (6-12-6): 33–35 dB R_w,
- secondary glazed windows (6-100-6), 35–40 dB R_w,
- secondary glazed windows (4-200-4), 40–45 dB R_w.

Against aircraft noise, good quality, well-sealed secondary glazing gives about 3 dB improvement over sealed units. It is only properly designed secondary glazing with wide air gaps and good seals, that gives enhanced sound insulation against external traffic and aircraft noise.

Burglary

Windows should:
- be glazed with appropriate materials firmly fixed in position,
- be constructed from materials of suitable strength and sizes,
- be securely fixed in position in the surrounding wall,
- have a minimum of opening lights, except where they form a means of escape,
- have as large a pane size as possible,

and, where opening lights are required,
- be fitted with locks and concealed or tamper-proof fixings and hinges.

Means of escape

Windows are used in certain circumstances as a means of emergency egress, particularly in ordinary houses and in loft conversions. Such windows need to have opening lights of minimum clear space 850 mm × 500 mm and are subject to appropriate positioning and to the installation of safety devices that prevent accidental opening. Where windows are

Garvin SL. *Insulating glazing units.* Digest 453. Bracknell, IHS BRE Press, 2000

British Standards Institution. BS 8233: 1999 *Sound insulation and noise reduction for buildings. Code of practice*

used as means of escape there may be a conflict with other safety considerations.

Window safety

The various hazards that are posed by window openings can be classified as given in Box 5.4.

In 1967 the UK Government produced a Design Bulletin *Safety in the home* which makes the following recommendations for windows installed above the ground floor, particularly in local authority property:

- If opening lights in any windows above ground floor level are lower than 800 mm from the floor, additional protection is needed, eg a guard rail.
- Above third floor level, the minimum internal height to the sill should be increased to no less than 1140 mm, and at this level windows should be designed so that they can be cleaned (safely) from the inside.
- 550 mm is the maximum allowable reach through an open window to clean glazing (Figure 5.26).
- Side-hung casements should have easy clean hinges with a clearance of about 95 mm.
- Fastenings to casement windows and pivot windows above ground floor level should limit the initial opening to 100 mm and provide continuous control of window movement (Figure 5.27).
- Windows that are reversible for cleaning should have locking bolts to secure them in the open

Figure 5.26 Maximum sideways reach for cleaning purposes should not exceed 550 mm

position; these locks should be out of the reach of children.

In *Safety and security in the home*, Sinnot considers that the safety of openable windows must be considered principally for children. Falls from windows, Sinnot estimated, were responsible for 1% of all fatal accidents involving children. To make openable windows as safe as possible the following advice was given:

- permanently restrict the distance the window can open to a maximum of 100 mm,
- make the height to the bottom of the openable window well above that which a young child can reach, including when standing on a chair, eg 1350 mm to the opening mechanism,
- the inner sill or ledge should not provide a platform on which a child could stand or sit,
- windows should open to sufficient dimensions to allow emergency escape.

Box 5.4 Types of hazard posed by window openings

- Falls through open windows.
- Inability to escape in the event of fire due to locked, inoperable or restricted window openings.
- Collision with glazing, which could be a result of people colliding with open windows on circulation routes around buildings or falling against glass in buildings, and trapping of fingers and other body parts in window frames or hardware.
- Burglary as a result of break-ins through open or insecure window openings.

Key Point

Window safety is not just a matter of the window and associated hardware but concerns the design of the building in general.

Ministry of Housing and Local Government. *Safety in the home.* Design Bulletin 13. London, HMSO, 1967

Sinnot R. *Safety and security in the home.* London, Collins, 1985

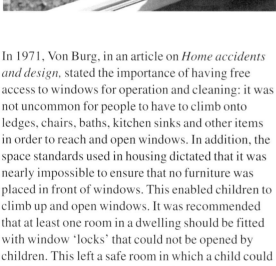

Figure 5.27 Safety features in softwood side-hung windows: **(a)** the limiter is engaged automatically on opening; **(b)** the casement opens to 90° to permit means of escape; **(c)** the casement opens to 180° to permit cleaning of the outside

In 1971, Von Burg, in an article on *Home accidents and design,* stated the importance of having free access to windows for operation and cleaning: it was not uncommon for people to have to climb onto ledges, chairs, baths, kitchen sinks and other items in order to reach and open windows. In addition, the space standards used in housing dictated that it was nearly impossible to ensure that no furniture was placed in front of windows. This enabled children to climb up and open windows. It was recommended that at least one room in a dwelling should be fitted with window 'locks' that could not be opened by children. This left a safe room in which a child could be left while an adult was otherwise engaged. However, if ventilation was required in this room then suitable hardware, guarding or window design should be fitted to reduce risks.

A government report by Tinker & Littlewood considered accidents involving falls from windows by children. Two-thirds of all the falls were from a bedroom window, half of which were the child's own. Almost two-thirds of the children were likely to have climbed onto furniture to reach the sill or window. Many had clambered onto furniture under the window that their parents had thought they were too young to reach, or had dragged furniture across the room specifically to be able to reach the window. Faulty or missing catches contributed to the accidents in one-third of the cases where windows were involved.

Von Burg E. Home accidents and design. *Build International* 1971: **October:** 268–270

Tinker A & Littlewood J. *Families in flats.* London, The Stationery Office, 1990

BRE *Information Paper* 17/93 summarises statistics on the occurrence of falls from windows using data from the Home Accident Surveillance System of 1985–1986. Fatal deaths are split into three age categories:

- 0–14 years = 20% of fatalities,
- 15–64 years = 60%,
- 65 + years = 20%.

A number of recommendations are given in IP17/93 on the design of windows in homes which are summarised in Box 5.5.

Window safety is covered by the Building Regulations for each part of the UK:

- **England & Wales:** AD N,
- **Scotland:** Section 4 of the Technical Handbooks, Domestic and Non-Domestic,
- **Northern Ireland:** Technical Booklet V.

Box 5.5 Safety recommendations for the design of windows

Placement of windows
Windows should be located so as to minimise accident risk, paying attention to the surroundings, likely position of furniture, the window itself and the window sill. All these features are used by children when they are playing.

Sills
Sills should be designed to discourage climbing or using them as seats or standing platforms; they could be sloped or eliminated completely.

Guarding
Guarding should be considered, especially if the sill height is below 800 mm.

Hardware
Windows on the first floor and above should be fitted with restrictors that allow no more than 100 mm opening.

Height above the floor
The height of the lowest part of the window opening should prevent occupants who overbalance or fall in the room from falling out of the window. In blocks of flats, from the second floor upwards, the lowest part of the window should not be less than 1100 mm from the floor.

Webber GMB & Aizlewood CE. *Falls from domestic windows.* Information Paper IP17/93. Bracknell, IHS BRE Press, 1993

However, this is mainly concerned with aspects of window cleaning and not child safety. BS 8213-1: gives good practice guidance for windows above the ground floor and recommends that safety restrictors should be fitted to accessible opening windows where children or adults are at risk of falling out. Table 1 in BS 8213-1 summarises different window configurations, the associated risks and means of managing those risks.

Doors

The prime purpose of a door is to:
- ease ingress and egress for occupants and goods.

External doors, as part of the external envelope of buildings, in theory, should conform as nearly as possible to the performance requirements of the remainder of the envelope. However, the extra requirements of access, and the wear and tear that this involves, have inevitably tended to make them more of a challenge to the designer than that of the wall in which they are situated. This leads to the probably inevitable compromise between lightness in weight, for ease in opening, and robustness to resist unauthorised entry.

Alternative materials for both leaves and frames, include wood, steel and aluminium, and now the practically all-glass door. Hinging is not always appropriate for some designs, and pivots are used, sometimes incorporating hydraulically operated closers.

Infiltration of wind and rain through open doors can be minimised by specifying a revolving doorset (Figure 5.28), smaller ones hand-operated, but larger versions power-operated. Remember, however, that alternative flat-threshold access will need to be provided for wheelchair users, parents with push-chairs, etc. There is guidance on

---

British Standards Institution. BS 8213-1: 2004 *Windows doors and rooflights. Design for safety in use and during cleaning of windows, including door-height windows and roof windows. Code of practice*
Stirling C. *Level external thresholds: reducing moisture penetration and thermal bridging.* Good Building Guide 47. Bracknell, IHS BRE Press, 2001

Figure 5.28 Revolving doors such as this assist in conservation of heat in well-used entrances

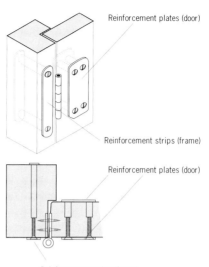

Reinforcement plates (door)

Reinforcement strips (frame)

Reinforcement plates (door)

Reinforcement strips (frame)

Section through door and door frame

Figure 5.29 Reinforcement strips and plates for hinges

designing level external thresholds in BRE's *Good Building Guide 47*.

A powered entrance door should be provided wherever possible to all buildings to which the public have access. They are particularly appropriate for use by the disabled, and information on them can be found in chapter 12, *Access to buildings*

All doors providing access to a building should be robust and capable of withstanding rough treatment. Doors in concealed positions should be able to resist attack by tools (eg hammers, tyre levers, etc.). In the domestic situation, 44 mm thick solid timber doors offer better protection than hollow-cored doors, and steel doors even better. Reinforcement plates for timber doors (Figure 5.29) may be worthwhile where the risk of unauthorised entry is high.

Final exit doors

Front doors are the main entry and exit routes from a house and are normally the main escape route in an emergency. Such doors should be provided with:
- a mortice deadlock operated from either side only by a key,
- a rim automatic deadlock,
- a door chain or limiter,
- a door viewer,

- a letter plate, if a suitable position is not available near the main entrance door. It should be positioned at least 400 mm from any locks to prevent manipulation of the locks.

Additional bolts should not be fitted to the main entrance door since they may delay escape of the occupants in an emergency.

Sliding patio doors are a common point of entry for intruders. Unless precautions are taken during fitting, it is comparatively easy to lever them out of their tracks.

Further information on security components can be found in LPS 1175-5.3.

Weathertightness

Laboratory measurements of water penetration through doors, using the same test methods as for windows have shown doors to be generally much less resistant to leakage than windows. It is difficult to design an inward-opening external door for an

Loss Prevention Certification Board (LPCB).
Requirements and testing of burglary resistance building components, strongpoints and security enclosures. LPS 1175-5.3. Available from www.RedBookLive.com

exposed situation which will meet a high standard of resistance to leakage without making it difficult to open, or without providing additional protection such as a canopy or even a porch. Outward-opening doors are much easier to make weatherproof, but even these may need the extra protection of a porch in the most exposed situations.

Alternative means of reducing the water load on the area in front of the door are by providing:

- shelter (eg a porch),
- a drainage channel or gutter in front of the threshold,
- permeable paving.

Thermal insulation

Thermal insulation values for opaque and glazed doors are effectively identical to those for windows ($U = 2.0$ W/m^{2}K).

Noise and sound insulation

It is difficult to improve the performance of external doors to more than around 30 dB, and where this figure needs to be improved on, consideration should be given to fitting a lobby. The use of two doors either side of a lobby, should, depending on the absorbency of the surrounding wall surfaces and how well the walls are constructed, give around 45–50 dB. A lobby will also be a useful additional precaution against heat loss.

Regulations and standards

Building Regulations

The combined dead, imposed and wind loads that must be transmitted to the ground safely, without undue movement. The structural requirements for walls are covered in the Building Regulations for each part of the UK:

- **England & Wales:** AD A,
- **Scotland:** Section 1 of the Technical Handbooks, Domestic and Non-Domestic,
- **Northern Ireland:** Technical Booklet D.

For resistance to weather and to ground moisture:

- **England & Wales:** AD C,
- **Scotland:** Section 3 of the Technical Handbooks, Domestic and Non-Domestic,
- **Northern Ireland:** Technical Booklet C.

The requirements that the building shall be designed and constructed so that there are appropriate provisions for the early warning of fire, and appropriate means of escape in case of fire from the building to a place of safety outside the building capable of being safely and effectively used at all material times, are covered in the Building Regulations for each part of the UK:

- **England & Wales:** AD B,
- **Scotland:** Section 2 of the Technical Handbooks, Domestic and Non-Domestic,
- **Northern Ireland:** Technical Booklet E.

Other regulations

Where housing developments are proposed, the NHBC or Zurich Building Guarantees requirements may also have to be met. These have specific requirements for the design and construction of walls.

In the UK, there is no general statutory requirement for a particular daylighting or sunlight level. However, the HSE Workplace Regulations require that:

'where practicable, the lighting of workplaces should be by natural light'.

To be fit for human habitation, the 1985 Housing Act stipulates that a dwelling in England and Wales should have adequate provision for lighting. Guidance on the application of this requirement is contained in DCLG's Housing Fitness Standard. However, this guidance is advisory; local housing authorities are asked to:

'have regard to it when applying the standard but must form their opinion in the light of all relevant circumstances'.

Health & Safety Executive (HSE)

Workplace (Health, Safety and Welfare) Regulations 1992
Health and safety (Display Screen Equipment) Regulations 1992

Section 8 of the Annex deals with lighting, states:

'Habitable rooms should have sufficient natural lighting to enable normal domestic activities to be undertaken without strain during the main hours of daylight without requiring artificial light, unless the day is particularly overcast. ...As a general guide the area of glazing in a habitable room should not be less than one tenth of the floor area, and some part of the window should normally be at least 1.75 m above floor level'.

There are proposals to replace the Fitness Standard by a Housing Health and Safety Rating System. This does not have a numerical criterion for good daylighting provision.

Standards and codes of practice

Relevant Standards and Codes of practice are listed in chapter 13 *References and further reading*.

6 Floors and ceilings

Performance requirements

The range of functions to be taken into account in the design of floors is not as extensive as in the case of those elements forming part of the external envelope of a building. Nevertheless, there are several (eg control of noise and wear), that can be of crucial importance for floors. So far as ceilings are concerned, they play a role in determining the fire resistance of a floor, and the topmost ceiling in a building under the roof also plays a role in determining thermal losses for the whole building and also additional sound insulation, particularly where the roof covering is lightweight.

Floors of a building can either be supported by the ground or can span across a space. Floors supported by the ground depend for their structural integrity on the adequacy of consolidation of any hardcore and selection of the correct grade of any thermal insulation.

Suspended floors can take many forms. Where beams are used to support decks, solid sections are used over the shorter spans, and progressively lighter and structurally more efficient solutions as the span increases: from perforation or hollow coring of beams and infill decks to web construction over the largest spans. As in the case of roofs, much attention has been given to the development of the most efficient sections of beams, often using a combination of different profiles, sections and reinforcement techniques to exploit the characteristics of the chosen materials.

Strength and stability

Floors usually contribute significantly to the overall strength and stability of the structure (Figure 6.1). Nevertheless, it is the correct assessment of the residual strength and serviceability of floors that is vital to their continued safe use in buildings.

Even smaller, domestic-scale buildings of loadbearing brick often rely on the floors to provide lateral restraint to the external walls.

However, all floors need to be sufficiently strong to carry the self weight of the structure together with imposed loads; for example those due to furniture, equipment or the occupants of the building.

Current requirements as far as the structure of floors is concerned are embodied in the accompanying documents of the national building regulations. Taking Requirement A1 of Schedule 1

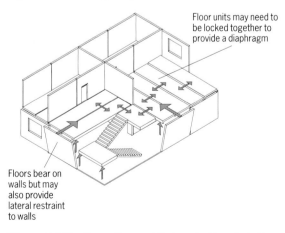

Floor units may need to be locked together to provide a diaphragm

Floors bear on walls but may also provide lateral restraint to walls

Figure 6.1 The floor often contributes significantly to the overall strength and stability of the structure of a building

of the Building Regulations (England & Wales) 2000 as an example:

(**1**) The building shall be constructed so that the combined dead, imposed and wind loads are sustained and transmitted by it to the ground
   (**a**) safely; and
   (**b**) without causing such deflection or deformation of any part of the building, or such movement of the ground, as will impair the stability of any part of another building.
(**2**) In assessing whether a building complies with sub-paragraph (1) regard shall be had to the imposed and wind loads to which it is likely to be subjected in the ordinary course of its use for the purpose for which it is intended.

Structural design of floors for buildings is also covered by the main British Standard codes of practice for the various materials:
- steel to BS 5950, various parts,
- concrete to BS 8110, Parts 1–3,
- timber to BS 5268, various parts.

Loadings

The main function of a floor is to support the loads placed on it during the life of the building. With some types of building, loadings are easy to predict and unlikely to increase with time. With others, particularly with industrial buildings, change of occupancy or developments in production may require floor strengths to be increased from those originally considered adequate. There is a dilemma here for designers. Strengthening an existing floor is normally inconvenient, difficult and expensive. On the other hand, design for improbably high loadings to allow for change will result in excessive cost of the floor construction itself, of the supporting framework or walling, and of the foundations.

Design values of floor loads for current use are specified in BS 6399-1.

Dead loads include the weight of the structure of the floor, calculated from the values given in BS 648 or, better still, from known actual weights. The load calculations should include all tanks, etc. filled to capacity (in fact, all permanent construction including services of a permanent nature).

Imposed loads are all those loads arising from the use of the building including movable partitions but excluding partitions of a permanent nature; wind loads are also excluded. There are reduction factors to take account of in the likelihood that all floors will not be loaded to the maximum values simultaneously, but these are not always applicable.

The structural codes over the years have given uniformly distributed loads and point loads for:
- residential buildings,
- institutional buildings,
- offices, banks, etc.,
- halls, libraries and theatres, etc.,
- shops,
- workshops and factories,
- warehouses, and
- garages and car parks.

The values for uniformly distributed loads, for example, have ranged from 1.5 kN/m^2 for bedrooms and 2 kN/m^2 for most other domestic-scale accommodation excluding circulation areas, up to 7.5 kN/m^2 for boiler rooms and such like. Ceiling structures with access provide for a uniformly distributed load of 0.25 kN/m^2 with an additional point load, though ceilings without access do not need such allowance.

Problems from local distress in a floor are likely but they will be small with, for example, a solid reinforced concrete floor which is capable of effective lateral distribution of loads. However, the effect of concentrated loads might be more serious where the floor is built up from a series of separate adjacent beams. The extent of load-sharing between beams depends on the character and extent of the connections between them formed, for example, by

British Standards Institution
BS 5950: *Structural use of steelwork in building* (various parts)
BS 8110: *Structural use of concrete* (Parts 1–3)
BS 5268: *Structural use of timber* (various parts)

British Standards Institution
BS 6399-1: 1996 *Loading for buildings. Code of practice for dead and imposed loads*
BS 648: 1964 *Schedule of weights of building materials*

grouting or by a surface layer such as a screed or other decking. In a timber floor, the load is shared between joists by the action of the floorboarding.

Lateral support and restraint

Lateral support is often required to be provided by floors to walls, and such requirements are set out for new buildings; for example, external and compartment or separating walls greater than 3 m long will require lateral support by every floor forming a junction with the supported wall; and all internal loadbearing walls of whatever length will require support from the floor at the top of each storey.

Straps may not be needed if the floor has an adequate bearing on the wall (eg 90 mm in the cases both of the bearing of timber joists and of the bearing of concrete joists).

Load testing

Load testing of a structure involves applying test loads to an existing structure to determine whether it is satisfactory. They may need to be employed to check compliance where the design is based on a performance specification.

Load testing is expensive and is used only infrequently in structural appraisal.

Impact resistance

The importance of impact resistance, both for the floor as a whole as well as the flooring, depends on particular circumstances; but it is generally the case that the solidity of the deck or the bedding substrate within the floor will be one of the most important factors in the resistance of the finish to indentation.

Impact resistance is not usually the most critical factor in the performance of floorings, and it is probably unrealistic to expect that under accidental impacts the flooring will not sustain some damage, though the floor itself should not be in danger of collapse.

Indentation resistance and recovery

Key Point
If a floor finish indents under load, it may not be a problem if the finish recovers when the load is removed.

Materials vary greatly in their recovery after loading. Surface texture and pattern may help to disguise indentations that do not fully recover.

Resistance to indentation (poinçonnement) has been considered by the European Union of Agrément (UEAtc) in relation to thin floorings used in domestic-scale buildings (UEAtc Method of Assessment and Test No 2). Tests are designed to provide assessment of the effects of dynamic and static furniture loads, falling objects, and the action of sharply pointed footwear.

Dimensional stability
Movements

Expansion and contraction of any part of the building fabric subjected to variations of moisture content or temperature will have the potential to cause problems if not accommodated in the design. As a general rule, all common building materials will be subject to thermal expansion and contraction. So far as materials used in floors and flooring are concerned, it is the larger components which need most consideration, especially where the building is only intermittently heated. Fortunately, though, movements will be much smaller than in roofs, for instance, because the internal environment will be that much less susceptible to external factors such as weather.

Moisture movement is mainly a property of porous materials; thus, concretes have reversible movements in the range 0.02–0.2%, and softwoods sawn tangentially across the grain in the range 0.6–2.6%, and sawn radially in the range 0.45–2.0%. Concretes also have irreversible drying and carbonation shrinkage in the range 0.02–0.1%. In most real situations there is some restraint, though rarely complete restraint, afforded to materials undergoing such movements.

Most substrates under floorings are affected to a greater or lesser degree, and the effects of movement can be sufficient to cause rippling in floorings if insufficient precautions are taken to prevent it.

European Union of Agrément (UEAtc). Directive for the assessment of floorings. Method of Assessment and Test No 2. Paris, UEAtc, 1970

Where two differing materials are joined, differential movement can occur, which usually exacerbates the problem. The best practice is to try to accommodate movements at the smallest and most elemental level, provided this does not prejudice the structural function.

In small structures such as detached domestic houses, it is often possible to omit explicit movement design and depend on restraint from other elements to accommodate movement loads, though floors in larger span buildings with higher loadings are another matter entirely. In timber floors, the relatively high moisture movement across the grain can lead to appreciable shrinkage in the section, causing gaps to open under skirtings.

Progressive movement may conceivably occur at the edges of floors adjacent to external walls, mainly as a result of ratcheting. Where movement in a material resulting from contraction (eg drying shrinkage) causes cracking, debris can fill the crack. Expansion cannot then close the crack because of the debris, in turn causing the material, and perhaps adjacent materials or elements, to move. Successive contraction and expansion movements, with more debris accumulating in the cracks, results in ratcheting.

Vibrations

Long-span lightweight floors are especially susceptible to vibrations which can be detected by building occupants. Although vibrations produced by people walking on the floor can prove annoying to other users, they do not create most concern. It is when floors are used for dancing or keep-fit classes where people are moving in unison, giving rise to resonance in the floor (which raises potential safety rather than serviceability issues) that the greatest problems may be encountered. The phenomenon has been generally appreciated for many centuries, demonstrated by the military instruction to break step when marching over bridges.

Floor designs can be analysed using various different methods to see whether they would have acceptable vibration characteristics in service. Movement can be minimised by building in bearings tightly (encastred bearings). This is a specialised

field but some further information is available in BRE's *Information Paper* IP 17/83.

Movement joints

Floors need to be provided with movement joints which reflect the positions of movement joints throughout the whole structure, and such joints should continue through the floor finish, and normally also provide continuity in other aspects of the performance of the floor such as its fire resistance.

Deflections

The amount of deflection that can be tolerated in a floor may depend more on the judgement of the occupants than on the actual construction of the floor and the effects of deflections on the integrity of finishes.

Thermal properties

The rate of heat loss through a floor situated next to the ground varies with its size and shape. Various requirements for thermal insulation in terms of thermal transmittance (U-values) are given in the accompanying documents of the national building regulations:

- **England and Wales:** AD L1 and L2,
- **Scotland:** Section 6 of the Technical Handbooks, Domestic and Non-Domestic,
- **Northern Ireland:** Technical Booklet F.

Indeed, the economic and other criteria that may be used to meet the values given in the national building regulations will vary according to building type, fuel used, its relative cost, and a host of other factors.

Key Point
What is important is where that insulation goes within the floor, that it is laid consistently, filling every space and avoiding thermal bridges, and that it is located on the correct side of any item of construction which functions as a vapour control layer (vcl).

Jeary AP. *Determining the probable dynamic response of suspended floors.* Information Paper IP 17/83. Bracknell, IHS BRE Press, 1983

The thermal insulation value of materials reduces with increased moisture content. It follows that materials that do not absorb water are needed where prolonged wetting, eg through dampness rising from below (from high water tables), is inevitable.

The insulation value of the floor can be degraded by thermal bridges where high thermal transmission materials penetrate layers of low thermal transmission material, such as may happen at thresholds. Thermal losses due to cold bridges are often ignored in calculations, especially where thin sections are involved, but these and other components and materials such as concrete floor beams become more important as thermal insulation standards increase. Some thermal bridges also have serious implications because they produce inside surface temperatures below the dewpoint of the air, leading to selective condensation on parts of the flooring. Suitable designs can overcome or reduce thermal bridging to acceptable levels.

In suspended ground floors, air movement into and within the void and especially through layers of low-density insulation material, can reduce the thermal efficiency of the floor considerably. Sealing at joints and around areas where services penetrate the insulation is important. Undesigned air movement within the floor void may also carry water vapour to areas where condensation can cause problems. U-values, however, can be approximated (see BRE's *Information Paper* IP 3/90).

Where the thermal insulation layer is intended to provide support for a screed or flooring, it will need to have certain minimum structural and other properties such as:
● resistance to loads,
● dimensional stability,
● chemical compatibility with the layers above and below.

The position of thermal insulation will determine the thermal properties of a ground floor. Where the insulation is placed as a sandwich between the base and screed in a solid floor, a quick warm-up is

achieved as only the screed is heated. However, once the heating is switched off the temperature will fall rapidly as the screed has little heat capacity. If the insulation is placed below the slab then, because of the large heat capacity of the concrete, the floor will be slow to respond to the heating being switched on but slow to cool when it is switched off.

Because most thermal insulating materials used in floors have some resistance to moisture vapour, but considerably less than that of a dpm, it is necessary with some constructions to provide additional moisture barriers to control constructional water. These are often referred to, rather loosely, as vcls, and are nearly always of polyethylene sheet.

Control of dampness/condensation, and waterproofness

Not all floors are subject to dampness problems, but those in contact with the ground in basements or ground floors will clearly be most at risk. Moreover, the risk should not be discounted altogether in the case of suspended floors, particularly at bearings on external walls. The form of any potential failure from dampness depends on the flooring material and the adhesive. Flexible sheet materials like PVC, linoleum and rubber commonly curl or blister, while tiles made from similar materials curl or tent at their edges. The basic trouble in most cases is some expansion of the flooring material accompanied by loss of adhesion between it and the base. Often, any moisture expansion is aggravated by stretching of the flooring material due to traffic after adhesion has been lost. Many adhesives soften in the presence of moisture. This adverse effect on flooring and adhesive is often made worse by the high pH of the moisture because it contains alkalis derived from the cement in the concrete base or screed to which it is fixed. Timber block and strip floors fail by disruption brought about by expansion of the wood following moisture take-up. Carpets and other textile floorings may become debonded and ruck because the adhesive has softened. In addition, there is the possibility that some fibres or backings will rot.

Anderson BR. *The U-value of ground floors: application to building regulations.* Information Paper IP 3/90. Bracknell, IHS BRE Press, 1990

Sources of moisture

There are four main possible sources of moisture in floors to consider:

- excess constructional water,
- ingress of water from the outside,
- condensation,
- water spillages and services leaks.

Excess constructional water

Water is added to a screed or concrete mix to make it workable (ie more water than is required for hydration of the cement). The water for workability usually amounts to between half and two-thirds of the water used for mixing. This excess must be allowed to dry out before fixing impervious floorings such as PVC sheet, or moisture-sensitive floorings such as wood or textile, as both of these categories can be affected. If moisture-sensitive flooring is used, an extra surface damp-proof membrane (dpm) will be needed (Figure 6.2).

Water moves through concrete partly by capillary action but mainly by diffusion. (Water moving through concrete is a slow process.) Unless the atmospheric humidity is high, that is to say in excess of 90% relative humidity (RH), water will evaporate from the surface of a base as fast as it reaches it. Moisture is never visible once the initial surface water has dried off. The result is that the surface may look dry even though there is still a considerable amount of water below the surface.

The rate of drying, and therefore the time to dry a concrete base or screed before floorings can safely be fixed depends on a number of factors. These include the mix proportions, amount of mixing water added, the temperature of the floor, the relative humidity of the air and the thickness of the base or screed. A rule of thumb method often quoted for forward planning purposes is to allow a day for each millimetre of thickness for screeds. This has worked well for thicknesses up to 75 mm, although many sand and cement screeds laid semi-dry will be sufficiently dry well within the predicted times. Thus, a sand and cement screed 50 mm thick laid on a dpm usually dries within 4–6 weeks. Proprietary screed systems are available which dry considerably more quickly than conventional screeds. The maximum RH value for screeds, etc. for laying moisture-susceptible flooring is 75% RH.

The rule of thumb method of predicting drying times does not apply to thick concrete slabs. Commonly, slabs 150 mm thick with a dpm immediately below, take between 6 and 12 months to dry, but there have been cases where they have not been dry even after 18 months.

Where it is impracticable to allow sufficient time for the concrete base to dry out, a dpm or vcl must be laid between the wet construction and the sensitive flooring (Figure 6.3). Conventionally, this is achieved by providing a dpm on top of the slab and covering it with a screed which, because it is unbonded, should be a minimum of 50 mm thick. The application of an epoxy bonding agent (which also acts as a dpm) to the surface of the base enables bonded screeds less than 50 mm to be laid.

Proprietary surface dpms based on solventless epoxy resins can be applied to the surfaces of screeds or concrete bases, or, where the old substrate is not in good condition, over a moisture-tolerant skim coat of underlayment. They are most often used where specifiers have misjudged how long a construction will take to dry and there is a need to lay the flooring on a base which is still wet.

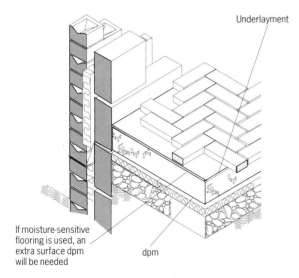

Underlayment

If moisture-sensitive flooring is used, an extra surface dpm will be needed

dpm

Figure 6.2 Where construction times are too short to allow adequate drying time, precautions may need to be taken to avoid construction water affecting any moisture-sensitive flooring

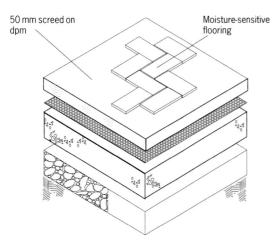

50 mm screed on dpm

Moisture-sensitive flooring

Figure 6.3 If the dpm is placed between base and screed, only the screed needs to be dry

Power-floated slabs, having a dense and therefore relatively impervious surface, dry out more slowly than trowelled slabs. Hygrometer readings from instruments left in position for at least 72 hours should provide a good indication of moisture conditions, even of power-floated slabs.

Water ingress from the outside
Water, which can potentially move from the ground through the base hardcore and concrete, or which can be conducted from wet external leaves of walls to the edges of sensitive floor construction, must be prevented from doing so by effective damp-proofing. This is especially crucial in basement floors with high water tables where tanking sufficient to withstand the hydrostatic pressure is required. In the case of a basement, it may be necessary to ensure that sufficient mass is available in the building above to prevent flotation of the basement.

If the surface of the screed or base is dry when flooring is laid, moisture from below the foundations can still rise after many months and the flooring could fail a year or two later.

Condensation
Condensation will occur when the surface temperature of the floor is below the dewpoint temperature of the atmosphere for a sustained period of time. The dewpoint will vary according to the air temperature and the relative humidity. Condensation occurs when the relative humidity of the air in direct contact with the cold surface rises to 100%.

The two most common situations in which condensation occurs on a floor are:
● where the floor is adjacent to an exterior perimeter wall and there is a loss of heat from the floor to the outside via a cold bridge,
● where the floor has a high thermal capacity and the floor temperature is unable to follow rapid changes to the air temperature which often falls below the dewpoint. This phenomenon particularly affects floors in warehouses in weather conditions where a cold spell is followed by a warm front.

It is not normally necessary to provide a vcl in floors to prevent moisture ingress downwards, though it may be required for special reasons; for example, to prevent water vapour entering from beneath the construction to affect sensitive materials.

For a material to qualify as a vcl it should have a vapour resistance greater than 200 MNs/gm.

Plastics films are the most usual materials for forming a vcl in a floor construction. Joints in a flexible sheet vcl should be kept to a minimum. Where they occur, they should either be overlapped by a minimum of 150 mm and taped, or sealed with an appropriate sealant, and should be made over a solid backing. Tears and splits should always be repaired with jointing or sealing as above. Penetrations by services should be kept to a minimum and carefully sealed at interfaces. Draughts of moisture-laden air through gaps in vcls are more significant than normal still air diffusion through materials; it is therefore of greater importance to avoid holes in the vcl than to take elaborate precautions for sealing laps in the layer.

Materials for dpms
Although concrete bases of good quality laid directly on the ground can be relatively impervious to the passage of 'liquid' water, they cannot be expected to stop all moisture rising from the ground. It is therefore necessary to provide protection from damp. The moisture barrier usually consists of a membrane either laid under the slab or sandwiched

between the slab and a screed. Undoubtedly, the ideal membrane is completely impervious to moisture, either as water or water vapour.

The degree of protection needed depends on a number of factors, eg the moisture sensitivity of the flooring, the fixing adhesive and site conditions.

Materials that can provide effective surface dpms, if properly laid, include:

- epoxy resins,
- pitch/epoxy resin mixtures,
- polyurethanes,
- mastic asphalt, not less than 12 mm thick,
- pitchmastic.

Concrete bases containing proprietary 'waterproofers' are not an acceptable substitute for a properly laid dpm.

Waterproofness of floorings

In certain circumstances, floors need to be completely waterproof, especially where the materials from which they are made are at risk of deterioration. Such circumstances arise, for example, in the bathrooms and kitchens of dwellings, or in manufacturing areas where water for cleaning is used on a large scale. There is no criterion which can be applied generally, and each case must be treated on its merits.

Air permeability

Air permeability is defined and discussed in general terms in chapter 1. So far as airtightness of ground floors is concerned, it is the perimeters of both suspended and solid floors that are most at risk. Where a suspended floor needs to be ventilated underneath to ensure freedom from condensation it will be the upper surface that needs to form the barrier. While intermediate floors for the most part are not in the external envelope, it is important to prevent air from cavities leaking into buildings around intermediate floor supports. Also, where such floors project from the building face to form balconies they can be a source of air leakage.

Comfort and safety

Comfort and safety are two functions which may be required of flooring. They are less amenable to

quantitative measurement and specification than other functions, though they are not of less importance. Indeed, aspects of safety should be of major concern to all specifiers of flooring.

Warmth to touch

Although a floor can be made to play a relatively small part in the heat losses from a building by suitable choice of thermal insulation, it is often the rate at which bare feet or thinly shod feet lose heat to the floor which can be of more immediate importance to the building user.

If a floor can be found to be too cold, it is reasonable to assume that it could also become too hot. Excessive heat in a floor might affect people and might also lead to degradation of the materials forming the flooring. In no case, should the temperature reached by the flooring exceed 27 °C; some materials are affected at lower temperatures.

The European Union of Agrément (UEAtc) has suggested a formula (Method of Assessment and Test No 2(25)) for considering the appropriate thermal comfort requirements of a flooring material:

$$\sqrt{\frac{k}{pc}}$$

where:
k = thermal conductivity (W/m °C),
p = bulk density (kg/m^3) and
c = specific heat (J/kg °C) of the material.

A test method is available to determine the quantity of heat lost in W(Jm^2) at the end of one minute and at the end of 10 minutes.

Resilience

Many floor finishes have a characteristic feel which is rather difficult to define and, hence, equally

European Union of Agrément (UEAtc). Directive for the assessment of floorings. Method of Assessment and Test No 2(25). Paris, UEAtc, 1970

difficult to select objectively when offered as a choice. Subjective views often predominate, as in the case of, for example, sales staff preferring a hard floor to a deep pile carpet which makes them feel tired. There may be problems with textile coverings; for example certain types of open-weave barrier matting without metal armouring can present wheelchair users with problems.

Accidents and safety

The contribution of the floor finish to accidents is an important issue. This is true not only for industrial locations, where the risks may be more obvious, but also in the home.

Responsibility for the safety of people engaged in building operations is not only confined to the contractor, the specifier also shares the responsibility in the sense that they must consider safety aspects of installing the items they are specifying, consider the risks, and decide whether to select alternatives. The specifier is also obliged under current legislation to provide information about health and safety issues for all items specified.

Compliance with all relevant items of safety legislation is outside the terms of reference of this book, but reference may be made to the various health and safety regulations and to the *Construction safety handbook*.

When considering floorings in buildings, it is important to consider the slipperiness characteristics of new materials and the lighting needs for all potential users, especially people who are physically or visually disabled.

Where features for stairs need to be checked, for example for conformance with the various national building regulations, the relevant detailed criteria will be found in the British Standards codes of practice (BS 5395: Parts 1–3) for the design of straight, helical and spiral stairs and industrial-type stairs, permanent ladders and walkways.

The size of the goings is the most critical factor in the safety of stairs. If the goings are large enough, the chances of a slip or overstep leading to a fall are dramatically reduced.

Nosings also have a part to play in increasing safety. Prefabricated nosings are available in a wide range of profiles and finishes to suit the materials forming the main part of the tread surface. The aim is to provide the right compromise between slip resistance and tripping hazard for both upward and downward travel. Full details can be found in BRE's *Information Paper IP 15/03*.

One further risk area to be considered is the sizes of gaps in flooring grilles. There is some difference of opinion on this matter. It can be argued that warnings are sufficient for those floors surrounding industrial processes where the public are not admitted. In public areas, particularly at entrances to buildings, consideration should be given to limiting the mesh clear gap size to 5 mm. This should allow snow, rainwater and most small stones picked up on moulded rubber shoe soles to fall through, while at the same time preventing narrow heels and umbrella ferrules from penetrating the gap.

Slip resistance

Slip resistance is a significant component in the avoidance of accidents, but it is also one of the most complex attributes of floorings in buildings to assess.

Most floorings when dry have adequate slip resistance, but many smooth floorings can become dangerously slippery when wet. In housing, where there is a tendency for floors to be wet sometimes, as in kitchens and food preparation areas, there needs to be a trade-off between ease of cleaning and slip-resistance. Uneven or slippery floors, however, make every activity potentially dangerous.

Durability and slip-resistant performance (the chief safety factors) have to be considered in

Davies VJ & Tomasin K. *Construction safety handbook.* 2nd edition. London, Thomas Telford, 1996. 320 pp
British Standards Institution. BS 5395: *Stairs, ladders and walkways*
   Part 1: 2000 *Code of practice for the design, construction and maintenance of straight stairs and winders*
   Part 2: 1984 *Code of practice for the design of helical and spiral stairs*
   Part 3: 1985 *Code of practice for the design of industrial type stairs, permanent ladders and walkways*

Roys M & Wright M. *Proprietary nosings for non-domestic stairs.* Information Paper IP 15/03. Bracknell, IHS BRE Press, 2003

relation to the particular use to which the floor will be subjected; for example, rubber flooring, which has good non-slip qualities when dry, is slippery when wet and therefore quite unsuitable for washing or cooking areas.

Adequate performance with respect to slipperiness is, however, also bound up with maintenance, an aspect which it is impossible for the specifier to control.

A coefficient of friction between the foot and the floor of 0.40 represents a safe value for walking on level surfaces. Few materials fall below this value when new, but in time a floor may become slippery through the polishing action of traffic, the presence of water or oil, or the use of too much wax polish. Any evaluation of slipperiness must therefore take account of these factors and is best based on experience as well as on measurement. The equipment normally used for tests is the portable skid tester developed by the Road Research Laboratory and described in BS 7976 Parts 1–3.

Specifiers have to balance the different requirements of a floor covering: cost, wear qualities, appearance, as well as slip resistance. Sometimes a slip resistance of less than 0.40 is acceptable, but only if certain conditions are met. For example, in dwellings the number of people using a floor is low so the chance that someone will need this high slip resistance value is low. However, in a shopping mall there are more people using the floor, increasing the chance of a slip so a high value is more likely to be required. The management of floor cleaning also has an effect on the value required; if the cleaning always takes place when there are no users present, or behind a barrier, lower slip resistance values in the wet state may be acceptable. Similar arguments follow for the correct size and placement of entrance matting, which can stop transfer of water from the outside onto the flooring.

Steps and stairs
Staircases provide one of the most hazardous situations in a building. So far as the horizontal

surfaces (ie treads) are concerned, the following points need to be carefully considered.

Changes of level and of material need to be made obvious to users, and this generally means good lighting conditions, a contrast in floorings either side of a change of level, or a strip of contrasting appearance such as a nosing.

At a change of level, flooring such as a textile which exhibits a strong or dazzling pattern, and which can camouflage the actual situation of the step or nosing to the user, should be avoided.

If a single step is isolated from any other change of level or occurs in an unexpected place, it is likely to cause a fall (Figure 6.4). Where a change of level is expected, eg at thresholds or entrances to buildings, it is not dangerous for the able-bodied.

AD B, Clause 5.25 of the Building Regulations (England & Wales) recommends that the floorings of all escape routes should be chosen to minimise their slipperiness when wet.

Chapter 12 gives specific information on staircase dimensions most suitable for the ambulant disabled.

Figure 6.4 Changes of level in unexpected places could lead to people tripping or falling

British Standards Institution. BS 7976: 2002 *Pendulum testers. Part 1: Specification. Part 2: Method of operation. Part 3: Method of calibration*

Radon and methane (see also chapter 2)

Soil gases including radon, methane and other landfill gases can enter buildings due to a combination of pressure differences between the soil and the inside of the building, and building characteristics. The risk applies to any part of the building in contact with the ground (eg ground floors, wall foundations, basement walls). Gas can enter through cracks and gaps in and around all types of floor construction (solid or suspended), around service penetrations and through construction joints and wall cavities.

The level of protection required within a new building will depend on:
● for methane or landfill gas: concentrations measured on site,
● for radon: geographical location.

The main method of protection for new buildings is to provide a gastight membrane across the full footprint of the building in contact with the ground. For radon, this might be further enhanced with provision for future subfloor ventilation by including a ventilated subfloor void or radon sump. For other gases such as methane, a permeable layer may be required beneath the airtight barrier to allow for natural ventilation, or where concentrations are very high, future mechanical ventilation.

Aldehydes and volatile organic compounds

Some binders employed in the manufacture of wood-based products and boards used for flooring contain aldehydes, formaldehyde in particular. The vapour is an irritant and, in sufficient concentrations in the internal atmospheres of buildings, can cause personal discomfort. When the problem became apparent the specifications of the boards were amended so new boards conforming to the latest specifications should not present a problem.

Commission of the European Communities, Directorate General for Science. Research and Development. Guidelines for ventilation requirements in buildings. *European Concerted Action: Indoor air quality and its impact on Man. Report No 11.* Luxembourg, Office for Publications of the European Communities, 1992

Some products emit other volatile organic compounds (VOCs) which, in sufficient concentrations, may be injurious to health. They include solvent naphtha and related compounds from dpm materials, and plasticiser from degraded PVC floor coverings. To enable specifiers to select suitable products, information is available about the types and amounts of VOCs emitted. (*European Concerted Action: Indoor air quality and its impact on Man, Report No 11*).

Levelness and accuracy

Any floor must be reasonably flat and level for safety reasons, and so that furniture standing on the floor is visibly plumb and level. The limits for local variations for newly built, nominally flat floors include three categories: up to 3 mm, up to 5 mm, and up to 10 mm under a 2 m straight edge laid directly on the floor. Although these figures refer to concrete floors, levelling screeds and in-situ flooring generally, there seems no reason why they cannot be applied to other floors as appropriate. The middle category is the one most often used in domestic and office buildings.

In housing, small changes in floor level represent a hazard and will need to be eliminated or avoided where possible, particularly if young families or elderly people are likely future occupants (see the section on safety earlier in this chapter).

Hygiene

Certain floors are required not to harbour dirt and germs, and should not allow water to penetrate below the surface. Hot welding of, for example, flexible PVC and linoleum sheets with rods of compatible material, given adequate standards of workmanship, can provide a continuous surface which offers considerable advantages in hospitals and certain industrial process areas. To provide a virtually tanked floor area, sheet flooring material can be formed with radiused corners at the floor perimeter and welded at the corners of the upstands. The integrity of the weld is of paramount importance.

Thermosetting resin floorings of epoxy or polyurethane can provide good resistance to a wide range of chemicals and be laid in-situ to provide smooth hygienic surfaces. These materials are used in food and pharmaceutical process areas and clean rooms.

Fire and resistance to high temperatures
Non-combustible floors

> **Key Point**
> Non-combustibility does not of itself ensure fire resistance.

Metals, for example, lose strength rapidly when heated above a critical temperature; any steel or aluminium used structurally must be insulated, and structural steel should also be protected by the use of fire-protecting suspended ceilings. The fire resistance of solid reinforced concrete slab floors depends on the minimum thickness of the slab and on the thickness of cover to the reinforcement. With hollow floors, including precast concrete or clay units, it is the minimum total thickness of solid material which determines their fire resistance. The fire resistance of prestressed units, though influenced by many factors, depends mainly on the thickness of cover to the steel.

Compartment floors

Compartment floors, which are in essence needed in conjunction with walls to separate one kind of risk from another or where a portion of the building (a compartment) must be restricted in size, are required to have enhanced requirements for performance in fire.

The fire resistance of a floor is judged according to three attributes:
- loadbearing capacity,
- integrity,
- insulation.

A fire-resisting floor will also need to satisfy sound and perhaps thermal requirements. Also, it is important to note that the finished widths of timber joists to achieve adequate fire resistance must be a minimum of 38 mm; and there may be other kinds of problems such as calcium silicate board being insufficiently robust to carry extra weight. Furthermore, the underside of all floors exposed to

rooms and corridors must satisfy provisions for surface spread of flame.

Openings

Where a floor needs to be a compartment floor, there are strict controls on openings in order to ensure that the effectiveness of the compartment has not been compromised. There are various ways of ensuring this; for instance by means of placing the opening within a protected shaft, or by means of a fire damper or shutter. However, where the floor is not a compartment floor, that is to say it is simply an element of structure, less stringent controls are applied depending on what penetrates the floor.

A protected shaft also requires the use of materials of limited combustibility and this will effectively rule out the use of certain categories of flooring. AD B of the Building Regulations (England & Wales), though, allows combustible materials to be added to the upper surface of stairs except in the case of fire-fighting stairs.

Galleries and balconies

Galleries and balconies cannot be compartment floors since, by definition, they are incomplete floors. Fire and smoke simply bypass them. However, these floors need to be fire-resisting and so resist collapse during fires (Figure 6.5).

Ceilings

So far as structural fire protection is concerned, it is the whole floor construction, including the ceiling if any, which determines the performance. The ceiling, whether suspended or not, may be required to contribute to the overall fire resistance of the floor or to prevent premature collapse of any beams which support the floor. An alternative to this is that, where the floor itself achieves adequacy of fire resistance, all that the ceiling is required to meet is the surface spread of flame requirement. A ceiling may not need to contribute to the fire resistance of a concrete floor but it may provide all or most of the protection needed for timber floors.

BS 476: Parts 20 and 21 provide for fire test procedures for elements of construction. So far as floors are concerned, these include the following components making up the floor element:

Figure 6.5 The floors of balconies and galleries must be fire-resisting

- floors (boards, etc.),
- beams,
- suspended ceilings.

Suspended ceilings can contribute to the fire-resistance of a floor, or indeed may be needed to provide protection to the underside of a steel beam, but all ceilings are subject to control under building regulations for their surface spread of flame classification on the face exposed to the room below (Figure 6.6).

Overlaying light fittings with fire protection material breaches the IEE Wiring Regulations (BS 7671) and the national building regulations:

British Standards Institution

BS 476: *Fire tests on building materials and structures*
      Part 20: 1987 *Method for determination of the fire resistance of elements of construction (general principles)*
      Part 21: 1987 *Methods for determination of the fire resistance of loadbearing elements of construction*
BS 7671: 2001 *Requirements for electrical installations.*
IEE Wiring Regulations. 16th edition

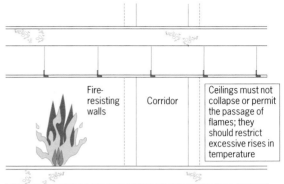

Figure 6.6 Suspended ceiling acting as a fire-resisting membrane

- **England and Wales:** AD P,
- **Scotland:** Section 2 of the Technical Handbooks, Domestic and Non-Domestic,
- **Northern Ireland:** Technical Booklet E.

Ceilings under roofs may also need to be airtight to reduce the amount of heat migrating into cold roof spaces.

Cavity barriers and fire stops

Cavity barriers within floor voids have an important influence on the growth and development of a fire within an enclosed space, though that is not their prime purpose under building regulations. For example, Clause 9.11d of AD B in the Building Regulations (England & Wales) states that in the case of a ground floor, cavity barriers will be needed where the height of a cavity beneath the floor is greater than 1 m or where crawl spaces are provided with access. In the case of suspended floors above ground level, a cavity within the floor void needs to be limited in size, just as with all other elements of structure.

Where services are required to penetrate compartment floors, the protection needs to be preserved.

Key Point

For further information on how to comply with the Regulatory Reform (Fire Safety) Order 2005, including the selection of approved products and services, see www.RedBookLive.com.

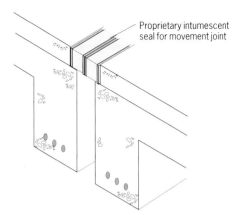

Proprietary intumescent seal for movement joint

Figure 6.7 A flexible fire-resisting seal in a movement joint

Movement joints

Flexible fire-resisting seals are available which can provide continuity in the fire performance of movement joints of compartment floors (Figure 6.7).

Floor finishes

In the UK, the rate of spread of fire on the upper surface of flooring is not provided for in building regulations since the upper surface of a floor is not significantly involved until a fire is well developed. Consequently, it does not play an important part in fire spread in the early stages of a fire. It is these early stages that are most relevant to the safety of occupants. Remembering, however, that building regulations are concerned with personal safety, there may be circumstances (eg with insurance) where designers concerned with the integrity of a building and its contents may need to take a different view.

The *Guide to fire precautions in premises used as hotels and boarding houses which require a fire certificate*, issued under the Fire Precautions Act 1971, calls for floor coverings to be included in the overall assessment of the suitability of surfaces to protected routes. (The same requirement also includes premises covered by the *Guide to fire precautions in existing places of entertainment and like premises*, and to factories, offices, shops and railway premises in the appropriate guides issued under the Fire Precautions Act). Where new textile coverings are being fitted, they should comply with BS 5287 as conforming to the low radius of fire spread (up to 35 mm) when tested in accordance with BS 4790.

Stair treads and landings within fire-fighting shafts are required to be constructed from materials of limited combustibility (Clause 9.2, Section 2 of BS 5588-5).

Floor heating

There are two basic kinds of underfloor heating laid in screed or slab:

- low pressure hot water in steel, copper or, more commonly, polypropylene pipes,
- electric cables laid either directly or in conduit.

In the case of hot-water-based systems, the temperatures reached in the pipes may be as high as 54 °C. In-situ floor finishes on newly laid floors which incorporate heating elements are more likely to curl and crack because of the increased drying shrinkage and the thermal gradient imposed across the section of the floor. Clay, concrete and terrazzo tiles usually present no problem provided that the base has been dried out by applying the underfloor heating before the tiles are laid. However, the heating must be switched off and the floor cooled while the tiles are being laid. Sheet and tile finishes of linoleum, cork, rubber and PVC, and the adhesives normally used with them, behave satisfactorily provided that the base is dry.

It is recommended (for a variety of reasons including indentation, shrinkage, and degradation of flooring and adhesives) that the surface temperature of a floor should not exceed 27 °C.

Great Britain Home Office. *Fire Precautions Act 1971. Guide to fire precautions in premises used as hotels and boarding houses which require a fire certificate.* London, The Stationery Office, 1991
Greenwood T. *Guide to fire precautions in existing places of entertainment and like premises.* 6th impression. London, The Stationery Office, 2000

British Standards Institution
BS 5287: 1988 *Specification for assessment and labelling of textile floor coverings tested to BS 4790*
BS 5588-5: 2004 *Fire precautions in the design, construction and the use of buildings. Code of practice for firefighting stairs and lifts*

Appearance and reflectivity

User satisfaction with floors depends largely on the condition of the top wearing surface. After installation, the surface should be clean and hygienic, and capable of being maintained in this condition. It should be suitable to receive a floor covering if one is not provided under the building contract. In areas where water may spill onto floors, both the effect of moisture on the floor and the ease with which the floor may be cleaned and dried, together with its subsequent rate of deterioration, will need to be considered.

Matters which should receive considerable attention are the needs of visually handicapped people for safe access and movement within buildings (RNIB's *Building sight*); these matters include appropriate contrasts at changes of level.

Appearance

The appearance of a floor finish is often considered to be of paramount importance, both in its initial selection and its acceptability for continuing use. This is probably true irrespective of situation and building type with the possible exception of those buildings used for certain industrial and storage purposes. The visual acceptability of the various kinds of floor finishes, in the final analysis, is a matter for individual taste and judgement. However, some of the factors that affect appearance, the quality and distribution of lighting in interiors, and the effects they might have on the choice of various materials can be identified (Figure 6.8).

Reflectivity

The reflectance of the flooring will affect the amount and distribution of light in a space. Reflectance values will always need to be considered when choosing floor coverings.

The CIBSE *Code for lighting* recommends an average floor reflectance of between 0.20 and 0.40. It recognises that in some industrial buildings a

Barker P, Berrick J & Wilson R. *Building sight.* London, RNIB and The Stationery Office, 1995

Chartered Institute of Building Services Engineers. *Code for lighting.* London, CIBSE, 2002. 130 pp + 2004 addendum

Figure 6.8 The highly reflective composition of this floor finish improves the level of lighting in this sports hall

Box 6.1 Typical reflectance values for floorings	
Portland cement screeds, cream PVC tiles and light-coloured carpets	0.45
Maple, birch and beech floorings	0.35
Light oak, marbled PVC tiles, medium shades of carpet	0.25
Other hardwoods and cork	0.20
Red and brown quarries, dark tiles and carpets	0.10

reflectance of 0.20 may be difficult to achieve; in these cases steps should be taken to keep the floors clean so that the average reflectance is maintained at 0.10 or above. Reflectance values for some floorings are given in Box 6.1.

Sound insulation

> **Key Point**
> It is crucial that floors provide adequate sound insulation against noise generated both above and below them.

Sound insulation of floors against internally generated noise comprises two aspects that have to be measured separately (Figure 6.9):
- airborne sound,
- impact sound.

There is a discussion of airborne sound in chapter 5 *External walls, windows and doors.* Airborne sound

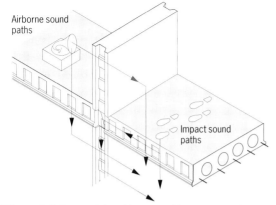

Figure 6.9 Some of the airborne and impact sound transmission paths

insulation of floors depends on the design, the type of material used, the mass per unit area for concrete floors or the degree of isolation achieved in lightweight floors.

Some general 'rules of thumb' for floors are listed in Box 6.2.

Types of separating floor

This section is based on AD E of the England & Wales Building Regulations. Similar requirements will be found in the Scotland and Nn Ireland Regulations:

Box 6.2 Rules of thumb for sound insulation in floors

- The more massive solid concrete floors are, the higher their airborne sound insulation is likely to be.
- Independent ceilings generally improve the airborne sound insulation of floors more than ceilings suspended from floors.
- Ceilings fixed to floors with effective resilient mounts generally improve the airborne sound insulation of floors more than ceilings having the same mass per unit area that are fixed rigidly to the floor.
- Well-designed lightweight floors with ceilings providing a high degree of isolation can have good airborne sound insulation.
- Flanking sound transmission limits the airborne sound insulation between rooms.

BRE. *Specifying dwellings with enhanced sound insulation: a guide.* BR 406. Bracknell, IHS BRE Press, 2000

- **Scotland:** Section 5 of the Technical Handbooks, Domestic and Non-Domestic,
- **Northern Ireland:** Technical Booklet G.

Separating floor systems must be used with the appropriate flanking structure (see BRE's report *Specifying dwellings with enhanced sound insulation* and AD E).

Insulation against impact sound is increased by the use of appropriate floating floors which help to isolate the surface that is walked on from the supporting base floor.

AD E describes three main separating flooring systems that are illustrated in Figure 6.10.

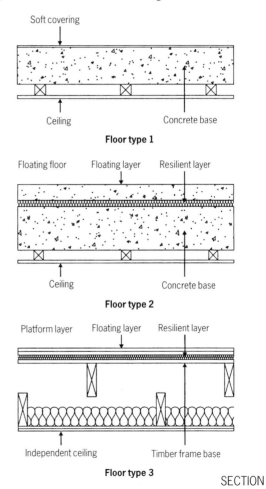

Figure 6.10 Separating floor types in AD E. Reproduced from Approved Document E of the Building Regulations (England & Wales) under the terms of the Click-Use Licence

Box 6.3 Some examples of floor constructions designed to meet building regulation sound insulation requirements

Floor type 2: Concrete base with ceiling and floating floor

The resistance to airborne and impact sound depends on the mass per unit area of the concrete base, as well as the mass per unit area and isolation of the floating layer (screed). The floating floor reduces impact sound at source. The floor slab can be either cast-in-situ or precast concrete. A ceiling finish is optional.

The resilient layer should be laid so that it is continuous over the floor surface, all joints should be tight butted with an upstand at the perimeter to isolate the floating layer from the surrounding construction (walls, etc.). The skirting should have a 3 mm gap from the floor surface.

The floating layer should be at least 65 mm cement–sand screed with 20–50 mm wire mesh to protect the resilient layer while the screed is being laid.

Services should not be taken through this type of floor but if unavoidable, the enclosure must have a mass of at least 15 kg/m^2 above and below the floor. Leave a 5 mm gap between the enclosure and the floating floor screed. Pipes should be isolated from the screed with a resilient material and in order to avoid noise being transmitted to the structure, there should also be a clearance around the pipe where it passes through it, suitably fire stopped.

Flanking requirements

Where the flanking walls are of cavity construction, the mass of the leaf adjoining the floor should be 120 kg/m^2 (excluding any finish) unless it is an external wall having openings of at least 20% of its area in each room, in which case there is no minimum requirement. However, flanking walls of mass 120 kg/m^2 give better results. The floor base must pass through the leaf and the cavity should not be bridged.

Where the flanking wall is of solid construction and less than 375 kg/m^2 (including any finish) the floor base must pass through the wall or be built in. If the wall is more than 375 kg/m^2, either the floor or the wall can pass through. For a floor base where the beams are parallel to the wall, the first joint should be 300 mm from the cavity face of the wall leaf.

Where there is also a separating floor, the requirement for a minimum mass per unit area of 120 kg/m^2 excluding finish should always apply, irrespective of the presence or absence of openings.

Floor type 3: Timber-frame base with ceiling for installation and platform floor

The resistance to airborne and impact sound depends on the structural floor base and the isolation of the platform floor and the ceiling. The platform floor reduces impact sound at source.

The floating layer should be 18 mm tongued and grooved (t+g) timber or wood-based board. All joints should be glued and spot bonded to a substrate of 19 mm plasterboard.

The floor base should be at least 12 mm thick timber or wood-based board. The floor and independent ceiling should be supported by separate staggered joists. The space between the structural floor and the ceiling should be a minimum 250 mm with a clearance of at least 50 mm between the bottom of the joists supporting the structural floor and the upper surface of the ceiling.

The ceiling should be no less than two layers of plasterboard with staggered joints and an overall thickness at least 30 mm with 100 mm quilt (density 10 kg/m^3) in the ceiling void.

Partitions should be taken through the floating layer and supported off the main structural floor layer (timber frame base) which is positioned on top of the joists. As for Floor type 2, avoid taking services through the system.

Flanking requirements

This floor should be used with timber frame flanking walls. The gap between the floating layer and the wall should be sealed with a resilient strip of the same mineral wool used as the resilient layer or by accurately sized semi-rigid materials such as EPS or extruded polystyrene. A 3 mm gap should be left between the skirting and floating layer. If this is sealed, flexible mastic should be used. Air paths between the floor base and the wall should be blocked, including the space between joists. The junction of the ceiling and the wall lining should be sealed with tape or caulking.

It is best to avoid taking services through this type of enhanced sound insulation floor. If it is essential for ducts or pipes to penetrate the floor they must be in an enclosure both above and below the floor. This enclosure must have a mass of at least 15 kg/m^2. The enclosure should be taken down to the top surface of the floating layer. A nominal gap should be left, sealed with acrylic caulking or neoprene. To prevent noise from pipes or ducts being transmitted to the timber structure, there should be a clearance where they pass through the structure and this should be sealed with flexible fire stopping. See the Gas Safety (Installation and use) Regulations 1998 for ducts containing gas pipes.

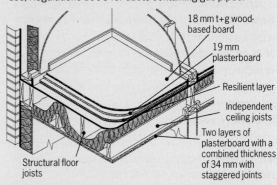

18 mm t+g wood-based board

19 mm plasterboard

Resilient layer

Independent ceiling joists

Two layers of plasterboard with a combined thickness of 34 mm with staggered joints

Structural floor joists

Timber-frame floor with independent ceiling for installation in timber-frame construction

Impacts on the surface of a floor can be transmitted through the structure and ceiling into the room below. They can be minimised by:
● using a floating floor, or
● for Type 1 floors, using a soft floor covering fixed or glued to the concrete floor.

Two examples of floors that comply with AD E requirements for sound insulation are described in Box 6.3. Other examples are given in *Specifying dwellings with enhanced sound insulation.*

Durability
Floor finishes may be considered to provide for the following main functions:
● protection of the structural floor,
● better appearance,
● increased comfort and safety.

The relative importance of these functions varies according to circumstances and budgets. However, one of the most important attributes is the maintenance of the functions over time; in other words, the finishes must be durable. Because of its importance, durability has often been regarded as a basic property of a floor finish, but it represents only the length of time that the chosen properties persist and depends as much on the conditions of use as on the properties of the finish. Many factors (see list in Box 6.4) influence the life of a floor finish; wear is perhaps the most influential of all because:
● it is unavoidable in normal use,
● it applies to all kinds of flooring,
● in many cases, it can be obvious to users when it occurs.

Expected life of floorings
The service life of a floor covering depends as much on the conditions in the building and the degree of use as on the inherent properties of the material and the techniques adopted in laying it. However, even bearing in mind the comparative ease of replacement of many thin floorings, building users should have a reasonable expectation of a minimum

BRE. *Specifying dwellings with enhanced sound insulation: a guide.* BR 406. Bracknell, IHS BRE Press, 2000

Box 6.4 Factors influencing the life of a floor finish
● Wear,
● Water and other liquids,
● Indenting loads and impacts,
● Sunlight,
● Insects,
● Moulds and fungi,
● High temperature,
as well as
● the fundamental properties of the floor finish materials and adhesives used,
● the compatibility of the floor finish materials with other parts of the floor structure, and
● the floor structure's behaviour in use.

service life for a properly selected flooring, taking all circumstances into account.

Resistance to wear
All floor surfaces wear to some degree when subjected to foot or wheeled traffic. There may also be other factors including the movement of furniture, which may scrape or cut the surface, and the movement of heavy loads such as in warehouses. However, using the definition of wear as the progressive loss from the surface of a body brought about by mechanical action, it is not just the quantitative loss of material which is important but also the qualitative assessment of the condition of the worn surface. Where changes in appearance are involved, the assessment becomes more and more subjective.

Indentation arises from the effects of furniture on relatively soft floorings, impacts from articles dropped onto the surface, and damage caused by footwear, whether by small heels or protruding nails.

Avoidance of wear. Wear can be prevented only by denying access. Nevertheless, there are various housekeeping measures, albeit outside the immediate control of the designer, that can be employed to reduce wear (eg the avoidance of unprotected metal on chair legs and trolley wheels). Also, the effects of wear on some floorings can be mitigated by good maintenance, for example, by applying seals and polishes to timber, cork and linoleum flooring. Good cleaning procedures and barrier matting will keep grit (which is often implicated in abrasion) to a minimum.

Floorings for the heaviest wear situations. In buildings such as shopping malls, railway and airport concourses, the volume of pedestrian traffic can be intense and large cleaning machines are often used to cover the vast areas involved in the shortest times. Only robust floorings of the highest quality laid to a good standard give good service. Materials which have performed best are relatively thick ceramic tiles, high quality decorative terrazzo concrete tiles with hard aggregates, and hard-wearing natural stones like granite and quartzite. High quality bedding and jointing with all these floorings is essential for good service. High quality flexible PVC and thermosetting resin screeds have been used less often, but usually with good results.

Resistance to water

> **Key Point**
> The effects of moisture probably cause more damage to floor finishes than any other agency, even abrasion.

However, in contrast to a commonly held view, moisture rising from the ground through a ground-bearing slab above a defective dpm is not one of the most frequent sources of dampness in floors. Excess construction water has become more of a problem with 'fast track' construction.

Osmosis. This is a phenomenon which affects some kinds of flooring, but it remains a rare and unpredictable event. It is described in more detail in the feature panel on page 60 of BRE's *Floors and flooring*.

Pye PW & Harrison HW. *BRE Building Elements: Floors and flooring: Performance, diagnosis, maintenance, repair and the avoidance of defects.* 2003 edition. BR 460. Bracknell, IHS BRE Press, 2003

Resistance to chemicals and effluents
There is a risk in all kinds of premises that chemical compounds will be accidentally spilled on floor surfaces. In housing these typically may include food products, pharmaceuticals and cleaning products. The products may stain the flooring, or cause its softening or other forms of deterioration. In other buildings such as factories and laboratories, special chemical products may require to be resisted.

Resistance to ultraviolet and sunlight
Many flooring materials fade on prolonged exposure to sunlight, including some woods and cork, and some plastics become yellow; not all materials, though, are affected. Whether colour changes are important depends on circumstances. Material specifications do change, and inspection of old samples is not necessarily a good guide to future performance. It may be possible to carry out laboratory assessments. Manufacturers' advice can also be sought.

Insects
Two types of problem occur in floors as a result of insect action: degradation of materials, and infestation and harbourage.

The degradation of timber and some timber-based board materials can result from the action of wood-boring beetles. *Anobium punctatum* is the most common species encountered in the UK. Though damage is potentially of structural significance, the risk of significant damage in floor structures is very low because the normal designs and conditions generally militate against initiation and development of attack.

Mould and fungus
Wet rot decay of timber and timber-based board materials can take place only where these are maintained in persistently damp conditions.

Surface moulds can be found on external and internal building surfaces, including ceilings, usually preceded by persistent condensation or some other form of wetting. On internal surfaces, as well as being unsightly, mould is thought to cause respiratory problems in susceptible individuals.

Adhesives

It is important that the correct adhesive is specified for floorings, and the manufacturer's instructions should be followed.

Suspended floors and ceilings

Suspended timber
Ground floors

In housing, and small cellular-type buildings, three types of timber ground floor are commonly used and are categorised as follows:

- single span (ie fully spanning) where perimeter walls provide support,
- multi-span (ie partially spanning) where sleeper walls or beams reduce effective spans,
- ground-bearing (ie non-spanning) where the floor deck is fixed to battens resting or partially set into a slab or screed of concrete or other solid material.

Upper floors

Timber upper floors for simple cellular buildings usually comprise joists spanning between walls, with stairwells trimmed to run either with or across the direction of span. Often, these joists have intermediate support provided by a beam or internal wall.

Performance

The main criterion for performance of the traditional suspended timber floor has been stiffness (lack of which gives rise to deflection) rather than strength (lack of which leads to collapse). The tabulated sizes given, for example, in AD A of the Building Regulations (England & Wales) 2000 will provide more than sufficient strength and stiffness for normal use in domestic construction.

However, even if the joists are of adequate section for the loads to be borne, there remains the problem of rotation. Joists normally need to be blocked at bearings in shorter spans, and intermediate points also need to be strutted at least:

- once at mid-span in spans of 2.5–4.5 m,
- twice in spans over 4.5 m.

Suspended in-situ concrete slabs

The variety of designs for concrete suspended floors is numerous, and, in practice, floors can be found that are a combination of techniques. For example, precast beams can be used that have projections from the top surface, designed to provide for composite action with in-situ structural concrete toppings.

Most floors serve also as a diaphragm and assist in the redistribution of horizontal (wind) loads on walls in one direction to walls running in an opposite direction. Connections between walls and floors are therefore important. Since many in-situ concrete floors will be found to span in two directions, lateral restraint can be taken care of by the bearings, and further strapping will not usually be present.

Choice of materials for structure

Concrete mix design can be complex, with many permutations of cement type, fine and coarse aggregates, plasticisers, retarders, pigments, air-entraining agents and waterproofers.

The dimensional stability of reinforced concrete floor decks may sometimes be a cause of concern. Dimensional changes in the decks can cause damage to the supporting structure and, in some circumstances, give rise to distortion in the deck itself. The most important causes of excessive deformation are:

- thermal expansion and contraction,
- drying shrinkage,
- elastic deformation due to self weight and imposed loading,
- creep due to prolonged loading.

Exclusion of damp

Dampness is a potential problem with this type of floor at ground level, including situations where permanent shuttering that has been left in place within the subfloor void. In a heated building, though, a suspended concrete ground floor is warmer than the ground and, therefore, conditions discourage the transfer of water vapour from the ground, via the void, to the underside of the floor where it could condense. Transient conditions can exist where moisture transfer does occur. It is therefore prudent to provide these floors with an integral dpm, particularly where moisture-sensitive floorings are to be applied. The provision of a dpm beneath a screed also cuts down the drying time as construction water in the concrete slab does not need to be taken into account. Only the screed, in such a case, needs to be dried.

Fire

Solid concrete floors generally have good fire resistance, provided reinforcement is adequately covered. Problems are more likely to arise where localised perforations occur in the floor, for example where pipes pass through a floor or where cracks exist between the floor edge and a wall.

Precast concrete beam and block, slab and plank

Precast concrete beam and block floors are widely used (Figure 6.11). Reinforced concrete plank floors tend to be used in building types other than housing, and may be made either from dense concrete or from autoclaved aerated concrete (see chapter 11, *Modern methods of construction*). Many precast beams and planks are prestressed. In many cases the planks are hollow.

Beam and solid block floors consist of inverted tee beams set at a distance that will allow them to accommodate precast infill blocks. In positions where increased loads on a floor are expected, and therefore extra local strength is required in the floor (eg under internal partitions built off the floor), twin or triple beams adjacent to each other are appropriate.

The floor is assembled when the walls are at dpc level. The beams are set on the inner leaf of the external wall and should not protrude into the cavity. They must be placed onto a dpc to prevent both moisture and soluble salts moving from the brickwork and ground into the beams. It is especially important to prevent the ingress of salts which can increase the rate of corrosion of the reinforcement.

Once in position, the top of the floor is treated with a cement-based grout to lock the beams and blocks together. When the building is watertight the floor is usually finished by laying a sand and cement screed or a floating chipboard floor. However, to meet increased thermal insulation requirements, it has become common to place a layer of insulation over the beam and block before laying a screed or chipboard.

Many designs of precast floor require some additional in-situ work, either to provide continuity of reinforcement between bays or to provide additional compression concrete on the upper portion of the floor deck.

In the direction of span, lateral restraint requirements for the adjoining wall can be met by the amount of bearing provided for the floor. However, in the non-span direction, strapping may be needed where the ground-to-floor dimension is considerable (Figure 6.12).

Figure 6.11 Precast beam and block floors for housing have increased considerably in popularity since the 1980s

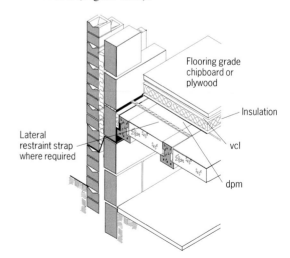

Flooring grade chipboard or plywood

Insulation

vcl

Lateral restraint strap where required

dpm

Figure 6.12 Strapping is sometimes needed over a precast floor to provide lateral restraint in the non-span direction

Other kinds of precast concrete floors

There is a great variety of proprietary designs of precast concrete floor units for use in floors other than for housing. Many of these designs rely on in-situ concrete toppings to provide continuity between units and to complete the floor. Many designs have protruding stirrup or stud shear reinforcement around which the topping is cast.

The surfaces of beam and block floors are often uneven. This problem can be prevented by levelling the beam and block surface with a thin layer of mortar. A 1:6 cement:sand mortar has adequate strength, and the normal thickness requirements for a levelling screed do not apply. Dynamic rocking of the floor might be prevented if the surface regularity of the base does not exceed 5 mm under a 3 m straight edge before insulation is laid.

Exclusion of damp

The various national building regulations require a dpm to be provided under a precast concrete floor if the ground level beneath the floor is below the lowest level of the surrounding ground and the ground beneath the floor will not be effectively drained.

A dpc must be placed below any beams or planks, at their bearings on supporting walls, to prevent moisture rising up the walls. This dpc also serves, rather more importantly, to prevent soluble salts reaching the components of the floor; salts will induce corrosion of the reinforcement.

Fire

Most beam and block floors, depending on the materials used, can be expected to achieve at least a half-hour fire resistance. The main exception to this are those which have used plastics infill blocks, particularly in ground floors where there is no requirement for fire resistance.

Most precast concrete floors with at least 50 mm in-situ concrete toppings can be expected to achieve one-hour fire resistance, although some designs can achieve more than this with no additional protection. Additional performance may come from replacing the ceiling finish.

Steel sheet and steel beams with brick and concrete infill

Where steel decking is used as permanent shuttering next to the ground, no further damp-proofing arrangements are normally required. Neither is ventilation of the underfloor void normally provided. In consequence, the underside of the deck may be subjected to long-term corrosion risk from the relatively high humidities if it is inadequately protected. BS EN ISO 12944 refers to steelwork in dry interiors and in wet or damp conditions. Ventilation of the void will reduce the risk of accumulation of moisture, but will not remove it altogether.

Platform and other access floors

A platform floor, complete with its supporting deck, is required to provide the full range of performance characteristics in addition to special requirements relating to demountability. These floors normally provide a system of load-bearing fixed or removable floor panels supported by adjustable pedestals or jacks placed at the corners of the panels to provide an underfloor void for the housing and distribution of services. In full access systems, most or all of the panels are removable. Partial access systems have runs of removable panels or individual traps, or both.

The platform and its supports will be required to possess certain properties over and above those that are normally required of floor finishes, including, for example, structural strength and demountability.

Since the suspension system for platform floors is usually in metal, there will normally be requirements for earth bonding in accordance with the requirements of the IEE Wiring Regulations (BS 7671).

British Standards Institution
BS EN ISO 12944: 1998 *Paints and varnishes. Corrosion protection of steel structures by protective paint systems* (8 Parts)
BS 7671: 2001 *Requirements for electrical installations. IEE Wiring Regulations.* 16th edition

Solid floors

Concrete ground-bearing floors: insulated above the structure

In-situ concrete ground-bearing slabs are the most common base used for floors in all types of buildings: houses, offices, factories, hospitals, warehouses, etc. The slab may be finished directly as the wearing surface or to receive:

● a bonded, unbonded, or floating screed before receiving other floorings,
● a granolithic or cementitious polymer topping,
● thermosetting resins or paints,
● mastic asphalt flooring,
● sheet and tile floorings, either with or without smoothing compound,
● ceramic tiles, terrazzo tiles, or slab flooring, either bedded in mortar or fixed with thin bed adhesives,
● prefabricated timber sheets (eg chipboard, plywood, etc.), either fixed on timber battens or laid directly on thermal insulation,
● wood block.

The quality and thickness of the concrete base will depend on the use to which the floor is put.

Hardcore

The principal uses of hardcore are as a make-up material to provide a level base on which to cast a ground floor slab, to raise levels, to reduce the capillary rise of ground moisture, and to provide a dry firm base on which work can proceed or to carry construction traffic. There are a number of factors that need to be taken into account in the selection of materials for use as hardcore. Ideally, the materials should be granular, and drain and consolidate readily (Figure 6.13). They should be chemically inert and not affected by water.

Three types of failure can occur to ground-bearing slabs from the use of unsuitable hardcore material, or from poor or unsatisfactory techniques in installing it. The three types are:

● inadequate compaction of the hardcore leading to settlement of the overlying slab,
● expansion and loss of strength of the concrete base caused by soluble sulfate attack on the

Figure 6.13 Blinded hardcore ready for laying the dpm and casting the concrete

concrete, leading to expansion, cracking and heave of the slab,
● expansion of the hardcore material leading to uplift and cracking of the slab.

Damp-proof membranes (dpms)

In all cases, a dpm will be required to comply with the national building regulations:

● **England and Wales:** AD C,
● **Scotland:** Section 3 of the Technical Handbooks, Domestic and Non-Domestic,
● **Northern Ireland:** Technical Booklet C.

This may be positioned above or below the concrete slab. In addition, thermal insulation in the form of foamed or expanded plastics sheets, glass fibre or foamed glass may be incorporated. As well as being required to prevent moisture from the ground reaching the inside of the building, the dpm may be required to fulfil other functions: preventing the interaction of ground contaminants with the concrete, stopping interstitial condensation and retaining constructional water. It may be placed in different positions. To prevent interaction with ground contaminants the membrane must be placed below the slab. The time required for the residual constructional water to evaporate should be taken into consideration when deciding on the position of the dpm. Box 6.5 lists materials suitable for dpms.

Concretes

The concrete for bases to receive other floorings is specified by strength. Grades based on characteristic

Box 6.5 Materials suitable for dpms

Sandwiched or buried dpms	**Surface dpms**
● Hot-applied pitch	● Epoxy resins
● Hot-applied bitumen	● Pitch/epoxy
● Polyethylene sheet	resin mixtures
● Polyethylene sheet	● Polyurethanes
backed with bitumen	● Mastic asphalt, not
● Bitumen sheet	less than 12 mm thick
● Bitumen/rubber	● Pitchmastic
emulsions	
● Cold-applied bitumen	
● Cold coal tar emulsions	
● Epoxy resin (also called	
epoxy bonders)	

strength of 30 N/mm^2 or 35 N/mm^2 are recommended; the higher strength may be required for structural reasons. Clause C4 of AD C of the Building Regulations (England & Wales) 2000 requires the minimum quality of the concrete to be at least mix ST2 of BS 5328-1, or, if there is any embedded steel, mix ST4.

Protection against radon and methane

Sealing cracks around the perimeter of solid concrete floors should reduce the seepage of radon from the ground. Sealing does remain a viable option for radon levels up to 400–500 Bq/m^3 if it is properly executed. There is more information in the *BRE guide to radon remedial measures in existing dwellings* and in chapter 2.

Services

Penetrations by services through the main area of a slab should be reduced to a minimum during the construction of the building. Drainage and service ducts, and service entries, should be confined to perimeters, wherever possible.

BRE. *BRE guide to radon remedial measures in existing dwellings.* BR 250. Bracknell, IHS BRE Press, 1993
Harrison HW & Trotman PM. *BRE Building Elements: Foundations, basements and external works: Performance, diagnosis, maintenance, repair and the avoidance of defects.* BR 440. Bracknell, IHS BRE Press, 2002

Effects of clay heave

Heave occurs when clay, that has dried down and shrunk (eg by tree roots drawing water from the ground), returns to its original volume on removal of the cause of the shrinkage (eg the tree). This expansion can be rapid, and perhaps destructive.

Two common methods for attempting to provide for movements in the clay subsoil beneath a ground-bearing slab are:
● a compressible layer between the slab and the hardcore,
● reinforcement in the slab.

Thermal properties and ventilation

Floors of hardcore and concrete laid on the ground do not lose heat in the same way as walls and roofs. The ground acts to some extent as an accumulator of heat, and in fact most of the heat loss occurs within a short distance of the edges adjacent to the external walls. Consequently, the average heat flow is dependent on the size and shape of the floor (there is more guidance in BRE's *Foundations, basements and external works*). Nevertheless, it is expected that in future most solid floors will require thermal insulation.

Excess construction moisture

Excess constructional water in concrete bases and screeds must be allowed to dry out before moisture-sensitive floorings (carpets, PVC, timber, etc.) can be laid. A sand and cement screed 50 mm thick laid on a dpm generally needs to be left about six weeks, but thicker slabs require much longer times

Proprietary screed systems are available which dry considerably more quickly than conventional screeds, but these are of little value if laid onto concrete bases that are still wet and protection will be needed.

The vcl used should be, for example, of polyethylene of not less than 125 µm (500 gauge) and preferably of 250 µm (1000 gauge). For bases laid directly on the ground the vcl is in addition to the dpm.

Key Point
Dpms must be continuous, whether above or below the concrete base, and be **linked** to the dpc in the walls.

Polyethylene below the base should be at least 300 µm (1200 gauge), or 250 µm (1000 gauge) if the product has a 3rd-party certificate or is to the Packaging and Industrial Films Association standard. Where a sheet material dpm is being laid below a replacement floor slab, the joints between sheets must present an impervious barrier to water entering the slab. The preferred method of forming the joint is to overlap the sheets by at least 150 mm and stick the joint with double-sided pressure-sensitive tape (Figure 6.14).

If welts are to be used to join sheets, they should be constructed in a three-stage operation (Figure 6.15). The welt should be held flat until the slab or screed is placed.

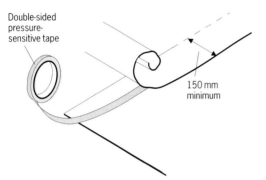

Figure 6.14 Sticking the overlap of a dpm

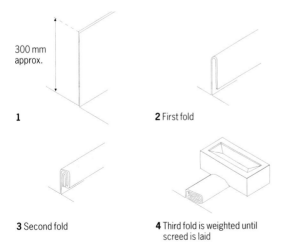

Figure 6.15 Forming a welt in a dpm (in 4 stages)

Condensation
Solid ground floors are at risk of condensation at their perimeters. Many floors have a cold bridge at the floor edge due to its proximity to low external temperatures and the fact that dense concrete has poor insulating properties. Raft construction is particularly at risk.

Spillages and leaks
Spillages of water from industrial or other processes, and leaks from embedded water pipes should be taken into account in the choice of solutions.

Water Regulations require pipes to be laid in ducts with removable covers.

Durability
Solid concrete ground-bearing floors should not deteriorate if they originally have adequate thickness, mix and compaction, and are laid on a prepared base of graded, compacted and inert hardcore.

Soluble sulfates leading to sulfate attack
Sulfate attack can occur on the underside of a concrete ground-bearing slab when fill below the slab contains sulfate salts and other contaminants, and the slab is not isolated from the fill by a dpm.

Expansion of hardcore
Although sulfate attack is the most common cause of distortion, cracking and upward movement of concrete floors, a small percentage of problems showing these symptoms have been caused by expansion of the hardcore. In these cases of expansion the strength of the concrete remains unaffected, and sideways expansion of the slab does not occur unless it is accompanied by sulfate attack. However, expansion of the hardcore sideways beneath the floor can cause distress to walls and ground beams below floor level.

The following hardcore materials have been known to cause problems due to swelling:
● steel slag or old blastfurnace slag,
● old broken re-cycled concrete,
● hardcore containing clay,
● hardcore containing pyrites.

Concrete groundbearing floors: insulated below or at the edge of the structure

The main difference between this kind of floor and that described in the previous section is that the operational sequence when the building is being built is varied to permit the placing of thermal insulation underneath the slab (Figure 6.16) and, sometimes, vertically at its perimeter.

It is important to avoid the risk of condensation occurring at thermal bridges by making sure that the continuity of insulation is unbroken, especially at the floor/external wall interface and at the interface of the floor with an internal wall occurring within 1 m of an external wall. Thermal bridges may also occur where changes of level between dwellings are made as stepped-and-staggered separating walls.

Rafts

Raft foundations are mainly used for small buildings on difficult sites such as made-up ground or over ground liable to subsidence from mining operations. In order to move as a whole, the raft is reinforced, often with deep downstand beams (toes) at the perimeter and under load-bearing walls.

There is a risk of condensation occurring at thermal bridges with this kind of floor. Where the toe of the raft is shallow, there is insufficient depth

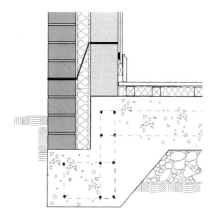

Figure 6.17 Construction should allow for the insulation to be started as low as possible; in other words from a deep toe in the raft

to allow the insulation in the wall to prevent heat loss through the perimeter of the slab. The construction shown in Figure 6.17 illustrates a method for reducing the loss to a minimum.

Services

Services penetrating the raft will need to be flexible and isolated by sleeves from the edges of the concrete, since movements in rafts must be expected to occur.

Screeds, underlays and underlayments

These materials are laid over the structural floor to provide suitable level surfaces for the flooring.

Some flooring materials, for example mastic asphalt, may be used as an underlay for other kinds of floorings. There can be advantages in using such an underlay as an alternative to waiting for cementitious screeds to dry before laying floorings.

Dense sand and cement screeds: bonded and unbonded

Dense sand and cement screeds are by far the most common form of providing a smooth and level base for the subsequent laying of floorings.

Screeds may be:
● fully bonded to the base,

Figure 6.16 Thermal insulation overlaid with a polyethylene dpm

● unbonded (ie not bonded to the base),
● floating.

For fully bonded screeds, the surface of the concrete base should be scabbled or shotblasted to remove laitance and to expose the aggregate. The optimum thickness of the screed is between 25 mm and 40 mm. For thicknesses greater than 40 mm, there is an increased risk of the screed debonding from the base.

All screeds laid directly over dpms (except for epoxy-bonding dpms) must be considered as unbonded. The minimum thickness should be 50 mm but, for certain kinds of floors, the screed may need to be thicker.

Proprietary screeds based on cementitious systems or calcium sulfate are available. Some of these are trowel-applied, others are self-levelling. Many proprietary screeds are designed to be laid more thinly than conventional screeds.

The standard appropriate to cementitious mixes described in this section is BS 8204-1.

Services
Services are best routed away from screeds if at all possible; for example, it may be possible to use hollow skirtings to accommodate service runs, thus avoiding the floor altogether.

Where building services are to be installed under screeds they should preferably be laid in ducts within the base structure. If there is no alternative but to lay them in the screed itself, the overall depth of the screed must be increased by the depth of the particular service accommodated. Otherwise, the screed will fail.

Control of dampness and condensation
The dpm in a ground-bearing floor slab and screed system can be laid in a variety of positions. In the case of a fully bonded screed the dpm must be laid under the slab, although for an existing screed a surface dpm can be used. The exception to this is the use of an epoxy dpm, which can act as a bonding agent and be laid as a sandwich.

Accuracy
Criteria for flatness of screeds are given in BS 8203 and BS 8204-1. The normal criterion used for levelness is a maximum deviation of 5 mm in a 2 m long straight edge laid in contact with the finish, though this can be reduced to 3 mm where greater accuracy is required. There is also a utility standard, where surface flatness is not critical, of ±10 mm.

For levelness, an overall deviation of ±15 mm for normal purposes is suggested for large floors. With reference to levelness, British Standard codes refer to a greater accuracy being required with respect to door thresholds and similar.

Further aspects of workmanship are covered in BS 8000-9.

Floating screeds: sand and cement
Floating screeds are used mainly to enable installation of one of two types of insulation:
● thermal insulation,
● impact sound insulation.

The required properties of these materials forming the layer immediately below the screed are very different.

Sand and cement screeds laid immediately over compressible layers like expanded polystyrene or glass fibre are classed as floating screeds, and should have thicknesses of not less than 65 mm for lightly loaded floors and 75 mm for more heavily loaded floors (see BS 8204-1).

With floating screeds used for thermal insulation purposes, it is important that the insulation board or mat should be sufficiently dense to support the likely loads when the floor is in use. For floating screeds used for impact sound insulation purposes, the board or mat should be sufficiently resilient to prevent impacts being transmitted to the structural

British Standards Institution
BS 8203: 2001 *Code of practice for installation of resilient floor coverings*
BS 8204-1: 2003 *Screeds, bases and in situ floorings. Concrete bases and cement sand levelling screeds to receive floorings. Code of practice*
BS 8000-9: 2003 *Workmanship on building sites. Cementitious levelling screeds and wearing screeds. Code of practice*

floor beneath, checking with the manufacturer that any expanded polystyrene is not too stiff for this application.

When detailing floating screeds, it is important that contact between the screed or flooring surface and any other part of the fabric by which impact sounds can be transmitted is avoided. This particularly applies to all edges of the screed; for example, small gaps should be left under skirtings . In addition, all penetrations of the surface (eg those caused by central heating pipes) should be suitably sleeved.

Levelling and smoothing underlayments

Levelling and smoothing underlayments are needed where a base is of insufficient quality to accept an intentionally thin flooring without the defects showing through (Figure 6.18).

Self-levelling underlayments are based on mixtures of powder and water which need a minimum of trowelling to achieve a smooth surface without bubbles. Other underlayments, based on mixtures of powder and natural rubber latex or synthetic emulsion, require more trowelling to achieve a smooth surface. All types are laid in thicknesses of just a few millimetres and therefore cannot be expected to remove major imperfections in the substrate.

The compounds generally produce a hard smooth surface which covers minor imperfections likely to show through thin coverings. They can also be used to provide slightly absorbent surfaces to allow better adhesion for fixing materials laid on other relatively hard non-absorbent surfaces.

Figure 6.18 Pouring an underlayment

Lightweight screeds

Lightweight screeds are used for many reasons including achieving better thermal insulation or simply saving weight. Since the body of a lightweight screed is less robust than a solid sand and cement screed, providing an extra and more durable topping to resist the in-service loads is necessary.

There are two basic types of lightweight cement-based screed:
- lightweight aggregate,
- aerated concrete.

Lightweight aggregates absorb moisture in the mixing process and high water:cement ratios are required to obtain workable mixes. As with dense screeds, therefore, the drying times tend to be prolonged. On the other hand, aerated screeds, containing much less water per unit volume, tend to dry more quickly.

Because of the reduced impact resistance when compared with dense sand and cement screeds, it is necessary to provide a thin topping of sand and cement. This may confuse later inspection. Neither type can be used as the wearing surface of a floor.

Matwells, duct covers and structural movement joints

Matwells are set into floorings in order to accommodate the thickness of mats or other forms of flooring used in entrances to prevent wetness and detritus from the exterior being brought into the interior of a building. Ducts and duct covers enable building services to be installed within floors to provide access for repairs and replacement of pipes and conduit.

To prevent disruption and unintended gaps in floor coverings, structural movement joints in coverings are made to coincide with movement joints in the floor structure below.

Structural movement joints in screeds are positioned to coincide with structural movement joints in the base and are normally anchored within the screed. Each side of the joint will be armoured with a metal angle fixed to the base. Alternatively, the armouring may consist of a strip of metal anchored back into the screed with wire ties. Metals may be stainless or galvanised steel or brass. Occasionally, the armouring may consist of plastics strips. The sides of structural movement joints too are vulnerable to scuffing if inadequately fixed. The level of finish of the joint and any possible cover strip will depend on the flooring.

It is important that structural movement joints remain free to allow the adjacent parts of buildings to move towards or away from each other, and their components to expand and contract.

Jointless floor finishes

This section deals with floorings that are formed in-situ from materials laid in molten, liquid or plastic form. The thicker types are normally trowelled and the thinner ones painted or poured to cover comparatively large areas without joints.

Since the finished surfaces of the materials are formed in-situ, the quality of workmanship is crucial to the acceptability and durability of the flooring.

British Standards Institution
BS 8204-2: 2003 *Screeds, bases and in situ floorings. Concrete wearing surfaces. Code of practice*
BS 8000-2: 1990 *Workmanship on building sites. Code of practice for concrete work. Section 2.1 Mixing and transporting concrete. Section 2.2 Sitework with in situ and precast concrete*
BS 8000-9: 2003 *Workmanship on building sites. Cementitious levelling screeds and wearing screeds. Code of practice*

Concrete wearing surfaces
Modern mix design and finishing can produce concretes with excellent abrasion-resistance suitable for the heaviest traffic. Table 6.1, adapted from BS 8204-2, relates the mix design and finishing technique for a concrete floor to the use to which the floor is intended to be put.

Aspects of workmanship are covered in BS 8000-2, Sections 2.1 and 2.2. BS 8000-9 also applies.

Polymer-modified cementitious screeds
These screeds are commonly used where thin toppings of around 8 to 10 mm are required to be laid onto concrete bases, although they can range in thickness up to 40 mm. They normally have good bond strength to the concrete base and good abrasion resistance. Chemical resistance is enhanced by the incorporation of polymers.

Although these floorings are most often laid as wearing surfaces, they can be used as screeds to receive other floorings. They are very useful in making up thicknesses of between 10 and 25 mm. Coves and skirtings can be formed in the materials.

Sand and cement screeds are not usually considered to be suitable as wearing surfaces. However, if they are laid slightly wetter than normal levelling screeds (ie not semi-dry), they can provide moderate abrasion resistance. The addition of small amounts of 10 mm single-sized aggregate to the mix will enhance the wear properties and reduce drying shrinkage problems.

Table 6.1 Selection of specification for a concrete floor according to use. Adapted from BS 8204-2			
Performance required	Use	Type of concrete and grade	Finishing process
Severe abrasion and impact	Very heavy duty engineering workshops, etc.	Special mixes, proprietary sprinkle finishes	Trowelling, depending on degree of compaction needed
Very high abrasion; steel-wheeled traffic and impact	Heavy duty industrial workshops, special commercial, etc.	High-strength toppings and other special mixes C60+ (granolithic concrete)	Trowelling two or more times followed by curing
High abrasion; steel or hard plastics wheeled traffic	Medium-duty industrial and commercial	Direct finished concrete C50	Trowelling two times followed by curing
Moderate abrasion; rubber tyred traffic	Light-duty industrial and commercial	Direct finished concrete C40	Trowelling two times followed by curing

In-situ terrazzo

In-situ terrazzo has been used successfully in cases where the quality of workmanship could be guaranteed, and where movements in the structure of the building could be sufficiently guarded against so as not to disrupt the terrazzo. It must be said, however, that many installations, even those in prestigious buildings, do suffer from cracking. This is one reason why terrazzo tiles, which are manufactured under controlled conditions, have taken the lion's share of the market.

Terrazzo is composed of a mixture of fine aggregates of marble or other decorative rock bound with one of a number of coloured cements. The finish is ground down after curing to expose the decorative aggregates. The terrazzo is normally laid on a 'green' cementitious screed to a thickness of not less than 15 mm and in bays of about 1 m^2 separated by strips of ebonite or plastics. Where aggregates in the mix exceed 10 mm, the thickness of the material is increased. Occasionally, it has been laid monolithically with the base which allows larger areas without joints.

The British Standard appropriate to the materials described in this section is BS 8204-4.

Synthetic resins

These floorings do provide an enhanced resistance to a wide range of chemicals and, being jointless, also offer a hygienic solution for buildings in the pharmaceutical and food industries.

There are two main types of in situ flooring based on synthetic resins and polymers used as binders.

Polymer-modified cementitious floorings

These comprise emulsified polymers and hydraulic cements which, when combined in certain proportions, produce the best characteristics of the resin and cement components. The main binding agent is the cement, the properties being modified by the polymer.

British Standards Institution
BS 8204-4: 2004 *Screeds, bases and in situ floorings.
Cementitious terrazzo wearing surfaces. Code of practice*

Resin-bound aggregate flooring

These are commonly syrups which cure (cross-link) by reaction of the chemical components. Hydraulic cements are rarely present. Curing happens after the addition and mixing of a catalyst or a cross-linking agent. This is commonly called the curing agent and its type depends on the resin system.

The resins used are epoxy resin, polyurethane, polyester and acrylic (methyl methacrylate). They are often referred to as thermosetting resins. Except where they are used as coatings, they are always mixed with substantial quantities of aggregates which include sand, quartz, carborundum and bauxite.

There are three basic types of resin-bound aggregate floorings.

- *Resin screeds.* These are relatively stiff mixes with the material being applied by trowel or sledge followed by power-trowelling (Figure 6.19).
- *Self-levelling resin floorings.* These are fluid materials, as their name implies. They contain much finer aggregate and provide smooth non-porous hygienic surfaces. Often laid in the range 1–3 mm thick, they provide excellent floors in clean rooms, pharmaceutical factories, food processing areas, etc.
- *Coatings and seals.* Floor paints are normally used over large surfaces as a relatively inexpensive means of improving appearance in the short term. Seals are applied to concrete floorings as a means of surface hardening and to

Figure 6.19 A sledge-applied epoxy resin screed flooring being laid (**a**) and the same flooring being power-trowelled (**b**). Courtesy of Flowcrete Systems

prevent dusting, again as a short-term measure. In many instances, the terms floor paint and floor seal are interchangeable.

Paints are normally one of the following main types:

❑ polymer based emulsions (eg acrylics),
❑ bitumen based emulsions (rarely encountered),
❑ resin based solvent mixes,
❑ oleoresinous paints,
❑ one and two-pack polyurethanes,
❑ two-pack epoxies,
❑ chlorinated rubber.

For concrete, seals or surface-hardening agents are normally one of the following principal types:

❑ sodium silicate solutions,
❑ silico-fluoride solutions,
❑ tung oil with resins,
❑ one and two-pack polyurethanes (usually pigmented),
❑ epoxies (usually pigmented).

Because of their small thickness, all paints and seals on floors will have a limited life before wearing through.

Mastic asphalt

Mastic asphalt has been successfully used as underlays for a variety of floorings such as rubber, linoleum, wood block, and cork and ceramic tiles. The grades are exactly the same as for finished floorings. Where used as an underlay for thin floorings, a levelling layer of latex may be required. Adhesives used for floorings must be compatible with the asphalt.

Mastic asphalt is prepared from mixtures of bituminous binders and inert minerals, usually in the form of aggregates. Jointless mastic asphalt flooring is normally laid in thicknesses from 15 to 50 mm.

The appropriate standards for asphalt floorings are: BS 8204-5, BS 6925 (Type F1076 for limestone

British Standards Institution

BS 8204-5: 2004 *Screeds, bases and in situ floorings. Mastic asphalt underlays and wearing surfaces. Code of practice*

BS 6925: 1988 *Specification for mastic asphalt for building and civil engineering (limestone aggregate)*

BS 1447: 1988 *Specification for mastic asphalt (limestone fine aggregate) for roads, footways and pavings in building*

aggregate or Type F1451 for coloured aggregate) and BS 1447. Material meeting BS 1447 should be used for waterproofing balconies and exposed roof-top access ways classified as floors.

Jointed resilient finishes

> **Key Point**
> The most common type of failure with all jointed resilient floorings is insufficient time being allowed for construction moisture in screeds and concrete bases to dry out.

None of these floorings should be laid on new construction until moisture condition has been established by the hygrometer method, and shown to be 75–80% RH (relative humidity) or less; and none should be laid in a ground-floor situation without the protection from moisture provided by an effective dpm.

BS 5442-1 describes adhesives for all sheet and tile floorings in this category.

Textile

Where textiles are stuck to the substrate, and they are effectively therefore part of the building contract, they are included in this section.

Textile floorings usually take the form of carpets of which there are many types and qualities. Materials used include wool and cellulose, and combinations of synthetic fibres such as acrylics, polypropylene and nylon for piles, with polypropylene, nylon and polyesters used for backings. Some carpets are reinforced with glass fibre. Most carpets are laid with an underlay or incorporate an underlay material in their construction.

Carpets laid as part of the building contract are normally fixed by gripper battens round the perimeter of the area to be covered; or they can be stuck which may give problems when they need to be replaced.

British Standards Institution. BS 5442-1: 1989

Classification of adhesives for construction. Classification of adhesives for use with flooring materials

Static electricity

Certain kinds of textile flooring, particularly those containing man-made fibres, in common with certain kinds of vinyl and other plastics, have been prone to problems caused by the build-up of static electricity from friction. The phenomenon occurs in conditions of low humidity, and can be uncomfortable to occupants and disastrous to microelectronics or to atmospheres where flammable gases can accumulate. This is mainly a problem with new carpets and is particularly evident during the first heating season. As the carpet becomes more soiled, conductivity improves and static is therefore less of a problem.

The problem can be ameliorated by incorporating conducting materials within the flooring.

Linoleum

Linoleum is prepared from a mixture of linseed oil, resins, cork and wood flour, traditionally adhering to a hessian backing, though other materials have been used as backings in more recent years. Heat treatment follows to create a tough material that is consistent throughout its structure.

The thicknesses available from stock are 2.4 and 3.2 mm, though other thicknesses may be available to special order.

Linoleum can be laid with butt joints or it can be solvent welded to form a virtually continuous surface. The appropriate standard is BS EN 548.

Cork

Cork, in tile or sheet (carpet) form, provides a flooring which has been used for many years, taking advantage of its warmth and resilience. However, there are negative properties too, particularly its susceptibility to wear if not suitably protected and to damaging cleaning methods.

Flexible PVC

Flexible PVC flooring (frequently referred to as vinyl) is appropriate for use in a wide variety of situations and can be obtained in several forms including those with resilient backing. Some grades can be used in heavy-wear situations and special grades are produced with enhanced slip resistance.

There are two types of unbacked flexible PVC flooring covered by BS EN 649:
- Type A, fully flexible,
- Type B, less fully flexible, but more flexible than those to BS EN 654 (formerly BS 3260).

Type A is available in sheets 2 m wide or occasionally in tile form, while Type B is available only in tile form (Type B is insufficiently flexible to roll up).

Flexible PVC can be laid with butt joints, or it can be hot-air welded to form a virtually continuous surface. With some materials joints can be solvent-welded. BS 8203 is relevant.

Semi-flexible PVC

These semi-flexible tiles are manufactured from with low levels of plasticisers. The resin being white, pigmentation is easier and light-coloured tiles are readily made.

They have been available in a range of thicknesses, usually 1.5, 2.0 and 2.5 mm. The relevant standard is BS EN 654 (formerly BS 3260).

Rubber

Rubber flooring is available in sheet or tile form, in varying degrees of hardness, and with various forms of surface profiles to enhance slip resistance. It may also be found laminated to a sponge backing, particularly for use in blocks of flats where it has been used to obtain improved impact sound insulation in separating floors. The appropriate standard is BS EN 1817.

British Standards Institution
BS EN 548: 2004 Resilient floor coverings. Specification for plain and decorative linoleum
BS EN 649: 1997 *Resilient floor coverings. Homogeneous and heterogeneous polyvinyl chloride floor coverings. Specification*
BS EN 654: 1997 *Resilient floor coverings. Semi-flexible polyvinyl chloride tiles. Specification*
BS 8203: 2001 *Code of practice for installation of resilient floor coverings*
BS EN 1817: 1998 *Resilient floor coverings. Specification for homogeneous and heterogeneous smooth rubber floor coverings*

A wide range of material specifications has been available comprising natural and synthetic rubbers with various fillers, pigments and curing agents.

Rubber flooring can normally be stuck to a flat screed using a contact adhesive or other recommended proprietary adhesive. Some thick tiles, with an undercut or studded profile to the back, are designed to be laid in a wet cementitious grout which fully bonds the rubber to the screed or power-floated slab; this type of rubber flooring is unaffected by moisture in the base. It can also be laid satisfactorily on metal plate floors or on timber boarded floors covered with a suitable underlay. Where smooth-backed rubber flooring is to be stuck to ground-bearing concrete floors with adhesive, there must be an effective dpm. BS 8203 is relevant.

Jointed hard finishes

All the floorings dealt with here are both hard and durable. They are commonly specified for use in public access areas such as concourses and include ceramic tile, brick, concrete flag, stone and its derivatives including terrazzo tile, composition block (mainly in the form of a cementitious binder with sawdust and wood flour) and metal.

Movement joints in hard finishes are extremely important and the recommendations in standards should be strictly observed.

Toughened and laminated sheet glass of appropriate thickness is occasionally used as a flooring for special effects such as concealed lighting within the floor or even as a completely transparent floor. Supports are similar in character to those used in platform floors described earlier, though additional measures will be needed in mountings to enable the glass to safely accommodate impacts. The glass does not retain its

gloss surface for long, owing to scratching caused by foot traffic.

Ceramic tiles and brick paviors

Ceramic floorings (tiles, quarries, faience, paviors and mosaics) are available in great variety. Laying techniques are covered by BS 5385, Parts 3 and 4. Aspects of workmanship are covered in Section 11.1 of BS 8000.

Concrete flags

When used internally, concrete flags offer many advantages compared with in-situ material, including being manufactured to a consistent quality and cured under controlled conditions. Consistent quality is achieved by hydraulic pressing and vacuum de-watering of the mix. BS 7263-1 is relevant.

The normal method of laying concrete flags is to fully bed them in mortar on a concrete base which has been well wetted to reduce suction.

There is no standard for laying these products internally but the recommendations given in BS 5385-5 provide a good guide, particularly with the semi-dry method.

Natural stone

Most stone floorings have been in the form of large thick flags although the more exotic stones such as marble have been sawn into thin slabs (eg of 12 or 20 mm thickness) and bedded on mortar on a concrete base. Flooring tiles are also made from natural marble aggregates or other stone aggregates commonly bound with polyester resins instead of Portland cement. Such resin bound tiles have many of the characteristics of natural marble, their

---

British Standards Institution
BS 8203: 2001 *Code of practice for installation of resilient floor coverings*
BS 5385: *Wall and floor tiling*
    Part 3: 1989 *Code of practice for the design and installation of ceramic floor tiles and mosaics*
    Part 4: 1992 *Code of practice for tiling and mosaics in specific conditions*

British Standards Institution
BS 8000-11.1: 1989 *Workmanship on building sites. Code of practice for wall and floor tiling. Ceramic tiles, terrazzo tiles and mosaics*
BS 5385: *Wall and floor tiling*
    Part 5: 1994 *Code of practice for the design and installation of terrazzo tile and slab, natural stone and composition block floorings*
BS 7263-1: 2001 *Precast concrete flags, kerbs, channels, edgings and quadrants. Precast, unreinforced concrete paving flags and complementary fittings. Requirements and test methods*

performance depending on the quality and relative amounts of marble in the matrix. Section 6 of BS 5385-5 is relevant.

Terrazzo tiles

Terrazzo tiles consist of similar mixes to in-situ terrazzo, namely crushed marble in a cementitious matrix. Consequently, floorings of these tiles share many characteristics with in-situ mixes. The most common size of tile is 300 × 300 × 28 mm: the upper half of the 28 mm thickness being terrazzo and the base being cement mortar. Overall thickness tends to vary with plan size. Hydraulic pressure is used to obtain a dense mix and the surface of the tile is ground after curing. The appropriate standard is BS 5385-5.5. Aspects of workmanship are covered in BS 8000-11.1.

Composition block

'Composition' flooring blocks consisting of sawdust, sand and pigment bound with Portland cement or calcium sulfate, and impregnated with linseed oil, unlike the wood blocks which they superficially resemble, are laid bedded in cement mortar. The appropriate standard is BS 5385-5.7.

Timber and timber products

Timber has provided the majority of decking for suspended floors over the ages, mainly softwood but sometimes indigenous hardwoods, and in most cases providing both deck and finish. Since WWII there has been a significant growth in the proportion of floorings formed from processed timber such as plywoods and chipboards.

There has been concern expressed about the profligate use of timber from non-renewable sources. This has meant that some of the hardwoods

traditionally used for floorings have declined in usage, and other species have been investigated.

The species of timbers suitable for flooring are too numerous to mention in detail in this book; Tables 1–11 in BS 8201 may be useful.

Small firm knots are normally acceptable in softwoods intended for use as the final flooring, but are not acceptable in hardwoods.

Wood and wood-based panel products are subjected to dimensional movement as moisture content changes, and usually provision must be made for this movement to take place without disrupting the flooring. Movement is normally accommodated at the perimeters of rooms where skirtings cover the edges.

The species of timber is often selected for its appearance, provided durability can be predicted.

Board and strip

In domestic construction, softwood tongued-and-grooved boarding has been favoured for suspended floors, with hardwood boards, traditionally of oak, used in other building types where wear is more severe. Fixing is invariably by nailing, often through the face of the board, although better class work is normally secret nailed through the tongues. The latter practice is much more labour intensive, for each board has to be cramped individually, rather than several boards at once. For the largest boards, screwing and pelleting is an alternative.

Safe maximum spans for softwood tongued-and-grooved boards are given in AD A of the Building Regulations (England & Wales) 2000. Maximum spans for hardwood boards will depend on the species.

Block

Wood blocks for flooring have been manufactured in many different sizes, with many different jointing techniques used: from tongued-and-grooved to metal tongues or wood dowels.

Block floors are stuck with a bitumen–rubber emulsion, an effective dpm being provided separately in the floor structure. If insufficient provision is made for expansion at perimeters, substantial upward displacement may occur, either as localised ridging or tenting or as uniform bowing

---

British Standards Institution

BS 5385: *Wall and floor tiling*
　　Part 5: 1994 *Code of practice for the design and installation of terrazzo tile and slab, natural stone and composition block floorings*
BS 8000-11.1: 1989 *Workmanship on building sites. Code of practice for wall and floor tiling. Ceramic tiles, terrazzo tiles and mosaics*
BS 8201: 1987 *Code of practice for flooring of timber, timber products and wood based panel products. Tables 1–11*

over the entire area. BS 8201 suggests providing an expansion gap of 10–12 mm at perimeters, but, for any areas larger than domestic room size, it recommends that sufficient space be provided between successive panels as well as at perimeters.

Parquet and mosaic

Parquet often has the appearance of wood blocks but the material is much thinner, down to a few millimetres, and is normally found glued and pinned to a boarded subfloor. Mosaic is simply the small sized units of similar thickness to parquet, but prefabricated and stuck to a backing for laying as a tiled finish.

Panel products

Particleboard (chipboard), plywood, and other wood-based panel products are not normally specified as finished floorings; nevertheless, they are often used in their unprotected state as deckings for suspended timber floors and for raised access floors, and as overlays and substrates for covering with other floorings after completion of the building contract. BRE's *Digest 323* gives further guidance.

The wood content of most of these materials is relatively high (eg 87% for chipboards). They do not, however, behave similarly to products entirely of natural wood (eg as to dimensional stability) and must be detailed accordingly.

Box 6.6 gives guidance on choosing a panel product for a particular substrate. Chipboard, OSB, OPC bonded particleboard and fibre building boards can be used as decks in suspended floors provided they satisfy the loading conditions.

For domestic floor loadings the thickness should be:

- 18/19 mm for joist spacings up to 450 mm,
- 22 mm for joist spacings between 451 mm and 600 mm.

For increased floor loadings with standard chipboards, the maximum spans will be smaller; for example, for 2 kN/m^2 it will be 450 mm. Alternatively, non-British Standard special grades may be available which give increased spans.

BRE. *Selecting wood-based panel products*. Digest 323. Bracknell, IHS BRE Press. 1992

Box 6.6 Choice of materials for substrates in new construction

For domestic floors with maximum uniformly distributed load (UDL) of 1.5 kN/m^2:
- Plywood to BS EN 636
- Particleboard to BS EN 312
- Oriented strand board (OSB) to BS EN 300 OSB/2 or OSB/3
- Medium density fibreboard (MDF) to BS EN 622-5 (MDF.LA)
- Fibreboard to BS EN 622-3 or BS EN 622-4 (MBH. LA1)
- Cement bonded particle board to BS EN 634

For non-domestic floors with maximum UDL of 2.5 kN/m^2:
- Plywood to BS EN 636
- Particleboard to BS EN 312
- OSB to BS EN 300 OSB/4
- Fibreboard to BS EN 622-3 or BS EN 622-4 (MBH. LA1)

Board quality must be chosen on the basis of likely location, especially in assessing risks of dampness (eg P5, P7, OSB/3 or OSB/4).

Special considerations apply to the use of board materials in non-domestic floors with maximum UDL above 2.5 kN/m^2.

For floating floors see BS EN 13810-1.
See also BRE's *Digest 394*.

British Standards Institution
BS EN 636: 2003 *Plywood. Specifications*
BS EN 312: 2003 *Particleboards. Specifications*
BS EN 300: 1997 *Oriented strand boards (OSB). Definitions, classification and specifications*
BS EN 622 *Fibreboards. Specifications.*
    Part 3: 2004 *Requirements for medium boards*
    Part 4: 1997 *Requirements for softboards*
    Part 5: 1997 *Requirements for dry process boards (MDF)*
BS EN 634 *Cement-bonded particle boards. Specification.*
    Part 1: 1995 *General requirements*
    Part 2: 1997 *Requirements for OPC bonded particleboards for use in dry, humid and exterior conditions*
BS EN 13810-1: 2002 *Wood-based panels. Floating floors. Performance specifications and requirements*

BRE. *Plywood*. Digest 394. 1994

Particleboard to BS EN 312 retains a high proportion of its initial strength after wetting. This property is due to the proportion of melamine in the composition. Where there is a risk of significant wetting, it is essential that the correct grade is used. Requirements for the protection of particleboard floors in potentially wet areas are given in BRE's *Defect Action Sheet 31*. Further information is given in BRE's *Information Paper IP 3/85* and *Digest 323*.

Where tongued-and-grooved (T&G) chipboards are continuously supported by the sub-base or underlay of insulation, it is essential that all the T&G joints are glued continuously; spot gluing is insufficient. It is now recommended for all situations. A PVA glue to BS EN 204 is satisfactory.

Regulations and standards

Building Regulations
Building regulations control the characteristics of floors in many ways.

Structural strength and stability
Arguably, some of the most important characteristics are those relating to structural strength and stability, particularly coupled with lateral restraint.
- **England and Wales:** AD A,
- **Scotland:** Section 1 of the Technical Handbooks, Domestic and Non-Domestic,
- **Northern Ireland:** Technical Handbook D.

Adequate consolidation of hardcore and correct compressive strength of thermal insulation are pinch points.

Fire safety
Fundamentally related to safety is the matter of the behaviour of floors and flooring in fire, particularly with regard to compartmentation and means of escape, and to ceilings in relation to surface spread of flame.
- **England and Wales:** AD B,
- **Scotland:** Section 2 of the Technical Handbooks, Domestic and Non-Domestic,
- **Northern Ireland:** Northern Ireland Technical Handbook E.

Conservation of fuel and power
Thermal properties are also controlled by regulations:
- **England and Wales:** AD L1 and L2,
- **Scotland:** Section 6 of the Technical Handbooks, Domestic and Non-Domestic,
- **Northern Ireland:** Northern Ireland Technical Handbook F.

Characteristics vital to performance, but in the opinion of BRE not always carried out effectively, include the elimination of thermal bridges (see the DTLR and Defra report *Limiting thermal bridging and air leakage*) and the achievement of adequate standards of airtightness.

Sound insulation
Sound insulation of floors between dwellings has taken on greater significance in recent years, and physical testing to demonstrate compliance will be needed in certain circumstances.
- **England and Wales:** AD E,
- **Scotland:** Section 5 of the Technical Handbooks, Domestic and Non-Domestic,
- **Northern Ireland:** Technical Handbook G.

BRE
Suspended timber floors: chipboard flooring — specification. Defect Action Sheet 31. 1983
Dinwoodie JM. *Wood chipboard: recommendations for use.* Information Paper IP 3/85. 1985
Selecting wood-based panel products. Digest 323. 1992

British Standards Institution. BS EN 204: 2001
Classification of thermoplastic wood adhesives for non-structural applications

DTLR and Defra. *Limiting thermal bridging and air leakage: robust construction details for dwellings and similar buildings.* London, The Stationery Office, 2001

Resistance to moisture

Building regulations also control the provision of dpcs and dpms.

- **England and Wales:** AD C,
- **Scotland:** Section 3 of the Technical Handbooks, Domestic and Non-Domestic,
- **Northern Ireland:** Technical Handbook C.

One vital provision sometimes overlooked is the adequacy of the link between adjacent dpcs and dpms.

Standards and codes of practice

The element of floors and flooring, as can be seen from the numbers of standards and codes cited in this chapter, is comprehensively covered, perhaps more so than any other element except building services.

The attention of readers is drawn to the rapidly changing situation reflecting the growing importance of European standards, particularly in the fields of structural strength, thermal requirements and airtightness.

7 Separating and compartment walls and partitions

Separating and compartment walls

The first part of this chapter covers separating and compartment walls in all building types and in all forms of construction (see Figure 7.9). The main performance criteria relate to sound insulation and fire resistance, though there may be cases where thermal insulation may also be relevant. However, deficiencies do exist in many cases: while it may be practicable to upgrade existing walls for increased performance, the most satisfactory solution is to design and build them correctly in the first place.

The term separating wall has been used as the generic term. Where a separating wall is in common ownership it is termed a party wall, but the various quasi-legal responsibilities of surveyors acting for owners of party walls in pursuance of the Party Wall etc. Act 1996 are not covered in this book.

AD E of the England & Wales Building Regulations groups separating walls into four main types (Figure 7.1). Sound insulation requirements for Scotland and Northern Ireland will be found in their national building regulations as follows:

- **Scotland:** Section 5 of the Technical Handbooks, Domestic and Non-Domestic,
- **Northern Ireland:** Technical Booklet G.

An introduction to enhanced sound insulation is given in chapter 5 *External walls, windows and doors* which includes reference to separating walls. To control all sound transmission, careful design and high standards of site supervision and workmanship are essential.

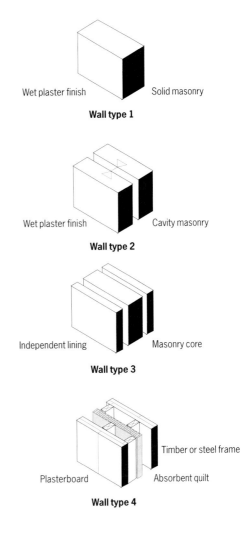

Wet plaster finish   Solid masonry

Wall type 1

Wet plaster finish   Cavity masonry

Wall type 2

Independent lining   Masonry core

Wall type 3

Timber or steel frame

Plasterboard   Absorbent quilt

Wall type 4

Figure 7.1 Types of separating wall

Masonry separating walls

Basic structure

Before the introduction of building regulations, most separating walls were of brick, with thickness depending on structural and fire requirements rather than on sound insulation requirements. However, well-built walls in wet-plastered dense aggregate block continue to provide adequate sound insulation in many domestic situations.

Under the influence of the London Constructional By-laws and certain local authority byelaws founded on the Model Byelaws, separating walls were often continued above roof level (Figure 7.2), but the need for this is anticipated to occur less frequently in the future. In a surprisingly high number of cases, though, there was no continuation of the wall into the roof spaces of terraces, leading to gross deficiencies in sound and fire performance.

For acoustics reasons, the separating wall will need to continue within the roof space, but not above it. Compartment walls may, however, need to continue above the roof finish.

Abutments

The separating wall should be joined to the inner leaf of the external wall in such a way that the separating wall contributes at least 50% of the bond at the junction.

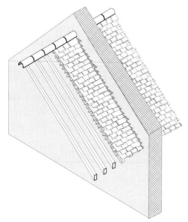

Figure 7.2 In Victorian and Edwardian houses, and even later houses, compartment walls were often continued above the roof slope

---

> **Key Point**
> There should be no gaps into any cavity in the external wall to provide flanking sound transmission routes, routes for fire spread or air penetration.

Openings and joints

There should be no openings as such, though in commercial premises it may be necessary to provide access, eg for means of escape.

In general, it is unwise to accommodate services within the thickness of masonry separating walls. Where chasing is necessary, it should be done according to prescribed rules. Avoid chasing and socket outlets back-to-back in separating walls (Figure 7.3) to the detriment of fire and sound insulation.

Main performance requirements
Choice of materials for structure

As already noted, brick was the universal solution for older buildings. Dense concrete block became popular in the drive for increased productivity in bricklaying when skills were scarce during the years immediately following the 1939–1945 war.

Figure 7.3 Electrical sockets set back-to-back in a separating wall should be avoided

These blocks (approaching 30 kg) now require two people to lift them so have been outlawed for health and safety reasons. Since the early 1990s, lightweight aggregate blocks with 75 mm cavities have become popular.

Strength and stability

Largely as a result of systematic testing, the strength and stability of loadbearing masonry walls can easily be calculated, and there are a number of examples of multistorey construction of 10 or more storeys.

Separating walls are often found to be inadequately bonded to front and rear walls. Wall-to-floor connections may be absent along spans which run parallel to the separating wall.

Dampness and condensation

Where a pitched roof abuts a separating wall in a stepped-and-staggered situation, part of the roof becomes an external gable, and the outer leaf of masonry becomes an inner leaf below roof level. Particular care must be taken to ensure that rainwater is prevented from reaching the interior.

Rising damp in separating walls will generally prompt the same technical considerations as for other internal walls, but additional problems may be encountered because the wall usually separates dwellings in different ownership and because floor levels are often stepped either side of the wall. Where access is available to one side only of a wall, the provision of a fully effective damp proof course (dpc) may be difficult to achieve and remedial work at a later date is almost impossible to achieve. Where there is a stepped wall between dwellings, the wall on the lower side may need to be considered as if it is part of a basement.

Fire

Soundly constructed masonry walls are usually able to provide fire resistance for periods longer than required by building regulations. This inherent resistance to fire is easily compromised, however, if any unsealed holes or gaps through a wall are left. Walls may rely for their stability on other parts of the structure, such as timber floors or roofs which are not so resistant to fire.

> **Key Point**
> Where a building contains several dwellings, each must be contained within fire-resisting compartment walls and floors.

A BRE survey of house design and construction revealed that fire-stopping at the top of separating walls is rarely effectively done. If the gap is not properly fire-stopped, fire can spread from one dwelling to the next (Figure 7.4).

Internal walls used for compartmentation (or perhaps earlier conversion) may not provide the required fire resistance because the supporting structure will fail before the wall. Sealing of all cavities that bypass the wall (eg floor and roof voids and around pipes in ducts) is often overlooked.

Standard fire resistance tests on fire-separating elements in buildings (eg walls) are undertaken on representative elements of construction of limited size. Modern buildings, however, can have elements that are many times larger and the real fire exposure can differ in timing, severity and extent. Careful engineering design and construction is needed to

> **Key Point**
> For further information on how to comply with the Regulatory Reform (Fire Safety) Order 2005, including the selection of approved products and services, see www.RedBookLive.com.

Figure 7.4 Lack of fire resistance at boxed eaves

ensure that thermal movements and restraining forces do not impair the stability, integrity and insulation of large assemblies.

The resistance of no-fines separating walls to the passage of smoke (as indeed for sound too) will depend on the integrity of the plastering.

Sound insulation performance

Physical characteristics that contribute to sound insulation

> **Key Point**
>
> The sound insulation of any single-leaf wall or floor built without gaps depends mainly on its mass.

According to the Mass Law there will be an increase of about 5 dB if the per unit area is doubled and there should be an increase of 6 dB for a doubling of frequency. However, sound insulation does not always increase as frequency increases. At a frequency known as the critical frequency, sound insulation can be reduced before increasing again. This is illustrated in Figure 7.5 which shows an idealised predicted Mass Law curve and the measured variation of the sound reduction index with frequency for a lightweight stud wall with panels on each side. The curve clearly shows the drop in sound insulation between 1000 Hz and

1600 Hz due to the critical frequency.

Walls and floors should be designed with critical frequencies below 100 Hz where possible because single figure values for sound insulation are determined from sound insulation in the frequency range 100–3150 Hz.

In principle, double-leaf walls can give better sound insulation than single-leaf walls. In practice, the sound insulation of a 280 mm plastered cavity masonry wall may be no better than that of a 240 mm plastered solid masonry wall because of the higher critical frequency of each leaf (around 200 Hz compared with 100 Hz for the 240 mm solid wall) and the bridging effect of the wall ties coupling them together. The coupling between the leaves of masonry cavity walls can be reduced by increasing the cavity width. Wall ties should be selected in accordance with the guidance in AD E.

Lightweight walls owe much of their sound insulation performance to isolation. The more effective the isolation, the better the sound insulation is likely to be. Hence the use of twin-leaf lightweight walls, resilient bars to fix plasterboard to and independent ceilings beneath floors.

Some general rules of thumb for walls are listed in Box 7.1.

Some examples of walls that comply with AD E requirements for sound insulation are described in Box 7.2.

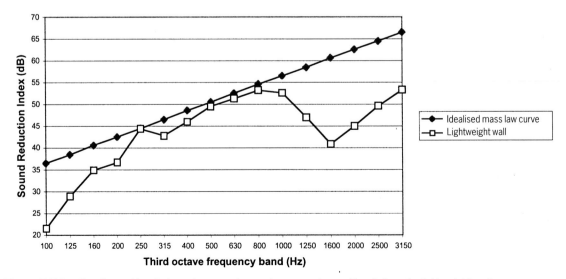

Figure 7.5 Predicted sound insulation using mass law and measured sound insulation of a lightweight wall

- The more massive solid masonry walls are, the higher their airborne sound insulation is likely to be.
- At low frequencies, massive masonry walls generally have higher airborne sound insulation than lightweight stud walls.
- Flanking sound transmission limits the airborne sound insulation between rooms.

Adequate sound insulation performance of separating walls may be achieved by:
- ensuring walls are of adequate mass,
- using only specified products,
- using wall ties with appropriate stiffness (eg butterfly ties) in cavity separating walls,
- not using lightweight flanking external walls,
- filling perpends and bed joints in masonry, particularly within floor zones and loft spaces (Figure 7.6),
- filling joints under hogging floor panels,
- not building joists into separating walls,
- not cutting deep chases back-to-back for services,
- ensuring adequate dry-packing under walls in concrete panel systems,
- ensuring lightweight linings in Wall type 3 do not touch the masonry core (see Figure 7.1).

Figure 7.6 Ensuring that bed joints and perpends in separating walls are filled properly will contribute to more effective sound insulation. In this photo, the perpends and bed joints below the joists are well filled but are inadequately filled above the joists

Where sound insulation of walls consisting of units of adequate mass proves to be deficient, the cause may be found in unfilled perpends within the wall construction; perhaps where the bricklayer has simply buttered the arrises of the units before laying (Figure 7.6).

Durability
Normally, there should be no problems with masonry separating walls unless they project through the roof. However, one factor which can and does affect masonry separating walls is movement from the ground which is sufficiently large to allow cracks to form in the masonry.

Framed separating walls

Basic structure
Two sets of steel or timber studs at 600 mm centres, with a cavity of at least 200 mm, clad on each face with plasterboard to a total thickness of 30 mm, and with an absorbent curtain or quilt incorporated in the cavity (Figure 7.7).

The quilt needs to be unfaced, possibly be reinforced, and should have a density of not less than 10 kg/m^3. If it is hung centrally in the cavity it should be a minimum of 25 mm thick, and if hung from one face it should be a minimum of 50 mm thick.

Steel-framed separating walls may also be used.

Accommodation of services
Ideally, services should not be incorporated in the separating wall, but where it is essential that electrical sockets are incorporated, they should not be fitted back-to-back, and should be backed by plasterboard of the correct thickness. Guidance is given in AD E and the *Robust Details Handbook*.

Main performance requirements and defects
Strength and stability
Strength depends on the adequacy of the studs, or frames, and how well they are nogged, or braced.

Robust Details. *Robust Details Handbook*. 2nd edition. Part E, Resistance to the passage of sound. Milton Keynes, Robust Details Ltd. Updated 2005

Box 7.2 Examples of masonry separating wall construction designed to meet building regulation sound insulation requirements

Wall type 1: Solid masonry

AD E states that the resistance to airborne sound depends mainly on the mass per unit area of the wall.

Enhanced sound insulation is achieved through a comparatively high mass (415 kg/m^2 combined block and plaster) and a 13 mm wet plastered finish to eliminate lining resonances and to ensure good air tightness. All masonry joints should be filled with mortar, including perpends, and the wall surface should be sealed with plaster, including hidden areas behind cupboards, bath panels, etc.

Flanking requirements

● Where possible, joists should run parallel to the separating wall. If timber floor joists are supported on the separating wall, hangers must be used. Concrete floors of mass at least 300 kg/m^2 may be taken through the wall.

● The plaster finish should be taken up into the roof void unless a sealed ceiling with a mass of 10 kg/m^2 is used.

● Concrete floors with a mass of at least 365 kg/m^2 may be taken through the wall. The cores of a hollow core concrete plank should be sealed with mortar or mineral wool at the junction of a separating wall.

● A flexible cavity closer must be provided where a separating wall abuts an external cavity , unless it has already been filled with cavity insulation. If the inner leaf of the cavity wall is constructed from lightweight block, it should abut the separating wall and be tied at no more than 300 mm centres vertically and the joint sealed with tape or caulking.

Wall type 2: Cavity masonry

AD E states the resistance to airborne sound depends on the mass per unit area of the leaves and on the degree of isolation achieved. The isolation is affected by connections (such as wall ties and foundations) between the wall leaves and by the cavity width.

For enhanced sound insulation minimise structural connections between the leaves and maximise the mass of the leaves.

The finish is 13 mm wet plaster. If dense blocks with a mass of 415 kg/m^2 including the plaster are used, the cavity should be not less than 50 mm. If lightweight blocks with a mass of 300 kg/m^2 are used then the cavity should be at least 75 mm.

Flanking requirements

● Where possible, joists should run parallel to the separating wall. If timber floor joists are supported on the separating wall, hangers must be used.

● Concrete floors, including suspended ground floors, should only be carried through to the inner face of each leaf and should not be continuous. Hollow cores of precast planks should be sealed with mortar or mineral wool at the junction of a separating wall.

● The plaster finish should be taken up into the roof void unless a sealed ceiling with a mass of 10 kg/m^2 is used.

● Wall ties should be butterfly type.

● A flexible cavity closer must be provided where a separating wall abuts an external cavity, unless it has already been filled with cavity insulation. If the inner leaf is of a lightweight framed construction it should abut the separating wall, be tied to it at no more than 300 mm centres vertically and have the joint sealed with tape or caulking.

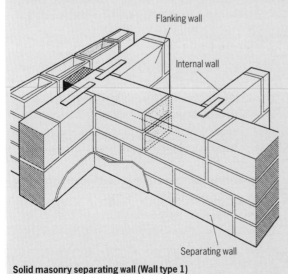

Flanking wall

Internal wall

Separating wall

Solid masonry separating wall (Wall type 1)

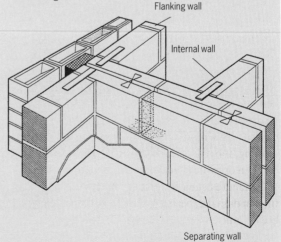

Flanking wall

Internal wall

Separating wall

Cavity masonry separating wall (Wall type 2)

Cont'd....

Box 7.2 Examples of masonry separating wall construction designed to meet building regulation sound insulation requirements (cont'd)

Wall type 3: Masonry between isolated panels

AD E states that the resistance to airborne sound depends the type and mass per unit area of the core and partly on the isolation and mass per unit area of the independent panels.

Enhanced sound insulation is mainly due to the separation of the panels from the masonry core. The lining support framework must be fixed only at the floor and ceiling.

Two types of independent lining can be used:

- Two sheets of plasterboard with a cellular core with a mass of at least 20 kg/m². Joints between panels should be taped.

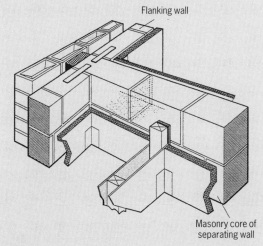

Flanking wall

Masonry core of separating wall

Masonry and panels separating wall (Wall type 3)

- Two sheets of plasterboard. Each should be at least 12.5 mm thick if they use a supporting framework and have a combined mass of at least 20 kg/m². If no framework is used, the total plasterboard thickness should be 30 mm thick with a mass of at least 25 kg/m².

Solid or cavity masonry walls may be used with a minimum cavity of 50 mm.

Flanking requirements

- The linings should be continued into the roof void unless a sealed 10 kg/m² ceiling is used.
- Timber floors carried on the masonry core must use joist hangers and there must be solid blocking between the joists.
- Concrete floors with a mass of at least 365 kg/m² may be taken through a solid wall. The cores of a hollow core concrete planks should be sealed with mortar or mineral wool at the junction of a separating wall. No floor should be taken through a cavity masonry core wall, except solid concrete ground-bearing floors slabs.
- The inner leaf of any external cavity wall should be lined internally with the same lining as used on the separating wall. Thermal insulation can be incorporated into the lining of the external wall, providing care is taken to prevent interstitial condensation. This can be useful in the case of dwellings with a substantial stagger on plan which can mean that a wall could be both a separating wall and an external wall.

Some separating wall leaves may not be adequately tied together as ties might interfere with sound insulation.

Dimensional stability, deflections, etc.

There may be some shrinkage of the timber studs as they dry down to their in-service moisture content, which could lead to cracking at the junction with the ceilings and with the external walls. It is unlikely to be extensive. If the smaller section timbers are dried down to 20% moisture content at the time of grading, the risk of subsequent shrinkage is reduced.

Twin or triple sheets of plasterboard will give sufficient resistance to both deflections and to impact damage, particularly if they have been laid to break joint.

Rising damp and condensation

Rising damp from a slab is unlikely to occur provided the damp proof membrane (dpm) is satisfactory. Condensation is a theoretical risk at the junction with the external wall if the thermal insulation is deficient, allowing a thermal bridge.

Thermal properties

The acoustic curtain goes some way to providing a reasonable standard of heat insulation on each side of the separating wall. There are no formal requirements.

Fire

Fire can spread between dwellings having framed separating walls where an effective fire separation has not been achieved. In some dwellings, the fire

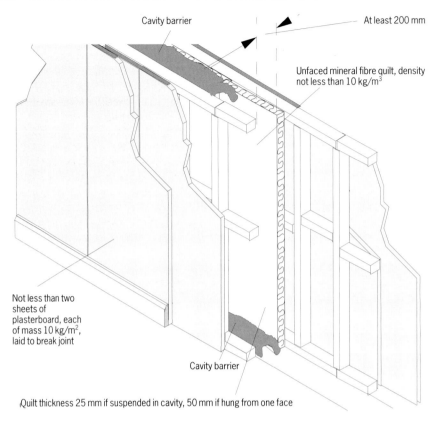

Cavity barrier

At least 200 mm

Unfaced mineral fibre quilt, density
not less than 10 kg/m³

Not less than two
sheets of
plasterboard, each
of mass 10 kg/m²,
laid to break joint

Cavity barrier

Quilt thickness 25 mm if suspended in cavity, 50 mm if hung from one face

Figure 7.7 Plasterboard on a timber or steel frame (Wall type 4)

separation is impaired where the fill between the
separating wall timber frames does not abut the roof.
Where the roof space separating walls are lined with
plasterboard, gaps have been noted at eaves and
purlin positions. Boxed eaves are particularly likely
to contain gaps which outflank the separating wall
(see Figure 7.4).

Cavity barriers will be needed at various places,
for example at the junctions with other elements of
construction, such as floors (Figure 7.8) and
external walls, for both fire and noise reasons.

One hour's fire resistance should be obtainable
from a framed separating wall built in accordance
with the manufacturer's instructions and faced with
various combinations of wallboard and plank.
Improved performance (one-and-a-half or two
hours) is also obtainable, but this may require the
use of boards of enhanced performance.

Figure 7.8 The diagonal bracing in the separating wall has
prevented correct installation of the cavity barrier at the
intermediate floor

Box 7.3 Example of a framed separating wall construction designed to meet building regulation sound insulation requirements

Wall type 4: Plasterboard panels on a frame of timber or metal (see Figure 7.7)

AD E states the resistance to airborne sound depends on the mass per unit area of the leaves, the isolation of the frames and the absorption in the cavity between the frames.

Enhanced sound insulation is mainly due to the isolation of the two plasterboard layers supported on independent studs with minimum connections and a sound absorbent quilt in the cavity.

- The independent frames, which may be light gauge steel or timber, should maintain a cavity of at least 200 mm between the inner faces of the plasterboard.
- The plasterboard lining to each side should be of two or more layers of plasterboard, each having a mass of at least 10 kg/m², with joints staggered.
- The sound absorbent in the cavity should be of unfaced mineral fibre batts or quilt, and should have a density of at least 10 kg/m³.
- Its thickness should be 25 mm if suspended in the cavity between the frames, 50 mm if fixed to one frame, or 25 mm per quilt if one is fixed to each frame.

Flanking requirements
- In the roof void, the separating wall construction can be carried through and the thickness of the cladding reduced to not less than 25 mm, or the cavity can be closed at

ceiling level without connecting the two frames rigidly together and one frame continued into the roof void with at least 25 mm thick cladding on both sides. In each case, the space between the frame and the roof finish should be sealed, eg with mineral wool.
- Intermediate timber- or steel-framed floors should not pass through the cavity wall. Air paths into the separating wall cavity should be blocked by carrying the cladding through the floor or using a solid edge to the floor. Ground floors should be a ground-bearing concrete slab, a suspended concrete slab or a suspended timber floor. Concrete ground floors may run through the separating wall. If a suspended concrete slab is used it must have a mass of at least 365 kg/m². If a suspended timber ground floor is used, the masonry foundation dwarf support walls should be built without gaps.
- The external wall inner leaf can be of either timber- or steel-frame construction. If there is a cavity masonry outer leaf, the cavity between the ends of the separating wall and the outer leaf should be stopped. The internal finish should be plasterboard of mass at least 10 kg/m².
- The inner leaf of the external wall can be steel or timber with plasterboard of mass at least 10 kg/m². Any cavity should be stopped between the outer masonry leaf and the ends of the separating wall.

Sound insulation

An example of framed wall construction that should achieve the minimum performance standard for airborne sound insulation is described in Box 7.3.

Stairway enclosures and protected shafts

> **Key Point**
> Stairways may need to be enclosed and protected since they are frequently designated as a means of escape in case of fire.

The enclosing walls of protected shafts may be constructed of any materials which meet the relevant fire-resistance criteria.

Where a stairway is contained in a protected shaft, and that shaft has a requirement of not more than one-hour fire resistance, glazed partitions

separating a corridor or lobby from the protected shaft may be used. They should be fitted with glazing having a half-hour's resistance, and can be used provided that the corridor or lobby is separated from the rest of the storey by construction also having a half-hour's resistance; that is to say two half-hour partitions between the protected shaft and the rest of the compartments (Figure 7.9).

> **Key Point**
> Permanent ventilation for smoke control purposes must be provided at the top of internal stairways serving more than one dwelling.

In many circumstances, compartment walls provide a separating wall function and therefore will need to provide adequate sound insulation.

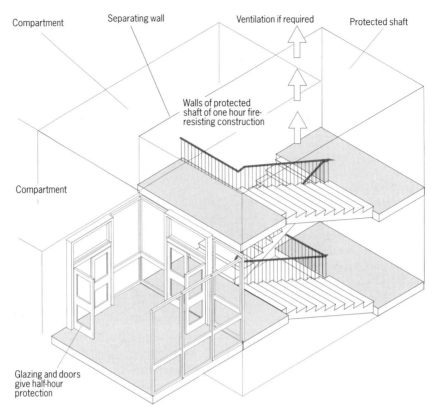

Compartment

Separating wall

Ventilation if required

Protected shaft

Walls of protected shaft of one hour fire-resisting construction

Compartment

Glazing and doors give half-hour protection

Figure 7.9 Glazed screens separating the lobby from the stairway in a protected shaft

Partitions

The second part of this chapter deals with all kinds of fixed partitions, whether of heavy construction (eg masonry) or light construction (eg stud and plasterboard), relocatable or movable. In essence, therefore, this section deals with internal walls that have no separating or compartmenting function. They may be of similar construction to separating or compartment walls.

Partitions constructed of brick or block may be either loadbearing or non-loadbearing. Imposed loads may be imparted mainly by floors, but also by tanks and WCs fixed to them.

Partitions have been built in a variety of materials over the years (eg hollow clay block and plaster slab reinforced with organic materials) either erected using normal bricklaying techniques or cast-in-situ. Modern methods of construction imply the use of

larger units to reduce site labour requirements both in assembly and in the finishing trades.

Internal walls for dwellings can be of any construction provided that they have a sound insulation of $\geq$ 40 dB R_w.

Masonry partitions

Basic structure

For many years, masonry loadbearing partitions have commonly been constructed in half-brick thicknesses laid in stretcher bond and they should continue to offer good performance if well built.

Abutments and corners

Partitions, especially non-loadbearing and less than 100 mm thick, cannot always be self-supporting. Masonry partitions can be provided with corner posts in timber to provide the necessary stability. In other cases, stability can be provided by full room height doorsets pinned to ceilings, with the blockwork held against horizontal displacement essentially by the architraves. Otherwise, returns are necessary.

Main performance requirements and defects
Strength and stability

Masonry partitions may be overstressed by superimposed loads (eg water tanks in the roof). They may be too thin in relation to their height, not tied to other walls at ends and not fixed to the floor or roof structure above.

In tests, single-leaf clay brickwork and concrete blockwork storey-height partition walls have been subjected to static vertical loads representative of those in cross-wall housing construction up to three storeys high. Simulated horizontal wind loading was applied to the end of the wall until failure occurred. Tests were carried out on walls both with and without a return at the loaded end and the results were compared. Walls with returns exhibited greater racking strengths than plain walls. Brickwork walls were generally stronger than the blockwork walls and exhibited lower strains.

Dimensional stability, deflections, etc.

Ideally, to maintain stability, partitions should be restrained at both top and sides, but they may be found restrained only at the top or the sides, and storey-height doorframes may prove useful additional support in this respect.

Rising damp and condensation

Partition walls should be at less risk of rising damp than external walls because of the normally drier environment around the foundations. However, where partitions are bonded to an external wall a discontinuity in damp-proofing can exist if the two dpcs are at different levels and are not linked. This condition is more likely if the partition and external wall were constructed at different times. As with external walls, the cause of rising damp is more likely to be bridging than a material failure or omission of the dpc.

Bridging of dpcs is commonly caused by plastering, but can also occur with solid floors and abutting external walls.

Fire

Loadbearing partitions will need to have fire resistance for a period of between a half and two hours, depending on the size of the building and location of partitions. Walls protecting means of escape only will require at least a half-hour protection. Where a partition supports a structural element with a designated fire resistance, the partition is required to meet at least the same criterion.

An unplastered half-brick loadbearing partition should give one-hour fire resistance. Sanded plasters give only marginal improvements, but lightweight plasters offer considerable improvement, for example by increasing the resistance of half-brick walls to six hours. However, they have to remain in place during a fire, and they may be too thin or lack adequate restraint, so that stability of the structure in fire might be jeopardised.

Services may pass through fire-resisting partitions without adequate fire stopping and there may be gaps at perimeters. Cavities in construction above a partition (eg a suspended ceiling), below a partition (eg a floor void) or to the sides of a partition (eg a wall cavity) may, in certain circumstances, provide a bypass route for fire in the absence of cavity barriers (Figure 7.10).

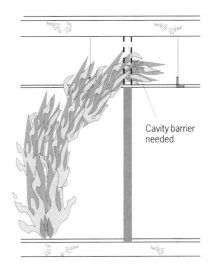

Cavity barrier needed

Figure 7.10 Partitions that do not reach the underside of the structural floor can provide routes for fire spread and sound transmission

Sound insulation

Partitions that do not meet the underside of the structural floor, for example those terminating at suspended ceilings, can be bypassed by sound travelling over the top of the partition within the suspended ceiling void. Masonry partitions may develop perimeter gaps, possibly resulting from structural movement or lack of bonding or tying.

> **Key Point**
> Well-designed lightweight stud walls with effective isolation between each side of the wall can have good airborne sound insulation, particularly at high frequencies.

Durability

Masonry partition walls commonly suffer from cracking, almost irrespective of material. In the case of non-loadbearing walls, this could be from building off a timber suspended floor that has yet to dry down to its in-service moisture content. If the wall is not bonded into the external wall, the crack may easily occur at the wallhead-to-ceiling joint. If it is bonded into the external wall, the crack may occur at any point in the height, depending where the bonding-in occurs. Cracking could also be from shrinkage of the material, especially lightweight concrete that has been allowed to become saturated on site.

In the case of loadbearing walls, one of the most likely causes of cracking is poor bonding of masonry. Also, there is the risk of cracking through shrinkage.

Framed partitions

This section deals with all kinds of framed partitions, including those that are static (eg ordinary timber studs with plasterboard facings) or movable (such as sliding and folding partitions) and those that are relocatable, sometimes called operable walls (eg those proprietary systems now widely used in office accommodation).

The 20th century saw a huge increase in the variety of solutions that are possible, particularly with movable partitions, some of which have excellent fire and acoustic performance.

Basic structure

Framed partitions may be loadbearing or non-loadbearing, just as with masonry partitions. The simplest form of framed partition is that which consists of nogged timber studs carried on floor plates, and covered each side with plasterboard which may or may not be skimmed.

Metal stud framed partitions, some of which may be demountable or relocatable but also faced with plasterboard, are a later development. Steel stud systems may have the advantage of rapid erection compared with, say, timber stud partitions cut on site. The framing for these systems, although formed from galvanized steel pressed metal sections, is in principle assembled the same way as traditional timber stud construction using horizontal head and sole plates, vertical 'studs' between these and horizontal cross-noggings at intermediate positions between the studs. Self-drilling self-tapping screws are used to connect the members together.

Yet other types might consist of prefabricated chipboard sheets faced with vinyl on the outer surface and a balancing layer, often paper mounted, on the inside.

Movable partitions

There are many types of movable partition. The simplest consist of lightweight accordian or concertina-type fibreboard, or steel cored vinyl-covered narrow slats hinged together by the coverings. More substantial systems consist of hinged leaves, similar to doors, connected in pairs or multiples and carried on overhead tracks and, for the heavier versions, also floor tracks. Fully glazed types are also available which are held in upper and lower clamping rails, but no stiles, where the safety glass alone acts as the leaf (Figure 7.11). Most of the types described here may be hand operated or mechanically driven and may or may not have pass doors and vision panels.

External corners

These are the most vulnerable areas of lightweight framed or cored partitions, and should be protected, for example by cover moulds or other forms of armouring.

Figure 7.11 Room height fully glazed operable wall

Accommodation of services

Partitions, even though nominally non-loadbearing, are frequently required to carry asymmetric hanging loads (eg bookshelves and sanitary fittings). Failure to allow adequately for live loads associated with these has been noted on BRE site investigations.

In the past, support for fixtures and fittings, particularly cantilevered sanitary fittings fixed on stud partitions, often relied on wood noggings between studs; this produced mixed results. In prefabricated paper-cored panels, support is provided by means of noggings driven into the cores.

If service pipes or wires are to be installed in stud partitions these must be installed before the plasterboard facing is fixed on both sides. Where the partitions are of the relocatable type, it is of course better to avoid siting service runs within or on their surfaces. Plasterboard is held in place using fixing strip and self-drilling self-tapping screws, the screw heads being concealed by clip fitting trim and skirting which is a push-fit over the fixing strip.

Main performance requirements and defects
Strength and stability
Stud partitions, irrespective of material, will need to be checked when under construction, particularly with respect to restraint at head and foot and at all returns.

Dimensional stability, deflections, etc.
Non-loadbearing partitions built off suspended timber floors depend for their stability on the stability of the floor. When this moves, the partition will move with it.

Rising damp and condensation
Dpc failure under a timber partition could result in rot in the sole plate and abutting studs, and to settlement of the partition.

Fire
Fire can have a devastating effect on lightweight framed partitions (Figure 7.12).

Timber stud partitions may have linings such as hardboard, plywood or medium board which do not satisfy fire resistance or surface spread of flame criteria (eg Class 0 where used adjacent to means of escape routes). Where plasterboard, or similar, is used it must have adequate thickness, sufficient plaster finish and noggings behind board joints.

Figure 7.12 A stud partition after a fire in an institutional building

Loadbearing stud framed partitions with studs must be a minimum of 44 mm wide, spaced at not more than 600 mm, covered with 12.5 mm plasterboard and have all joints taped and filled to give a half-hour's fire resistance. Similar performance can be obtained from partitions with metal studs. Doors and frames in partitions enclosing fire escape routes will normally require a minimum of a half-hour fire resistance.

An extra layer of 12.5 mm plasterboard to each side of the partition already described will increase the fire resistance to a full hour.

Relocatable and movable partitions will need to conform to fire resistance and surface spread of flame requirements, just as for any other partition in the same situation. Concertina-type partitions have little fire resistance, but other types are available which give up to two hours, depending on construction and glazing.

Sound insulation

Fixed timber stud partitions should be checked to ascertain that the correct facings have been used. Using low mass boarding can result in poor sound insulation.

Sliding/folding partitions generally have poor sound insulation because it is difficult to seal them effectively and often their mass per unit area is relatively low. Where the system is top-hung only, and the bottom is sealed by retractable seals, the condition of the floor surface will affect sound insulation, particularly where, for example, there are recessed or eroded joints in the flooring, allowing air gaps under the partition in the closed position to reduce sound insulation.

Durability

Timbers that are damp, but show no signs of decay, need to be dried out to an acceptable moisture content to avoid the risk of decay. Application of wood preservatives may be appropriate if drying time cannot be reduced to less than about six weeks.

Apart from impact damage, satisfactory service from movable partitions depends to a considerable degree on wear and tear in the suspension systems.

Some wood-based boards used for facing relocatable partitions may be treated with fire retardants which are hygroscopic in nature, leading to increases in moisture content and a risk of bowing if the moisture take-up is not uniform. Moisture contents of 20–22% have been measured by BRE on site. Whether this bowing is contained depends on the integrity of the fixings.

Regulations and standards

Building Regulations

The structural requirements for walls are covered in the Building Regulations for each part of the UK. The combined dead, imposed and wind loads must be transmitted o the ground safely, without undue movement:
- **England and Wales:** AD A,
- **Scotland:** Section 1 of the Technical Handbooks, Domestic and Non-Domestic,
- **Northern Ireland:** Technical Booklet D.

Resistance to moisture is covered by:
- **England and Wales:** AD C,
- **Scotland:** Section 3 of the Technical Handbooks, Domestic and Non-Domestic,
- **Northern Ireland:** Technical Booklet C.

The requirements that the building shall be designed and constructed so that appropriate provisions for the early warning of fire, and appropriate means of escape in case of fire from the building to a place of safety are covered by:
- **England and Wales:** AD B,
- **Scotland:** Section 2 of the Technical Handbooks, Domestic and Non-Domestic,
- **Northern Ireland:** Technical Booklet E.

The relevant accompanying documents for the Building Regulations relevant to sound insulation for each part of the UK are:
- **England and Wales:** AD E (revised 2003) which stipulates that separating walls and partitions in purpose-built houses and flats need to achieve a minimum value,
- **Scotland:** Section 5 of the Technical Handbooks, Domestic and Non-Domestic,
- **Northern Ireland:** Technical Booklet G.

Glazing in all new movable partitions in domestic accommodation, and in all movable partitions in other building types, must be safety glazing, complying with the relevant accompanying documents for the Building Regulations for each part of the UK:

- **England and Wales:** AD N requires safety glazing in doors up to a height of 1.5 m and in windows up to a height of 800 mm above finished floor level. Annealed glass, for example, in vision panels and pass doors, is permitted provided the smaller dimension of the pane is less than 250 mm,

- **Scotland:** Section 4 of the Technical Handbooks, Domestic and Non-Domestic,
- **Northern Ireland:** Technical Booklet V.

Fire doors have additional requirements.

Other regulations

Where housing developments are proposed, NHBC or Zurich Building Guarantees requirements may also have to be met.

8 Roofs

Roofs have to keep the weather out, maintain the comfort of the building occupants and ensure all this is done safely. Climate change with increasing wind and rainfall is likely to focus more attention on the design options (see BRE *Digest 486*). The designer has a wide choice of designs and freedom to create many different shapes. However, for the majority of buildings, the first consideration is between flat or pitched roofs, or a combination of the two. 'Green roofs', where plants grow as the top layer, are becoming more popular and claim to have environmental benefits.

Flat roofs are particularly suited to buildings which are irregular in plan and have large areas to cover (eg supermarkets and leisure centres). Conversely, they are popular for housing extensions. Bitumen, plastics and rubber sheet materials are used for the waterproofing layer. Design of insulation needs careful consideration to avoid condensation problems.

Pitched roofs include the traditional slate or tiled roof used on domestic and prestige buildings. They are usually limited to roofs that are rectangular in plan and can be found on small- to medium-sized industrial roofs. In general, the walls have to support a higher load from pitched roofs.

A third roof type, to be considered separately from the flat or pitched roof, is the profiled metal roof. This has grown in popularity recently,

particularly for covering industrial units such as warehouses, although some can be seen on office and housing developments.

Not considered here are roofs covered with fully supported metal such as lead, copper or stainless steel.

Design considerations

Factors including aesthetics, cost and planning requirements will all have to be assessed when making the choice between pitched (including angle of slope), or flat roof, or a combination of the two. All can suffer from premature failure if they are not designed, specified or constructed correctly.

A typical design life of 60 years for traditional slate or tiled pitched roofs has been used in the past, although many roofs last longer than this. It is a popular belief that pitched roofs last longer than flat roofs. However, this may not always be the case.

Flat roofs have in the past suffered from premature failures and short service life because relatively weak bituminous felts were laid over thermal insulation that was not dimensionally stable. The felts would rip or tear in as little as three years due to the movement of the insulation to which they were bonded. More stable insulation and much stronger materials have resulted in service lives which can approach 30 years (where properly installed) without indications of widespread failures. Failures are most likely to occur at parapet and other abutments and where services pass

Saunders GK. *Reducing the effects of climate change by roof design.* Digest 486. Bracknell, IHS BRE Press, 2004

> **Key Point**
> Roofs are dangerous places to work so designers should take account of safety and future maintenance requirements.

through the membrane. Control of the risk of condensation also needs more careful consideration.

Pitched roofs can also suffer from condensation problems if not constructed correctly, which results in a shortened service life. Also, some imported slates have not performed well in the UK climate. Pitched roofs rely on the fixings to keep tiles and slates in place. Verges are particularly vulnerable and end-of-life can be due to corrosion of the fixings. In this case, the tiles or slates can be removed and reset in place.

Roofs are dangerous places to work and the opportunities for accidents are frequent. Because of this roofs may not be inspected as often as they should be and preventative maintenance is often put off. The designer is ultimately responsible for ensuring that the design has taken account of safety and future maintenance requirements (see BRE's *Digest 493* and *Information Paper IP 7/04*).

Key performance requirements

Strength and stability

Every roof needs to be sufficiently strong to carry the self weight of the structure together with intermittent loads such as snow, wind or maintenance, without undue distortion or damage to the building and finishes. There is more guidance in BRE's *Information Paper IP 13/98*.

As discussed in chapter 3, the key environmental loads on roofs are as follows.
- **Wind,** influenced by wind speed, altitude, direction, height above ground level, gusts, topography and seasonal factors.

- **Snow,** influenced by the amount of snowfall, geographical location and altitude, as well as through drifting and the geometry of the roof itself.

BS 6399-2 governs the calculation of wind and snow loading on roofs.

Designing for wind resistance: some key issues
- Wind speeds in coastal regions tend to be greater than inland.
- Basic wind speed (measured at 10 m above ground) is adjusted by altitude and other factors (see Figure 3.4).
- Wind speed over reasonably level ground increases about 10% for every 100 m above sea level. About half the UK is below 200 m so wind speed-up due to altitude effects will be, at most, 20% above basic wind speed for these locations.
- Prevailing winds come from the west and south west, but BS 6399-2 calls for the design to take into account wind from any direction.
- The speed of wind is reduced by nearby buildings, allowing the effective height of buildings for calculations to be reduced.
- The critical calculation in respect of gusting winds is the gust which is just large enough to envelop the whole building or roof, or the whole component when designing a part such as a roofing panel, or calculating suction on an individual tile.
- Increase in wind loads due to gusts is greater for small buildings, and greater still for small components.
- Severe eddies and vortices will tend to affect corners, eaves and ridges, and any projections such as chimneys, dormers or tank rooms, creating both positive and negative pressures.
- For wind loads on roof-based photovoltaic or other forms of solar collector systems, see BRE's *Digest 489*.

BRE
Saunders G. *Safety considerations in designing roofs.* Digest 493. 2005
Saunders GK. *Designing roofs with safety in mind.* Information Paper IP 7/04. 2004
Menzies JB & Gulvanessian H. *Eurocode 1: The code for structural loading (2 Parts).* Information Paper IP 13/98. 1998

British Standards Institution. BS 6399-2: 1997 *Loading for buildings. Code of Practice for wind loads*
Blackmore P. *Wind loads on roof-based photovoltaic systems.* Digest 489. Bracknell, IHS BRE Press, 2004

Designing for snow loading: some key issues
- Density is increased by compaction, but loading is not unless rain saturates unmelted snow.
- Snow tends to accumulate at obstructions to form local drifts, and asymmetric loads can occur, particularly at abrupt changes in height greater than 1 m (eg dormers, parapets or chimneys; Figure 8.1) or through manual or wind snow clearing.

For small roofs, BS 6399-2 permits simplified design where:
- there is no access to the roof except for cleaning and maintenance,
- total roof area is below 200 m², or the roof is pitched and not wider than 10 m with no parapets,
- no other buildings are within 1.5 m,
- there are no abrupt changes of height greater than 1 m, or
- there are changes in height, but the lower area of roof is not greater than 35 m², and
- there are no other areas subject to asymmetrical snow loading.

In these circumstances, a uniform snow load based on geographical location, with a specified minimum value, can be used.

Outside these circumstances, the design snow load as required by BS 6399-2 requires regional variations in loading and redistribution of snow by wind to be taken into account. Load shape coefficients provide for local drifting of snow, and a snow load statistical factor provides estimates of loads with probabilities of exceedence. Roofs up to 30° pitch are assigned the same maximum uniform snow loading as a flat roof. For steeper pitches the asymmetrical loading is different. Outside the simplified design rules the following load conditions must be considered:
- uniformly distributed snow load,
- asymmetric snow load due to redistribution by wind,
- local drifting (see BRE's *Digest 439* for more information),
- partial loading due to snow removal.

Accidental and maintenance loads

During service, roofs will be subjected to irregular applications of accidental and maintenance loads, which can vary widely. Designers should take into account the typical problem areas which include:
- inadequate support for wide valley gutters that provide easy routes for maintenance,
- damage caused by the unprotected feet of ladders and mobile platforms used for access,
- damage by missiles where vandalism is a problem.

Energy conservation, vapour control and avoidance of condensation

All the materials of a roof, including the spaces and voids, contribute to thermal performance. Measured thermal performance, particularly of insulation, reduces with increased moisture content. Non-absorbent materials are preferable where prolonged wetting (eg through condensation) is probable. In most buildings, insulation is kept at an acceptable moisture content through protection from rain and designing to avoid the build-up of condensation. Water build-up can also be reduced by ventilating cavities adjacent to the insulation. Thermal losses due to thermal bridging (eg from concrete beams) is often ignored in standard calculations, but will become more important as thermal insulation standards increase. Thermal bridges can also lead to internal surfaces reaching dew point, leading to

Currie D. *Roof loads due to local drifting of snow.* Digest 439. Bracknell, IHS BRE Press, 1999

Figure 8.1 Drifting snow behind a parapet

selective condensation on, for example, ceilings (for more information, see BRE's Report *Thermal insulation: avoiding risks* and *Information Paper IP 1/06*. Unintended air movement (eg through failure to seal joints and service penetrations through ceilings) can carry water vapour to other areas.

Key design issues include:
● whether the location of the insulation makes the flat roof a cold deck (insulation below the deck), a warm deck (insulation above the deck) or a warm deck inverted roof (insulation on top of the waterproof layer),
● whether insulation is laid consistently (filling every space and avoiding thermal bridges),
● whether insulation is located to the correct side of any component which acts as a vapour control layer,
● whether insulation provides structural support to weatherproofing, in which case it will need adequate positive and negative load resistance, appropriate movement potential for weatherproof layer, chemical compatibility and appropriate fire resistance.

The need to minimise the risk of condensation is covered in the supporting documents of the national building regulations:
● **England and Wales:** AD F,
● **Scotland:** Section 3 of the Technical Handbooks, Domestic and Non-domestic,
● **Northern Ireland:** Technical Booklet C.

The Scotland and Northern Ireland regulations suggest that cold, level-deck roofs should be avoided.

If the relative humidity of the air within the building is above about 60%, there will be a risk of condensation on the underside of roofs on cold clear nights. The risk is particularly high for profiled sheet roofs (see later section in this chapter). BS 5250 provides the calculation for the risk of

BRE
Thermal insulation: avoiding risks. 2002 edition. BR 262
Ward TI. *Assessing the effects of thermal bridging at junctions and around openings*. Information Paper IP 1/06. 2006

> **Key Point**
> Cold level-deck roofs should be avoided.

condensation occurring, which should always be made for new flat roofs, and for warm deck pitched roofs. It should be noted that predictions suggest that climate change may exacerbate problems of condensation in buildings due to higher relative humidity. Higher winter temperatures combined with increased vapour pressures may give rise to more severe problems, particularly in roof spaces.

There are three main routes to reducing the risk of condensation:
● installing thermal insulation to reduce heat losses,
● installing a vapour control layer (vapour resistance greater than 200 MNs/g) on the warm side of thermal insulation,
● ensuring effective cross-ventilation of any cavity or void above the insulation.

Vapour control layers (vcls) are notoriously difficult to install without gaps at joints between sheets or interfaces with other parts of the structure. They are best fixed to continuous stiff sheet materials (but not fully bonded to avoid the risk of movements fracturing them at joints) with lapped glued joints between the vapour control sheets.

Control of solar heat and air temperature

As discussed in chapter 3, all construction materials are affected by heat gain through solar radiation. Highly reflective top layers help to reduce heat absorption, and light-coloured materials have high reflectance and therefore absorb less solar energy. Bright sheet metal materials can, however, experience significant solar heating. Different locations will tend to have different highest maximum temperatures, but all roofs are subject to rapid swings in temperature due to sudden showers on already hot surfaces. For this reason, repeated rapid reversals of temperature are included in tests of new materials. Similarly, material testing will

British Standards Institution. BS 5250: 2002 *Code of practice for the control of condensation in buildings*

cover resistance to frost damage and salt crystallisation.

Air permeability

Air permeability is defined and discussed in general terms in chapter 1.

> **Key Point**
> Although flat and pitched roofs offer quite differing geometries, the requirement for airtightness applies equally to both.

A continuous membrane over a flat roof will in principle be airtight, but it still has vulnerable edge joints. But the same is not necessarily true for pitched roofs covered with a flexible sarking and overlapping slates or tiles that can result in a leaky roof. If the roof has been designed as a warm roof, the airtight but vapour-permeable membrane will normally be on the slope under the slates or tiles. Although cold roofs are expected to become less common in future, they still may be needed in certain circumstances. Therefore, if the roof has been designed as a cold roof, the airtight but vapour-permeable membrane will be over the insulation at ceiling level to allow ventilation to take place to eaves and ridge. Sheet material overlaps will need to be properly sealed, as will all traps and service perforations into the roof void.

Protection against fire

Fire safety (particularly relevant to rooms in lofts), compartmentation and the need for cavity barriers in concealed spaces and control of internal and surface fire spread is covered in the supporting documents of the national building regulations:

- **England and Wales:** AD B,
- **Scotland:** Section 2 of the Technical Handbooks, Domestic and Non-domestic,
- **Northern Ireland:** Technical Booklet E.

British Standards Institution. BS 476-3: 2004 *Fire tests on building materials and structures. Classification and method of test for external fire exposure to roofs*

> **Key Point**
> For further information on how to comply with the Regulatory Reform (Fire Safety) Order 2005, including the selection of approved products and services, see www.RedBookLive.com.

Fire can attack a roof from outside or inside, and key design features to resist each origin are outlined as follows.

Key design considerations for fire from outside

- *Combustibility of materials* (including softening or melting temperatures, fixing and jointing methods and thermal properties of insulation and surrounding materials).
- *Fire performance of roof coverings* (resistance to penetration and performance in respect of flame spread, drawing on BS 476-3) based on distance from the relevant boundary as defined in the supporting documents of the national building regulations. Where separating or compartment walls are carried up only to the underside of the roof covering, the roof covering will need to be AA, AB or AC for a distance of 1.5 m either side of the wall (Figure 8.2a and 8.3). As long as the wall does not exceed 15 m in height, boarding or tiling battens can be carried over it provided they are fully bedded in mortar or other no less suitable material where they pass over the wall. Mineral wool quilt will suffice. Alternatively, the separating wall must extend 375 mm above the upper surface of the roof, measured vertically (Figure 8.2b).

Specific rules will apply to the location, area and spacing of roof lights, depending on their constituent materials and the fire performance, in particular for thermoplastics materials, such as PVAC (polyvinyl acetate), PVC-U and polycarbonate. Thermoplastics rooflights must not be placed over protected stairways. These rules are included in AD B of the Building Regulations (England & Wales) (Figure 8.4).

(a)   **(b)**

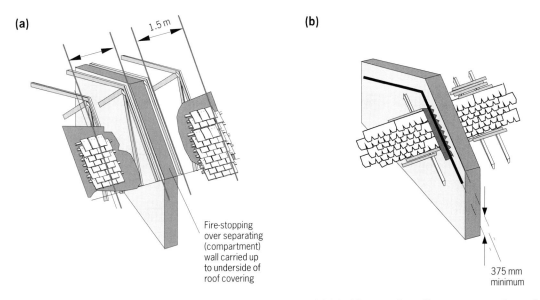

1.5 m

Fire-stopping
over separating
(compartment)
wall carried up
to underside of
roof covering

375 mm
minimum

Figure 8.2 Distances given in the Building Regulations (England & Wales) for covering adjacent to separating walls

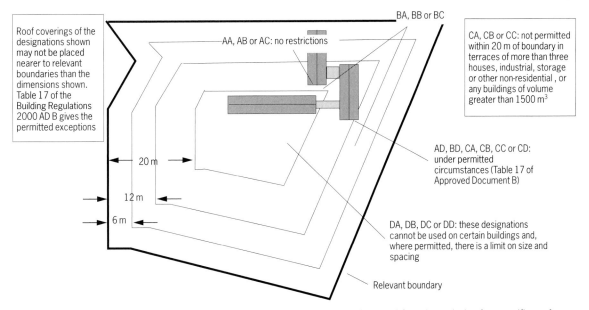

Roof coverings of the
designations shown
may not be placed
nearer to relevant
boundaries than the
dimensions shown.
Table 17 of the
Building Regulations
2000 AD B gives the
permitted exceptions

BA, BB or BC

AA, AB or AC: no restrictions

CA, CB or CC: not permitted
within 20 m of boundary in
terraces of more than three
houses, industrial, storage
or other non-residential , or
any buildings of volume
greater than 1500 m³

20 m

12 m

6 m

AD, BD, CA, CB, CC or CD:
under permitted
circumstances (Table 17 of
Approved Document B)

DA, DB, DC or DD: these designations
cannot be used on certain buildings and,
where permitted, there is a limit on size and
spacing

Relevant boundary

Figure 8.3 Permitted distances given in the Building Regulations (England & Wales) from boundaries for specific roof coverings of various performances

Key design considerations for fire from inside

- Where the roof forms part of an escape route, it must be constructed to achieve particular standards, just as a fire-separating floor must, as indicated by tests in BS 476: Parts 20–23. Where there is a fire-protecting ceiling, the construction as a whole is subject to test.

- The roof must satisfy for the specified time:
  - ❑ *loadbearing capacity* (freedom from collapse),
  - ❑ *integrity* (no holes through which fire can pass),
  - ❑ *insulation* (temperature rise contained).

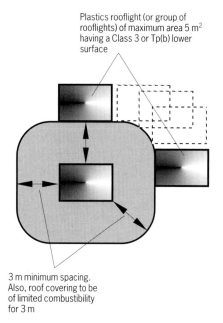

Plastics rooflight (or group of rooflights) of maximum area 5 m² having a Class 3 or Tp(b) lower surface

3 m minimum spacing. Also, roof covering to be of limited combustibility for 3 m

Figure 8.4 Criteria for distances between roof lights

Lightning protection

Roofs carry the terminals of any lightning protection system. If the system is to be effective, no part of the roof should be more than 10 m from a conductor, or 5 m for high-risk buildings. BS 6651 recommends that the risk of a building or part of a building being struck should be calculated. Figure 8.5 gives the number of lightning strikes to the ground by square kilometre per year. This is multiplied by the effective collection area of a structure to give the probable number of strikes to the structure per year. Weighting factors take account of:

● use of the structure (eg a school, hospital, office block, church),
● type of construction (eg thatched or metal roof, stone construction),

---

British Standards Institution
BS 476: 1987 *Fire tests on building materials and structures*
  Part 20: *Method for determination of the fire resistance of elements of construction (general principles)*
  Part 21: *Methods for determination of the fire resistance of loadbearing elements of construction*
  Part 22: *Methods for determination of the fire resistance of non-loadbearing elements of construction*
  Part 23: *Methods for determination of the contribution of components to the fire resistance of a structure*

> **Key Point**
> No part of the roof should be more than 10 m from a conductor, or 5 m for high-risk buildings.

● contents of the building or consequential effects (eg museum or art gallery, hospital, school, telephone exchange, power station),
● degree of isolation (a building on a flat plain or a tall building that is more than twice the height of surrounding buildings),
● type of terrain in which the building is situated (eg flat or mountainous).

A risk of 1 in 100,000 per year is the recommended threshold for protection.

If protection is required the following networks form the lightning protection system.

● *Air termination* (rod or horizontal conductor, normally placed at the eaves of flat roofs and linked to roof-mounted services such as air-conditioning units and to metal parts of the structural frame of a building).
● *Down conductors* (low impedance conductors taking the shortest possible route to earth, often positioned on chimneys and routed over sloping surfaces. Bends should be avoided as much as possible).
● *Earth electrodes* (metal rods, tubes, strips or plates, and indigenous earths such as piles or foundations).
● *Earth termination* (interconnection of the electrodes by a conductor where the earth electrodes network does not achieve a sufficiently low ohmic value).

More guidance can be found in BRE's *Digest 428*.

Daylighting, control of glare, etc.

The provision of daylighting through roofs generally relates only to larger span buildings where

---

British Standards Institution. BS 6651: 1999 *Code of practice for protection of structures against lightning*
BRE. *Protecting buildings against lightning.* Digest 428. Bracknell, IHS BRE Press, 1998

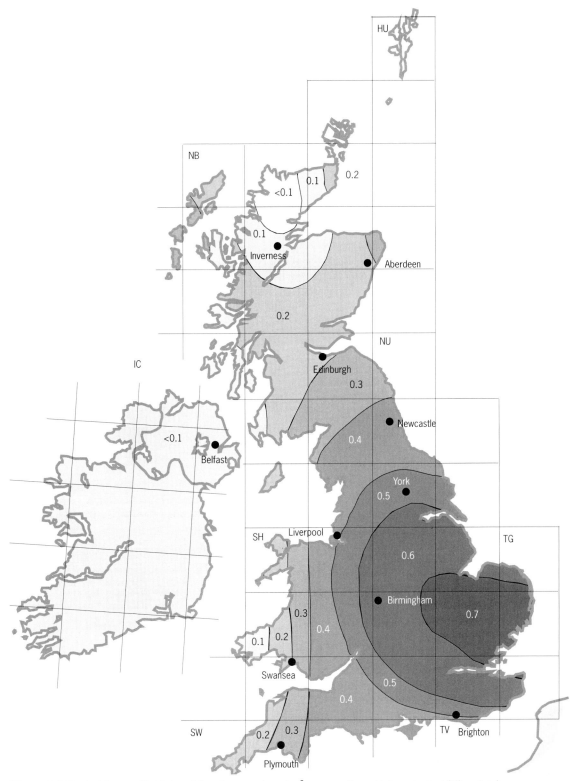

Figure 8.5 The incidence of lightning strikes to ground per km² per year (from Meteorological Office data)

it is not possible to provide adequate light through windows. Guidance on daylighting has been provided in chapter 2. As a brief reminder of the issues, the key factors are:

● *Daylight factor* (amount of daylight received under any roof lighting: the aim is to give an average daylight factor of about 5% for most tasks, where less than 2% daylighting factor is provided, the permanent use of artificial light may need to be considered).
● *Reflection* (glare or dazzle from glazed rooflights, or light-coloured roof surfaces: calculated using the relationship between the angles of incidence and reflection to derive the sun positions at which dazzle may be a problem).

The greatest dazzle problems tend to occur with surfaces facing within 90° of due south, sloping back at angles between 5° and 30° to the vertical.

Sloping glazing systems or on a smaller scale, rooflights of glass or plastics, may be used to improve daylighting to large-span buildings, or where the layout obstructs natural daylighting from windows. Many glazing systems can provide a fairly high light transmission, but it should be noted that most of the plastics will suffer weathering which will reduce their light transmittance significantly. They will also require safe access from inside and outside for cleaning (an HSE regulation requirement except for single dwellings but a requirement under CDM Regulations in any event) and may require guarding where they form part of a means of escape or a regular access route (see BRE's reports *Sloping glazing: understanding the risks* and *Highly glazed buildings* for more guidance).

Light tubes, also called light pipes, are a modern replacement for the traditional skylight. They are designed to transfer diffused light through a large diameter pipe (200 mm to approximately 500 mm

diameter stock sizes) through the ceiling. As with any penetration of the roof cover, due attention must be given to the flashing to ensure a watertight joint for flat roofs, a weathered detail for pitched roofs and avoidance of condensation.

Sound insulation

Roofs may be required to protect their occupants from noise just as much as walls. This book does not deal with protection of the outside from noise generated within the building, but rather with protection within from noise generated from the outside.

In addition to meeting the requirements for resistance to noise from adjacent dwellings, other aspects to be considered include:

● *Weather-generated noise* (eg drumming on thin metal sheets, whistling at ventilation slots, movements between metal to metal joints or from thermoplastics gutters, slack sarking membranes vibrating in windy conditions). BRE's *Information Paper 2/06* gives guidance on assessing the likely effect of rain noise from lightweight roofs and roof elements on the indoor ambient noise levels in rooms.
● *Noise travelling from room to room* (failures tend to be concentrated on missing elements of separating walls in roofs, unfilled joints in masonry, or holes around services and joists).

Durability and ease of maintenance

Durability and the service life of roofs are significantly affected by:

● poor workmanship,
● condensation,
● rain penetration,
● airborne and wind-driven pollutants,
● insects and other animals, rodents, bats and birds,
● mould, fungus, etc.,
● plants.

Even if the materials have been suitably specified, poor workmanship or installation, or insufficient or inappropriate maintenance will affect the service life. More details are given later under specific types of roof.

BRE
Kelly D, Garvin S & Murray I. *Sloping glazing: understanding the risks.* BR 471. 2004
Ridal J, Reid J & Garvin S. *Highly glazed buildings: assessing and managing the risks.* BR 482. 2005
Hopkins C. *Rain noise from glazed and lightweight roofing.* Information Paper IP 2/06. 2006

Pollutants

In terms of pollutants, the specific local environment is the key consideration. Industrial pollutants can be deposited on roofs by rain, snow or air. Wet deposition of pollutants tends to be a greater problem in areas with high rainfall (eg west coast of Scotland) and dry deposition in low rainfall areas (such as Eastern England). Dry deposition leads to damage when the material becomes wet and run-off occurs. Lower emissions of sulfur dioxide have reduced damage to calcareous stones and some metals, but increasing concentrations of oxides of nitrogen (linked to traffic emissions) continue to cause damage to metals and porous clay or concrete roof tiles.

Chlorides arising from marine and industrial sources affect metals such as iron, zinc, copper and aluminium and certain other relatively porous building materials such as roof tiles. Natural concentrations of chlorides reduce with distance from the coast. The highest concentrations are found where salt spray is present and more often on the landward side of the building where washing by prevailing rainfall is less likely.

Ozone concentrations are higher in rural areas than in urban areas. Ozone is implicated in the deterioration of polymeric materials, including plastics and paints. It does not affect inorganic materials directly, but contributes to converting sulfur and nitrogen oxides to their respective acids, which do affect inorganic materials.

Insect and other animal infestations

Insects degrade materials and can cause infestation and harbourage problems. Degradation from common furniture beetles is relatively rare in roofs, as design and conditions generally discourage initiation and development of attack. House longhorn beetles can cause severe damage but are confined to areas to the south west of London. Areas where treatment of timbers is mandatory to protect against House longhorn beetle are listed in the Building Regulations (England & Wales). The majority of other larger insects can be deterred by provision of 3–4 mm mesh over ventilation slots.

Infestations by other animals rarely cause severe structural damage, but may cause nuisance and

Figure 8.6 Tree growth on an inverted flat roof

fouling and have health and hygiene implications. Access normally occurs through external openings at eaves or occasionally by gnawing of timber fascias. Seagulls and other birds have been known to peck gaskets, putty and shiny roof membranes. Rabbits burrowing in green roofs have damaged membranes. Bats are protected under the Wildlife and Countryside Act. Guidance is available from English Nature.

Mould and fungal growth

Wet and dry rot are most likely to occur in fascias and bargeboards which are best protected through adequate weatherproofing and detailing to drainage installations. The risks of mould growths and fungal bloom are increasing, and can increase the temperature of the roof by reducing reflectance.

Plant growth

Most moss or lichen growth is harmless, but weed growth in gutters may be a problem if debris is not cleared regularly. Weed growth, and even tree seeds taking root, can be a problem for inverted flat roofs (Figure 8.6).

Roof design options

The following design options are based on good practice. The cold deck flat roof is not a recommended method but it is included here as it

English Nature. *Bats in buildings*. Free leaflet IN15.2. Peterborough, English Nature

> **Key Point**
> The cold deck flat roof is **not** a recommended method.

still remains a design option. However, as mentioned earlier, the method is not permitted in Scotland and Northern Ireland. Where there may be a limitation with a particular type of roof construction, this is indicated in the text.

Flat roofs

Flat roofs are generally considered to be those built below a pitch of 5° (BS 6229). The main elements are the structural deck, thermal insulation and waterproof membrane/covering.

Structural deck. Timber decks are usually found on housing and concrete and metal decks on blocks of flats and other developments. In timber decks, the simplest way of ensuring that basic strength and stability requirements are met is to ensure that members are sized in accordance with the appropriate national building regulations. In that way, deflections, which are frequently a source of later defects, can be taken care of. Concrete rarely deflects to the extent that problems ensue for the coverings.

Thermal insulation. The national building regulations and technical standards require roofs to be built so that heat loss is limited. For flat roofs the location of the insulation largely determines the type of insulation used. More detail is given in Box 8.1 and relevant British Standards are listed in chapter 13 *References and further reading.* The inclusion and location of the vapour control layer is critical to the avoidance of condensation (see BS 5250).

Waterproof membrane/covering. The choice for flat roofs is bitumen, plastics and rubber sheets or metal (see Box 8.3).

The structural deck, thermal insulation and waterproof membrane/covering can be configured in three ways to give what are known as warm deck sandwich, warm inverted and cold deck roofs (see Box 8.1; more guidance is given in BRE's *Good Building Guide 36*).

Details needing careful consideration are:
- *Abutments.* A generous fillet should be provided at abutments or changes of level for all kinds of flat roof coverings. Built-up membranes, for example, that are turned through a right angle without employing a fillet, are likely to crack at the angle even if subjected to only a very small strain.
- *Service perforations.* The aim in new design should be to keep the flat roof as free as possible from services; it is these interruptions in the waterproof layer that make it vulnerable to rain penetration. Heating system flues should be routed through external walls in preference to roofs wherever possible. Alternatively, outlets should be adequately detailed to allow for the effects of both movement and heat (Figures 8.7 and 8.8).

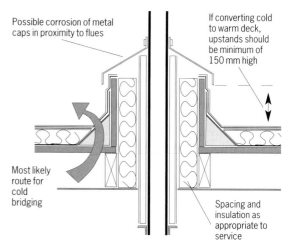

Figure 8.7 Service outlets need to be adequately capped over upstands

Labels in figure: Possible corrosion of metal caps in proximity to flues / If converting cold to warm deck, upstands should be minimum of 150 mm high / Most likely route for cold bridging / Spacing and insulation as appropriate to service

British Standards Institution
BS 6229: 2003 *Flat roofs with continuously supported coverings. Code of practice*
BS 5250: 2002 *Code of practice for control of condensation in buildings*

BRE. *Building a new felted flat roof.* Good Building Guide 36. 1999
British Standards Institution. BS 6399-3: 1988 *Loading for buildings. Code of practice for imposed roof loads*

Box 8.1 Configurations for flat roofs

Warm deck sandwich roof

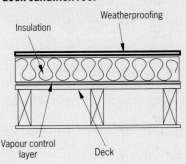

The warm deck roof is described from the external layer downwards. The choice of materials for each layer is:

- **Surface finish:** none, solar reflective paint, chippings, ballast, paviors;
- **Waterproof covering:** bitumen, plastics or rubber membranes, metal sheets, ie lead, copper zinc, etc., liquid-applied systems, mastic asphalt;
- **Thermal insulation:** foamed plastics, mineral fibre batts (not quilt), cellular glass, cork, composites;
- **Vapour control layer:** bitumen sheets, polyethylene;
- **Structural deck:** timber, timber-based sheets, profiled metal, concrete);
- **Cavity** (not ventilated);
- **Internal lining:** plasterboard suspended ceiling.

This type of roof construction can be used for most flat roof applications. If heavily trafficked the insulation will need to have a high compressive strength and the membrane will need to be covered with paving slabs.

Cold deck roof

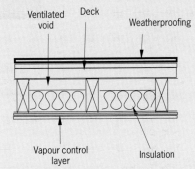

The cold deck roof is constructed with a choice of materials as follows:

- **Surface finish:** none, solar reflective finish, chippings, ballast;

Warm inverted roof

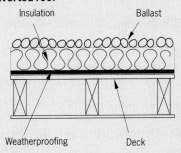

The warm inverted roof construction, with choice of materials as follows, is:

- **Surface finish:** ballast or paving slabs;
- **Geo-synthetic filter membrane;**
- **Thermal insulation:** expanded extruded polystyrene;
- **Optional separating membrane** (needed to ensure a reaction does not take place between the polystyrene insulation and some types of plastics membranes);
- **Waterproofing membrane:** bitumen, plastics, rubber, mastic asphalt, liquid-applied;
- **Structural deck:** concrete & screed, timber, profiled metal;
- **Cavity** (not ventilated);
- **Internal lining:** plasterboard, suspended ceiling.

With this type of roof, it is possible for condensation to form beneath the membrane when cold rainwater percolates below the insulation. This can be a problem for decks of low thermal mass (profiled metal, plywood) over high moisture-producing areas (eg swimming pools); there is more guidance in BRE *Digest 336*.

- **Waterproof membrane:** bitumen, plastics, rubber, mastic asphalt, liquid-applied;
- **Structural deck:** timber, timber sheets, profiled metal, concrete;
- **Ventilated cavity:** at least 50 mm deep;
- **Thermal insulation:** foamed plastics, mineral fibre;
- **Vapour control layer:** polyethylene sheet (this is almost impossible to seal);
- **Ceiling:** plasterboard).

This roof design is prone to condensation forming on the underside of the structural deck, especially where ventilation is restricted. It has been a common construction for domestic extensions but condensation problems have led to collapse of the deck where chipboard has been used. It is therefore not a recommended design because of the condensation risk.

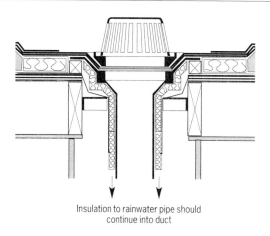

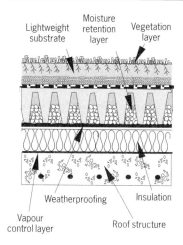

Lightweight substrate / Moisture retention layer / Vegetation layer / Weatherproofing / Insulation / Vapour control layer / Roof structure

Insulation to rainwater pipe should continue into duct

Figure 8.8 Rainwater outlets penetrating warm roofs should be thermally insulated

Figure 8.9 Structure of a green roof

● *Rainwater disposal.* For flat roofs falls are no longer recommended as there is no evidence that they have any effect on the life of the coverings. Ponding is less harmful to coverings than was at one time supposed; but if a leak occurs in a ponded area, water will enter the building in greater quantities than through an efficiently drained surface. So, although there is not the same strict requirement for adequate falls as there used to be in the days when oxidised bitumen felts were the only solutions available, a slight fall is a useful insurance.

Additional performance requirements include:
● *Providing safe access.* This may be for occasional maintenance, for means or escape or for access to other parts of the building. BS 6399-3 governs the requirements: additional loading, including impacts, should be allowed for in design, and guarding should be provided.

Green roofs

Colonisation by mosses and lichens on roofs is sometimes unavoidable. However, the 'green roof' is designed to accommodate and encourage plant growth both on pitched and flat roofs (Figure 8.9). Roof gardens have been built for many years but if trees and shrubs are to be used, then the structure

will need to support an increased load. In Germany, where the construction of green roofs is more common than in the UK, green roofs are divided into two categories: 'intensive' and 'extensive'.

The 'intensive' green roof is essentially a garden and is accessible to people. Planters are required for trees, a deep earth layer is needed for shrubs and artificial irrigation is necessary. Landscaping and paths all add to the weight so the structural implications of this need early consideration. Snow loading should not be forgotten.

The 'extensive' green roof is simpler in concept with the planting selected to have low management requirements and no artificial irrigation. The thin soil layer has only to support low-growing plants or turf. However, the structure will usually be more substantial than for a tiled or membrane-covered roof.

The reasons for selecting a green roof design include:
● visual appeal,
● controlling rainwater,
● improving the thermal performance, and
● providing a habitat for insects and other wildlife.

On greenfield sites they can, to some extent, replace lost habitat. There is more guidance in BRE's *Digest 486*.

BRE. *Swimming pool roofs: minimising the risk of condensation using warm-deck roofing.* Digest 336, 1988

Saunders GK. *Reducing the effects of climate change by roof design.* Digest 486. Bracknell, IHS BRE Press, 2004

Thermal performance

Thermal performance of lightweight inverted warm deck flat roofs can fall below expectations in certain locations, and allow risk of condensation to occur. Further guidance is available in BRE's *Information Paper 2/89*.

Pitched roofs

Key additional performance requirements

The design of pitched roofs is covered by BS 5534, BS 5268-3, BS 8103-3 and for cut roofs, there is design guidance in BRE's *Good Building Guide 52*.

A domestic-scale pitched roof is typically constructed from a structural timber framework either using factory-produced roof trusses (Figure 8.10), built in-situ (cut roofs; Figure 8.11) or in the case of larger roofs, steel and timber (Figure 8.12). These support a roof covering of tiles or slates mounted on timber battens, or profiled metal sheets fixed directly to purlins. A decision as to where the insulation should be placed and the type of slate or tile to be used will need to be made at an early stage. An underlay is used beneath the slates or tiles as a second line of weathering defence. Ventilation may need to be provided at some location within the roof construction depending on the need to control condensation (see BS 5250 and BRE's Report, *Thermal insulation: avoiding risks,* for more guidance). Guidance on ventilating thatched roofs is given in *Good Building Guide 32*.

Figure 8.10 Factory-produced roof trusses

> **Key Point**
> Ventilation may be needed somewhere within the roof, depending on the need to control condensation.

BRE

Beech JC & Saunders GK. *Thermal performance of lightweight inverted warm deck flat roofs*. Information Paper IP2/89. 1989
Site-cut pitched roofs. Part 1: Design, Part 2: Construction. Good Building Guide 52. 2002
Thermal insulation: avoiding risks. 2002 edition. BR 262
Ventilating thatched roofs. Good Building Guide 32. 1999

British Standards Institution

BS 5534: 2003 *Code of practice for slating and tiling (including shingles)*
BS 5268-3: 1998 *Structural use of timber. Code of practice for trussed rafter roofs*
BS 8103-3:1996 *Structural design of low-rise buildings. Code of practice for timber floors and roofs for housing*
BS 5250: 2002 *Code of practice for control of condensation in buildings*

The more complex the geometry of the structure of the pitched roof, the more likely it is that faults will ensue. Locations that might need special attention include:

● *Valleys*. Many new roofs involve the construction of a valley gutter, one method being to apply a metal lining to the valley and to cut the tiles to shape (Figure 8.13). The cutting is often clumsy and the appearance compromised. A better solution, involves the use of purpose-made valley tiles of various configurations. Valleys formed from these tiles may be swept or laced, either to a radius or to an acute angle. A double thickness of sarking felt is needed in valleys, lapped at least 600 mm over the centreline of the valley.

● *Overhanging eaves.* Overhanging eaves may be of 'closed' or 'open' configuration. These terms refer to whether the rafters are boxed in with fascia and soffit, or left exposed with the wall surface carried up to the underside of the roofing. In order to achieve adequate weather protection to

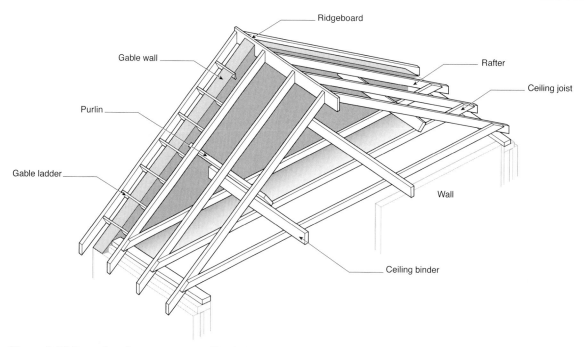

Figure 8.11 Examples of components used in a built in-situ cut roof

Figure 8.12 Steel beams increase the design options for a large roof

the wall below, BRE recommends an overhang at eaves of at least 300 mm (Figure 8.14).

- *Clipped eaves.* BRE does not recommend the use of so-called clipped or flush eaves. However, they are used. The reinforced BS 747 class 1F felts that were commonly used in this situation as an underlay disintegrate after a few years' exposure

> **Key Point**
> The use of clipped or flush eaves is **not** recommended.

to external conditions, leaving the detail vulnerable to rainwater blown back by the wind onto the head of the wall. The more satisfactory solutions, from the point of view of durability, for overhanging eaves as well as for clipped eaves are those in which the detail has the benefit of a strip of better quality material lapped under the sarking felt and dressed into the gutter.

- *Verges.* Mortar bedded verges should oversail the wall or bargeboard by 50 mm, and should be undercloaked with a durable and compatible material (Figure 8.15). Pointing between the undercloaking and the tile above should not exceed 35 mm in thickness.

British Standards Institution. BS 747: 2000 *Reinforced bitumen sheets for roofing. Specification*
Note: This Standard will be replaced by BS EN 13707

Figure 8.15 A good projection was given to this single lap tiled verge, and the tiles were neatly clipped

Figure 8.13 Work in progress on a valley. The supporting board can be seen on the left, though the sarking overlap and battening leave something to be desired

Figure 8.14 The shelter from rain falling nearly vertically which is provided by this projecting verge is clearly apparent. The protection will, however, vary with wind direction and intensity

- *Abutments.* Where a pitched roof abuts a separating wall in a stepped-and-staggered situation, part of the roof becomes an external gable, and the outer leaf of masonry becomes an inner leaf below roof level. Particular care is needed to ensure that rainwater is prevented from reaching the interior. Soakers for plain tiles should be a minimum width of 175 mm to give an upstand of 75 mm against the wall, and a lap of 100 mm under the tiles. A cover flashing then laps the soakers by 50 mm — more if it can be arranged. For interlocking tiles, a stepped flashing dressed over the tiles must be used; soakers are not feasible since they would interfere with the interlock.
- *Hips.* Hips in plain tiled roofs are most often formed with ridges bedded in mortar (Figure 8.16), and held at the foot on hip irons or hooks. It is common to clip each ridge tile to the hip rafter. Alternative methods of forming hips, giving neater appearance and probably better

Figure 8.16 The mortar is already breaking up in this comparatively new hip and the galleting is becoming loose

Figure 8.17 Dry laid verges, but the ridges are mortared. Rainwater run-off from the upper valley will be concentrated in the single trough of the tiles, and there is a risk that it will overshoot the lower gutter in all but the lightest of rainfalls

performance, include purpose-made hip tiles and bonneted hip tiles. Mitred hips in plain tiles should be laid over soakers lapping at least 100 mm to each side of the hip. Single lap tiles are in general not suitable for exposed mitres at hips.

● *Ridges.* A great variety of shapes of ridge tile is available. Traditional practice is to bed them solidly on mortar but even when so bedded, ridges can be subjected to loss of tiles in gales, particularly the end tiles facing edge-on to the wind. A 1:3 mortar mix can be used for bedding, preferably air entrained, but this is a comparatively strong mix and could lead to cracking. A high durability mortar is nevertheless required, and an air-entrained 1:0.5:4.5 cement:lime:sand mortar is an alternative. Dry fixing and clipping is a more positive method (Figure 8.17). Some fixings rely on there being a ridge board at the apex of the pitch, and noggings may be necessary where they are used with trussed rafters. These noggings need to be particularly securely fixed.

It is essential that adequate bracing and strapping is provided to roof trusses as the responsibility for the stability of the roof is not passed to trussed roof suppliers.

Traditionally, the insulation has been placed at ceiling level and the loft space ventilated at the eaves and also at the ridge. It is common practice in England to use a bituminous underlay (type 1F to BS 747), although proprietary underlays are frequently used and should be covered by third-party certification. In Scotland, the roof may be fully boarded immediately beneath the tiles, in which case counterbattens may be needed. These types of pitched roof are referred to as cold ventilated pitched roofs.

A more recent development has been the use of the cold sealed pitched roof. In this construction, there is no provision for ventilation of the loft space at all. However, the underlay used is a 'breather membrane' which allows water vapour readily to pass through it. It is common practice to provide a

British Standards Institution. BS 747: 2000 *Reinforced bitumen sheets for roofing. Specification*
Note: This Standard will be replaced by BS EN 13707

space beneath the tiles by the use of counterbattens. This space can be ventilated by the use of special tiles. Some 'breather membrane' manufacturers claim this is not necessary.

The use of the loft space has become more prevalent and 'the room in the roof' type of construction is used where the insulation is placed at rafter level. Counterbattens are used to provide a space between the insulation and the underlay which is a 'breather membrane' type.

Profiled metal roofs
Key additional performance requirements
Profiled metal roofs can provide a versatile, economic form of roof construction and have become popular both for domestic use and for covering large industrial buildings such as warehouses. They can either be assembled on site, supplied separately as an inner liner, thermal insulation and profiled outer sheet, or as metal skinned composite panels. Both of these systems rely on sealants to ensure end and side laps are watertight. The outer profiled sheets are often coated for additional protection and appearance.

Typical assemblies
Profiled steel or aluminium sheets are used in a number of assemblies as described in Box 8.2.

Design considerations
A comprehensive and flexible approach to roof design is essential to reduce the risks with moisture that may be associated with increased levels of thermal insulation.

- *External moisture.* Wind-driven rain and snow are less likely to penetrate joints if the roof pitch is greater than 15° or the joints are sealed. Avoid penetrations through the external skin of the roof; if this is not possible, adequately seal the roof around the penetration and provide insulation to reduce thermal bridging and subsequent condensation at the penetration/roof interface.
Detail around rooflights needs care to prevent moisture entering around the rooflight and to prevent condensation within sealed units.
Detailing to perimeter, internal or valley gutters

should be simple, and should reduce the risk of blocking and subsequent overloading or flooding of weirs and overflows. Internal metal gutters can become extremely cold when exposed to flowing rainwater or melting snow. Insulate internal gutters and associated penetrating pipework with closed-cell, moisture-resistant insulation, and seal joints between sections of the insulation with a vapour-resistant tape.

- *Internal moisture.* An effective internal vapour control layer is imperative to the success of systems incorporating a cavity above the insulation. This vapour control layer must extend over the full area of the roof and joints should be sealed with vapour-resisting tape. The vapour control layer may be a separate (polyethylene) layer or may be formed from an impermeable liner system.
In addition to this vapour control layer below the insulation, a permeable membrane may be needed above the insulation. This permeable membrane should direct any dripping condensate away to guttering. Effective detailing and subsequent installation of this membrane may be difficult around purlins and rooflights.
To reduce further the risk of moisture build-up within voids in the roof construction, any voids directly below the outer sheet should be ventilated, with inlets at eaves and ridge. The filler strips at the eaves and ridge should be perforated to provide the recommended ventilation.
Site-assembled, double-skin systems should be considered only where there is a low risk of moisture from internal processes or where extract ventilation is provided close to this moist process. Where the risk from internal processes is high, for example in swimming pools, steam or wet manufacturing, double-skin construction (with voids) needs careful consideration; if composite systems are used, they must be sealed at the joints between panels.

- *Ventilating air moisture.* There is a small risk of condensation, arising from external ventilating air, coalescing and subsequently running or dripping onto the underlying insulation. A permeable membrane can be installed to direct

Box 8.2 Profile metal roofs: typical assemblies

Single skin

An uninsulated profiled sheet is fixed directly to the purlins. This form of construction is generally used only over storage buildings, warehouses, animal housing and uninhabited buildings or as canopies over workspaces.

Single skin

Double skin

The most common type of profiled metal roof construction consists of a shallow profiled metal liner, a structural spacer system and an outer sheet that is normally of a deeper profile. Typically, the void between the two skins accommodates thermal insulation, usually mineral wool, and some form of vapour control on the warm, liner side of the insulation. A vapour permeable membrane can be placed above the insulation to direct any dripping condensate towards external guttering.

Used only in double-skin roofs, spacers create the cavity between the liner and external sheeting which accommodates insulation and appropriate membranes. They are structural elements and must be positioned correctly over the purlins or other structural elements and must be fixed securely to these elements.

Traditional zed spacers are normally made from 1.5 mm-thick galvanised steel and are supported on a plastics ferrule. They are generally suitable for insulation up to 100 mm thick. If the insulation is thicker than this, the zed spacer may become unstable so it is advisable to use bracket and rail systems. These are more stable over a greater range of depths and are designed to reduce the effects of thermal bridging. The rails are formed from structural grade steel and the brackets incorporate a thermal break and vapour seal.

Double skin with zed spacer

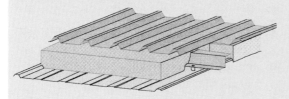

Bracket and rail

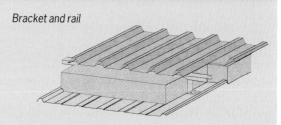

Composite systems

These combine a liner sheet, rigid insulation and an external profiled weather sheet forming an integrated composite panel. Most composite panels are factory-assembled and incorporate a rigid plastics insulant that is foamed to fully fill the void between the inner and outer sheet. Site-assembled composites consist of profiled inner and outer sheets and a separate pre-formed rigid insulant. This insulant may be plastics or mineral and, as it is rigid in nature and fully fills the void between the sheets, no structural spacer system or ventilation is required.

Composite

Boarded systems

These systems do not incorporate a metal liner but need a rigid insulation board supported below the external profiled sheet. The internal finish is not usually considered to provide vapour control so it may be simply the exposed face of the insulation board. These systems are uncommon and are only suitable where dry processes are enclosed and there is little risk from generated moisture penetrating the roof structure.

Boarded

this moisture towards gutters. If this risk is considered high and the detailing of a membrane impracticable, consider a site-assembled or prefabricated system incorporating low permeability, closed-cell insulation, which fully fills the void between the liner and outer sheet.

There is further guidance in BRE's *Roofs and roofing* and *Good Building Guide 43* and in LPS 1181-1 and BS 5427-1.

Installation

There are a number of Codes of Practice available which relate to the installation of the various materials found on roofs. These Codes of Practice deal with the material aspects and not with construction site management. It is important that the health and safety of the workers is considered early in the design stage to comply with the principles of Construction Design and Management Regulation 1994. For pitched roofs, workmanship is covered by BS 8000-6 and for flat roofs, BS 8000-4.

Key Point

The health and safety of workers should be considered early in the design stage.

BRE
Harrison HW. *BRE Building Elements. Roofs and roofing: Performance, diagnosis, maintenance, repair and the avoidance of defects.* BR 302. 2000 edition
Stirling C. *Insulated profiled metal roofing.* Good Building Guide 43. 2000

Loss Prevention Certification Board (LPCB).
Requirements and tests for built-up cladding and sandwich panel systems for use as part of the external element of buildings. LPS 1181-4.1. Available from www.RedBookLive.com

British Standards Institution
BS 5427-1: *1996 Code of practice for the use of profiled sheet for roof and wall cladding on buildings. Design*
BS 8000: *Workmanship on building sites*
    Part 4:1989 *Code of practice for waterproofing*
    Part 6: 1990 *Code of practice for slating and tiling of roofs and claddings*
BS 6229: 2003 *Flat roofs with continuously supported coverings. Code of practice*

However, these standards have, in part, become superseded by later standards which are listed in chapter 13, *References and further reading*.

A list of other relevant health and safety regulations are given in BS 6229. In addition, the Reporting of Injuries, Diseases and Dangerous

---

Box 8.3 Specification of materials

Bitumen sheets
At the time of writing, it is planned that BS 747 will be withdrawn and replaced by BS EN 13707. This new British Standard refers to many test methods, and CE marking will be accompanied by results from these test methods either on the product label or a data sheet. There is no system for identifying levels and classes included in BS EN 13707, so the user of the standard is left to decide what value they require for a given test result declared by the manufacturer, eg tensile properties will be in terms of N/50 mm width and % elongation.

BS 8747 is under development by the British Standards committee and is a specification and selection guide for reinforced bitumen membranes (RBMs) for the use of CE-marked products in the UK. This new standard proposes the SnPn classification system where S is for strength and P is for puncturability. It is intended to ensure that an S1P1 membrane has the same performance level as a type 3B to the old BS 747. The higher the number, the better performance of the membrane up to S5P5. This new standard should be issued by BSI before withdrawal of BS 747.

Plastic and rubber sheets
A product standard dealing with plastic and rubber sheets for roof waterproofing, BS EN 13956, has also recently been introduced.

Sarkings
The new European Standards also include a product standard called BS EN 13859-1. This covers products such as type 1F to BS 747 and products with low resistance to water vapour as defined in BS 5534, the code of practice for slating and tiling.

Slates and tiles
There are many types of products to choose from. See BS 5534 for a list of product standards.

Profiled metal sheets
The main product types are various types of steel or aluminium. BS 5427-1 is the relevant code of practice for the design and includes reference to all the product standards for the various materials used including fixings.

Occurrences Regulations 1995 (RIDDOR) should be noted and acted on if necessary.

Box 8.3 gives guidance on specifying materials.

Drainage

Considerable quantities of rainwater falls on roofs during storms. Required practice is for roofs to fall towards gutters, either at the perimeter or to valleys, where the water flows via the required number of downpipes into drains. The method of design and calculations is given in BS EN 12056-3. For short span roofs, BRE has developed a simplified method of working out the drainage system which is given in Box 8.4. A typical example is calculated in Box 8.5.

Downpipes should be positioned so that they are not needlessly complicated with the layout of gully and connection to the drains planned. Where overflow of the system may occur (typically during exceptional rainfall) on flat roofs with parapets and non-eaves gutters, the design should include an overflow or emergency outlet to reduce the risk of overspilling of rainwater into the building.

An alternative to simple gravity flow drainage is the syphonic roof drainage system. This uses a gravity-induced vacuum principle to create a syphonic action. The system requires fewer roof outlets and uses smaller horizontal pipes to carry the discharge. There is more guidance in HR Wallingford's report *Performance of syphonic drainage systems for roof gutters*.

Regulations and Standards

Fire safety

The main safety requirement for roofs is to ensure the external fire performance of roofs is such that they do not burn through quickly or that they do not allow the flames to spread across the roof quickly.

The requirements are given in the supporting documents of the national building regulations:
- **England and Wales:** AD B,
- **Scotland:** Section 2 of the Technical Handbooks, Domestic and Non-domestic,
- **Northern Ireland:** Technical Booklet E.

The external fire performance is defined as penetration and surface spread of flame and is assessed by the test methods in BS 476-3. The level of performance is designated by two letters: the top designation is AA, AB or AC. AD B recognises that not all roof coverings require testing and that some are given a rating without testing. This includes slates and tiles and flat roofs with a minimum cover of 12.5 mm stone chippings.

Conservation of fuel and power

There is a requirement for roofs to be built so that the heat loss from the building is limited. The requirement has been expressed as a U-value which is specified for a range of building types.
This means that any roof design must include thermal insulation. The requirements are given in the supporting documents of the national building regulations:

HMSO. *The Reporting of Injuries, Diseases and Dangerous Occurrences Regulations 1995*. Statutory Instrument 1995 No. 3163. London, The Stationery Office, 1995

British Standards Institution
BS 747: 2000 *Reinforced bitumen sheets for roofing. Specification*
Note: This Standard will be replaced by BS EN 13707.
BS EN 13707: 2004 *Flexible sheets for waterproofing. Reinforced bitumen sheets for roof waterproofing. Definitions and characteristics*
BS EN 13956: 2005 *Flexible sheets for waterproofing. Plastic and rubber sheets for roof waterproofing. Definitions and characteristics*
BS EN 13859-1: 2005 *Flexible sheets for waterproofing. Definitions and characteristics of underlays. Underlays of discontinuous roofing*
BS 5534: 2003 *Code of practice for slating and tiling (including shingles)*
BS 5427-1: 1996 *Code of practice for the use of profiled sheet for roof and wall cladding on buildings. Design*
BS EN 12056-3: 2000 *Gravity drainage systems inside buildings. Roof drainage, layout and calculation*

Hydraulics Research Ltd. *Performance of syphonic drainage systems for roof gutters*. Report SR 463. Hydraulics Research Ltd, Wallingford
British Standards Institution
BS 476-3: 2004 *Fire tests on building materials and structures. Classification and method of test for external fire exposure to roofs*

Box 8.4 How to calculate roof gutter size

Step 1

Find out how much rainwater will be discharged from the roof by calculating the *effective catchment area*, abbreviated to E, of each section of roof that will discharge to an eaves gutter as follows.

- For a flat roof, E is the relevant plan area of the roof (in m²).
- For the slope of a pitched roof, the plan area, A (in m²), plus half the elevation area, B (in m²), see Diagram (a).
- For a pitched roof abutting a wall, the plan area, A, plus half the elevation area, B, plus half the wall area, C, above the roof slope, see Diagram (b).

Step 2

Next work out the quantity of rainwater run-off in litres per second, abbreviated to Q, draining to each gutter. For this you need only a rule-of-thumb: Q equals E divided by 48. (This converts the effective catchment area into run-off in litres per second in a typical thunderstorm.)

This gives a figure for how much water each gutter has to deal with without overflowing. How well it will cope depends on how many downpipes there are, so this is the next decision.

Step 3

If there is only going to be one downpipe, the gutter will have to cope with the entire flow. If there will be a

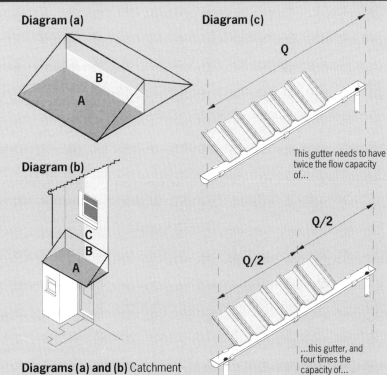

Diagram (a)

Diagram (b)

Diagrams (a) and (b) Catchment areas for calculating rainwater run-off

downpipe at each end, the gutter will only have to deal with half the flow (Q/2) at any point along its length. If the two downpipes are positioned at a quarter of the gutter's length from each end, the gutter only has to cope with Q/4, see Diagram (c). So in general, it is better to position downpipes in the centres of runs of guttering, rather than at the ends.

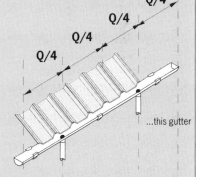

Diagram (c)

This gutter needs to have twice the flow capacity of...

...this gutter, and four times the capacity of...

...this gutter

Spacing between outlets to reduce water load on gutters

Step 4

Once you have decided how many downpipes there will be, you can use Table 8.1 to work out the size of standard eaves gutter needed.

Box 8.5 gives the calculation for a typical example.

Table 8.1 Flow capacities of standard eaves gutters (when level)

Size of gutter (mm)	Flow capacity (litres/sec)	
	True half-round	Nominal half-round
75	0.38	0.27
100	0.78	0.55
115	1.11	0.78
125	1.37	0.96

Box 8.5 Example of gutter sizes for a typical house

To illustrate how to use this calculation procedure, we can apply it to a typical example:
A house with a simple dual-pitched roof, eaves length 8 m, truss span 6 m and height 3 m.

Step 1
Plan area = $8 \times 6/2 = 24\,m^2$
Projected elevation area = $8 \times 3/2 = 12\,m^2$
So effective catchment area, E = $24 + 12 = 36\,m^2$

Step 2
Flow rate Q is $36/48 = 0.75$ litres/sec

Steps 3 and 4
If there is only one downpipe for each side of the roof, Table 8.1 shows that there would have to be a 115 mm size nominal half-round gutter to cope with the flow. But if there is a downpipe at each end of the eaves, so that the gutter only has to cope with half the flow rate (0.375 litres/sec), a 100 mm half-round gutter would be adequate. If the two downpipes were sited at the quarter points, the flow-rate would be halved again to 0.188 and a 75 mm gutter would cope.

- **England and Wales:** AD L1 and L2,
- **Scotland:** Section 6 of the Technical Handbooks, Domestic and Non-domestic,
- **Northern Ireland:** Technical Booklet F.

The design must be checked for condensation risk. Guidance is given in BS 5250 and BRE's report *Thermal insulation: Avoiding risks*). In some types of roof design some provision of ventilation will also be required.

The requirements are given in the supporting documents of the national building regulations:

British Standards Institution
BS 5250: 2002 *Code of practice for control of condensation in buildings*
BS 6399: *Loading for buildings*
    Part 1: 1996 *Code of practice for dead and imposed loads*
    Part 2: 1997 *Code of practice for wind loads*
    Part 3: 1988 *Code of practice for imposed roof loads*

BRE. *Thermal insulation: avoiding risks*. 2002 edition.
BR 262. Bracknell, IHS BRE Press

- **England and Wales:** AD F,
- **Scotland:** Section 3 of the Technical Handbooks, Domestic and Non-domestic,
- **Northern Ireland:** Technical Booklet C.

Structural requirements

The roof needs to be built on a supporting structure that will adequately transmit any loads through to the walls or other vertical supports. The loading on a roof is in the form of dead loads, imposed loads and live loads. Live loads include wind loading.

The requirements are given in the supporting documents of the national building regulations:

- **England and Wales:** AD A,
- **Scotland:** Section 1 of the Technical Handbooks, Domestic and Non-domestic,
- **Northern Ireland:** Technical Booklet D.

There is more guidance in BS 6399: Parts 1–3.

Other requirements

Design features on roofs may be subject to other building regulations. If a roof is used for recreational facilities where the public have access, ramps and doorways will need to be provided to ensure disabled access. The requirements are given in the supporting documents of the national building regulations:

- **England and Wales:** AD M,
- **Scotland:** Section 4 of the Technical Handbooks, Domestic and Non-domestic,
- **Northern Ireland:** Technical Booklet H.

Parapets or balustrades also need to comply with national design guidance:

- **England and Wales:** AD A,
- **Scotland:** Section 1 of the Technical Handbooks, Domestic and Non-domestic,
- **Northern Ireland:** Technical Booklet H.

Construction Design & Management (CDM) Regulations

The CDM Regulations were introduced in 1994 with a view to reducing the accidents associated with construction work. Since roofs are constructed at height they have been a major cause of accidents, some of which can be fatal.

The CDM principles are that the construction and maintenance must be considered at the design stage and a health and safety plan drawn up. However, it is considered by most designers that providing points to attach safety harnesses is sufficient. This is only applying a low-level strategy which, although an improvement, still requires more attention to be given to making the roof a safe place to work.

CDM also covers the construction phase and the maintenance phase, ie access to service items such as header tanks must be given proper consideration.

HSE. The Construction (Design and Management) (CDM) Regulations 1994. Available as a pdf from www.hse.gov.uk/pubns

9 Building services

Although the envelope of the building can provide the occupants with some protection from extremes of climate, both winter and summer, it is the servicing subsystems which now provide the fine tuning and correction of any imbalances in comfort levels. Designers would benefit from spending time examining the influence of one services subsystem on another in the buildings they are designing; for example, the inter-relationship of different forms of heating systems with thermal capacity, thermal insulation and ventilation provision, and the effects of extraneous air leakage. The performance of the whole building has to be viewed as a complex interaction of all its parts and all its subsystems, and designers need to recognise that what is in balance for one set of circumstances may not be the same for another.

There is not the slightest doubt that in the past few years the increasing sophistication and consequent demands of users for improved standards in all kinds of buildings, together with increased requirements for energy conservation, are driving an accelerating rate of change in the technological development of building services. As a consequence, building services plant is getting more and more sophisticated (Figure 9.1), and services are taking an

Figure 9.1 Part of the plant room for a small office building. The rate of change of technological development is increasing, and plant rooms can become congested

ever-increasing share of the total capital and lifetime costs of a building project.

Building services can broadly be categorised into three types:
- large self-contained elements with few connections to other services, such as lift installations,
- utility services, such as sanitary accommodation and kitchens, usually grouped into fairly well-defined areas of the building, with connections with hot and cold water systems, and to drainage and ventilation,
- environmental and security services, which by their nature extend on a significant scale

throughout a building; these include heating, ventilation, small power, standby power, lighting, sprinkler fire protection and communications systems.

Although cables are relatively small, and can be routed through buildings, the same is certainly not true of air-handling trunking. One of the most common design faults with regard to the installation of services is inadequate allocation of space; this fact is likely to constrain, if not to govern, the adaptation or provision of building services in the future, let alone the ease with which buildings can be completed in the first place.

Some servicing systems tend to create noise so provisions for their distribution may provide pathways for the transmission of sound between rooms. There may sometimes be requirements for background noise in some circumstances; for example, to enable conversations to be conducted in private.

Service installations are potential causes of accidents. Reduction of the risks calls partly for suitable shielding of, for instance, surfaces that are too hot to touch, but reliability and ease of operation will also contribute.

Energy rating of dwellings

An energy rating for a dwelling aims to provide a single expression of its energy efficiency. The rating gives households a means of comparing one dwelling with another, or the same dwelling before and after refurbishment. In particular, the use of energy ratings should enable householders to take energy efficiency into account on a more rational basis when buying or renting. For architects, ratings can be used as a design tool for optimising energy efficiency in new dwellings.

Two home energy rating schemes were launched in 1990:

● National Home Energy Rating (NHER) and
● Starpoint.

Both schemes rely on a computer program to calculate the annual energy requirements of the dwelling from a description of its construction and its heating system. Both also base their respective

scales for energy efficiency on estimated annual energy costs, assuming a standard pattern of heating and occupancy. The Government has developed a Standard Assessment Procedure (SAP) for calculating energy ratings to enable comparisons to be made between the two methods. Both schemes incorporate the SAP in their calculations.

The SAP provides a common basis for the energy rating of dwellings. All other approved energy labelling schemes are now obliged to include SAP ratings, and the ratings are referenced by the relevant parts of the Building Regulations.

The procedure is based on calculated annual energy costs for space and water heating, and assumes a standard occupancy pattern derived from the measured floor area of the dwelling and a standard heating pattern. The rating is normalised for floor areas so that the size of the dwelling does not strongly affect the results, which are expressed on a scale of 1–100: the higher the number, the better the standard. The method of calculating the rating is set out in the form of a worksheet, accompanied by a series of tables. A calculation may be carried out by completing, in sequence, the numbered boxes in the worksheet. The tables include data to support the assumptions of standard occupancy and heating patterns.

EcoHomes, the new homes version of BREEAM, has been discussed in chapter 1.

At the time of writing, the revision of Part L (due for publication 6 April 2006) is expected to introduce target CO_2 emission rates (TER) for new dwellings coupled with the need to demonstrate that the anticipated dwellings emission rate (DER) comforms to the target. Further information can be obtained at www.odpm.gov.uk.

The Code for Sustainable Homes is expected to become mandatory in the near future.

Energy rating of buildings other than dwellings

The Energy Performance of Buildings Directive is now in force. The procedure for demonstrating compliance with the Building Regulations for

---

European Parliament and the Council of the European Union. *Energy Performance of Buildings.* Directive 2002/91/EC. Available as a pdf from http://europa.eu.int

buildings other than dwellings is by calculating the annual energy use for a proposed building and comparing it with the energy use of a 'notional' building. Both calculations make use of standard sets of data for different activity areas and call on common databases of construction and service elements.

The National Calculation Methodology (NCM) allows the actual calculation to be carried out either by accredited simulation software or by a new simplified tool based on CEN standards, the Simplified Building Energy Model (SBEM). SBEM calculates monthly energy use and carbon dioxide emissions of a building given a description of its geometry, proposed use, and HVAC and lighting equipment.

Condensation

Condensation is usually seen as a completely unacceptable phenomenon in modern construction, particularly when it is accompanied by mould growth; the risks of its occurrence merit the closest attention on the part of designers.

At any temperature, air is capable of containing a limited amount of moisture as an invisible vapour; the warmer the air the more water vapour it can contain before it becomes saturated and liquid water may be deposited. The amount of water vapour present can be expressed as a vapour pressure of the mass of water per mass of air (kg/kg).

Two other important parameters of water vapour in the air are also important:
● relative humidity and
● dewpoint temperature.

The interrelationship of air temperature and moisture content can be graphically illustrated on a psychrometric chart which is a basic design tool for building engineers and designers.

Relative humidity

This is simply the ratio, normally expressed as a percentage, of the actual amount of water vapour present, to the amount that would be present if the air was saturated at the same temperature. Relative humidity is especially important as it is the parameter that determines the absorption of water by porous materials, governs the growth of moulds and other fungi and is the basis of many systems for measuring the water vapour content of air. Relative humidity depends on both the temperature and the vapour pressure of the air and may therefore be modified by changing either or both of these. A relative humidity of 100% is necessary for condensation, with deposition of liquid water, to occur, for example on the inside of single glazed window panes on winter mornings. However, moulds can grow on an internal wall surface when the relative humidity at the wall surface is only 80%.

Dewpoint temperature of the air

This is the temperature at which condensation of liquid water starts when air is cooled, at constant vapour pressure. It is important to realise that, although the dewpoint is expressed as a temperature, it depends only on the vapour pressure; warming or cooling the air makes no difference to the dewpoint.

The occurrence of condensation is a function of four factors:
● heating,
● ventilation,
● insulation, and
● occupant activity.

All four factors need to be in balance or under control to eliminate condensation (and, therefore, mould growth), ie reasonable heating and ventilation provision, the absence of thermal bridging in the building's materials and components, and the occupants avoiding excessive production of moisture. It is quite possible that correction or enhancement of any one of these factors subsequent to construction by placing too much reliance on the servicing subsystems, will not necessarily reduce the problem and might encourage a wasteful use of energy to correct it — clearly unacceptable these days. Such a case might be a poorly heated dwelling in which the occupants routinely dry their washing indoors.

A comprehensive examination of condensation and its prevention is included in BRE's book *Understanding dampness*.

Space heating and cooling

There has been a dramatic reduction in the use of open fires in the UK, in parallel with the growth in central heating, and the effects of the Clean Air Act. However, millions of existing dwellings still have chimneys, and where circumstances are appropriate, for example in rural areas, they are still used to permit the burning of wood, coal or peat in open fires. These fuels, including anthracite, can also be used in closed stoves, again where circumstances are appropriate. Whether or not using open fires, which usually means burning solid fuels, is desirable from an environmental standpoint is perhaps open to debate, but in particular circumstances such as for example public houses, there is an undoubted demand for the open fire in spite of the concomitant labour requirements.

The Building Regulation requirements for each part of the UK are:
- **England & Wales:** though not mandatory, there is a tacit assumption in AD L that most houses will have central heating, and that it must then conform to some minimum standards of control. Also, the Standard Assessment Procedure (SAP) rating, now part of the Building Regulations, gives more points for more efficient heating systems. In these cases, it may then be possible to relax some other requirement; for example, insulation in wall cavities.
- **Scotland:** Section 6 of the Technical Handbooks, Domestic and Non-Domestic.
- **Northern Ireland:** Technical Handbook F.

Thermostats for controlling air temperature have been based on a variety of operating principles (mainly pneumatic or electric) until electronic versions became available in the late 1970s. Electronic thermostats did not become popular until the 1980s but are now virtually universal.

Although carbon dioxide production varies with the efficiency of the boiler, it also varies significantly with the type of fuel used. A boiler

---

Trotman P, Sanders C & Harrison H. *Understanding dampness.* BR 466. Bracknell, IHS BRE Press, 2004

Box 9.1 Temperature limits for heating systems

The following may help in preparing criteria for a performance specification.
- No control of any heating appliance shall at any time be set at a temperature ≥ 60 °C if it is of a material which is a good diffuser of heat, or ≥ 65 °C if it is a poor one.
- Every non-sealed (open-vented) domestic heating appliance should (ideally) incorporate a control which shuts down the system at a water temperature of 95 °C. For a sealed system this value is set at 110 °C for operational reasons.

---

using LPG produces, on average, around 25% more carbon dioxide than a mains natural gas boiler, and one burning oil around 45% more.

There are many aspects of health and safety for the designer to consider in relation to heating systems. The temperatures of heating surfaces, and the possibility of contact by and sensitivity of a building's occupants to high temperature surfaces, is clearly an important consideration in selecting radiators. Where temperatures exceed 80 °C, some form of shielding or casing may be required. Low surface temperature radiators are available.

The risks of building occupants (especially small children in dwellings) suffering burns from heating equipment should be eliminated, or otherwise minimised as far as possible. Controls should not get too hot, and there should be no feeling of discomfort from contact with or close proximity to radiating surfaces (Box 9.1).

Durability
Assessment of the durability of heating installations is difficult because of the lack of reliable data from a continuously changing practice. BRE investigations dating from the early 1980s put the expected lives of cast iron boilers at around 25 years, with around 10–20 years for welded mild steel boilers. The life of oil storage tanks will depend on material and the degree of protection, if of steel, but around 20 years seems probable. Automatic stoking apparatus for solid fuelled boilers lasted around 12 years. For the most part, however, objective evidence on longevity seems to be in short supply, but recent anecdotal evidence seems to support these figures. In general,

though, ancillary controls and even the boilers themselves, governed largely by commercial considerations of the maintenance of a spares bank and obsolescence, would appear to have an economic life of about 10 years.

Individual installations

Heating systems for individual buildings include:

- boilers and (wet) radiator systems,
- underfloor heating (wet or electrical),
- electric storage systems,
- electric radiators,
- hot air.

Boiler systems can be fuelled by natural gas (currently in the majority of installations), liquified petroleum gas (LPG or bottled gas), oil or solid fuel. Mixed systems are feasible, but they are not usually recommended due to their impracticability and cost. A summary of designs for boilers and heating systems generally available is in BRE's *Good Repair Guide* 26, *Part 2*.

Combination boilers, or 'combi-boilers', that is to say boilers used for the supply of both direct domestic hot water and space heating, are in common use. In domestic situations, boilers can be fitted wherever it is most convenient; previously, the locations of boilers were constrained in a number of ways.

As an alternative to an open vented heating system which has been the most popular form of installation to date, a sealed system may be installed. The advantage of these pressurised (sealed) systems is that they can be used where there is no space for a feed-and-expansion tank or insufficient head room to provide the necessary pressure. Pressure may be supplied from the mains or from a pneumatic device (eg a pump applied to the pressure expansion vessel).

A sealed system requires:

- a boiler that is approved for this purpose,
- special safety controls,
- an expansion vessel that will accept the changing volume of water as the temperature changes.

Air in the pipework system is generally vented through automatic release valves. The heating circuit must not be permanently connected to the mains water supply.

Apart from the solid-fuelled type, boilers may be divided into those that take their air supply for combustion purposes from outside the habitable space (called room-sealed units), and which are the preferred type, and those that take their combustion air from within the habitable space (called open-flued types). Solid-fuelled boilers are exclusively of the open-flued kind. Electric boilers do not need a combustion air supply.

Boiler efficiencies in the 1990s increased significantly, in part due to the EU Boiler Efficiency Directive.

Gas boilers may be either condensing or, for the time being in some parts of the UK, non-condensing. A wide range of subtle variants is possible.

Condensing boilers

In new construction, the use of condensing boilers is now practically universal because of their significant improvement in performance

It has been shown that condensing boilers can help considerably to reduce energy consumed, thereby saving on fuel bills and producing less carbon dioxide than conventional boilers. Although slightly more expensive than their former standard counterparts, the extra cost of condensing boilers can be recovered in a very short time: within a year in large commercial applications and in 2–6 years in housing. A boiler using less fuel will also reduce its carbon dioxide emissions by an equivalent amount; consequently, condensing boilers help to mitigate climate change.

As well as being used in domestic situations, condensing boilers are also being installed in non-domestic buildings (Figure 9.2). Despite the low load factors on heating systems in office buildings, these boilers can reduce gas bills significantly and cost effectively. This is particularly true where condensing boilers are arranged to provide the base load in a multiple boiler installation.

A special requirement of condensing boilers should be considered early in the selection and design process. Condensate has to be drained from

BRE. *Improving energy efficiency. Part 2: Boilers and heating systems, draughtstripping.* Good Repair Guide 26. Bracknell, IHS BRE Press, 1999

Figure 9.2 A condensing boiler installation in an office building

the boilers and so a suitable drain location must be found. Normal plastics materials are needed for the drain connection (metal is not allowed); this should include a U-trap and perhaps a tun-dish (to provide visible indication of condensate flow or drain blockage). If the drain pipe is exposed, it must be protected from freezing. A continuous fall to the drain connection of at least $3°$ is recommended. The cost of providing a drain is usually small. The quantity of condensate produced will rarely exceed 50 ml/h for each kW of boiler input rating. (The theoretical maximum value is 150 ml/h per kW.) Actual volumes vary with operating temperatures.

Key Point

Remember that condensing boilers need a drain for condensate.

Thermostats

Thermostats, whether for controlling boiler temperature or room temperature, are crucial for the economical use of energy, and demand careful selection and siting. A room thermostat is a device for measuring the air temperature within a space and for switching space heating on and off. A single target temperature may be set by the user. A programmable room thermostat is a combined time-switch and room thermostat that allows the user to set different periods with different target temperatures for space heating.

Delayed start is a feature of a room thermostat which delays the chosen starting time for space heating according to the temperature measured inside or outside the building. Optimum start is a feature of a room thermostat to adjust the starting time for space heating according to the temperature measured inside or outside the building; it aims to heat the building to the required temperature by a chosen time.

Since a room thermostat is a device that operates a switch at a preset temperature, it could turn a boiler off when the temperature rises above a certain value and turn it on again when the temperature drops below that value. In practice, some thermostats have a dead band such that the switch-on point is slightly below the switch-off point. This means that the control is less accurate, though it does prevent rapid cycling and wear of the boiler, and can help to improve overall system efficiency.

There are two types of room thermostat:

● electromechanical and
● electronic.

The sensing part of the electromechanical type is either a bimetallic strip or a vapour capsule. In the electronic type the sensing part is usually a thermistor with the electronics setting the dead band and operating the switch.

Electronic thermostats can also be programmable, allowing different temperatures to be set at different times. Both types of thermostat can also have separate frost protection so that if the temperature drops below say $5°$ C, the heating will come on.

If a house heating regime is to be controlled by a single thermostat, there is no ideal location for it. Usually, it devolves into a choice between the hall or the main living room. In the hall, it will maintain the whole house at some reasonable temperature which has to be determined by trial and error to suit the occupants. If the living area temperature is more important, the thermostat can be placed there, but other parts of the house may then be too cold.

It is important that the thermostat receives no direct heat from, say, the sun or other sources, or be in a cold draught. In other words it needs to sense a

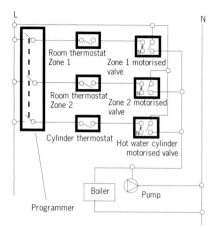

Figure 9.3 Fully pumped independent zone and hot water control

representative temperature for the whole house or room.

Separate upstairs and downstairs zone temperature controls are now required in new installations and a satisfactory way to do this is shown in Figure 9.3.

Motorised valves
Motorised valves do two things:
● open up water valves and
● operate electrical switches to energise connections to other devices.

In summary, motorised valves turn water flow on and off and they operate electrically. A two-port motorised valve controls water flow to a single destination while a three-port motorised valve controls water flow to two destinations (usually for space heating and hot water), and may be either a diverter valve (only one outlet open at a time) or a mid-position valve (either one, or both, outlets open at a time). Valve movement will also open or close switches which are used to control the boiler and pump. Automatic bypass valves control water flow. They are operated by the water pressure across them and are used to maintain minimum flowrates through boilers when alternative water paths are closed; for example, to maintain a water path when all thermostatic radiator valves (TRVs) are closed. Sensor units should be mounted horizontally rather than vertically to minimise this problem. For this

reason too, it is normal practice to install at least one radiator without a TRV.

There is more information on heating controls in BRE's book *Building services*.

A number of possible unwanted side-effects with boiler systems need the attention of the designer and are discussed in the following sections.

Noise and vibration
The pump is perhaps the main source of noise and vibration in small-scale central heating systems. While some noise and vibration at the source has to be accepted (eg with some kinds of boilers), proper mounting of the pump is essential in reducing noise. A partial but probably less effective solution is to insert a short length of flexible piping into the heating pipework to isolate the pump from other parts of the system.

When a central heating system fires up intermittently, the expansion and contraction of the pipes and radiators as they heat up and cool down is often accompanied by loud clicking and knocking noises. Invariably, a noise is due to friction at a point of restraint, and the sudden release of a restraint is accompanied by a loud report. It is not always easy to locate precisely these sources of noise, or to find effective remedies. It helps, therefore, if the pipes, where they pass through walls, joists, etc., are kept out of direct contact with the structure by means of sleeving or packing material; also, pipes should not be gripped too tightly by their supporting clips. Radiator brackets in particular may need attention, perhaps by the insertion of short pieces of PTFE tape or another lubricating medium between the radiators and suspension brackets.

Underfloor heating
Modern underfloor heating systems have a number of significant advantages over radiator systems, and for many applications underfloor heating represents the ideal solution. The low temperatures are well suited to condensing boiler operation. Two basic

Harrison H W & Trotman P M. *BRE Building Elements. Building services: Performance, diagnosis, maintenance, repair and the avoidance of defects.* BR 404. Bracknell, IHS BRE Press, 2000

kinds of underfloor heating laid in screed or slab have been used in recent years:

- low pressure hot water in steel or copper pipes, or, more commonly, polypropylene, and
- electric elements laid either directly or in conduit.

The main point to consider at the design stage is to make sure that any floor finish can tolerate the temperatures reached by the system and avoid pipe joints unless accessible.

Lightweight cementitious screeds are not appropriate for use with underfloor heating because of their insulation value. Ordinary cementitious screeds are appropriate for use with underfloor heating, but they can show excessive drying shrinkage caused by the heat drying out the material too quickly. Anhydrite screeds are appropriate for use with underfloor heating, but operating temperatures should not normally be allowed to exceed 50 °C. Although the screed can take higher temperatures, the finish may not.

Ceiling heating

For many years it was believed that radiant panel ceiling heating was more efficient than other forms of heating in achieving comfort levels, though there is little evidence to support this view. Ceiling heating is not appropriate for all circumstances, and there have been complaints, particularly from sedentary occupiers, of being uncomfortable. So far as the characteristics of the systems are concerned, either hot water or electric heating elements have been used in suspended ceilings, sandwiched between metal panels with thermal insulation quilts laid over the panels. Alternatively, electrically conducting membranes sandwiched between thin electrically-insulating sheets are now marketed.

Electric heating

An electric boiler can be used for domestic installations instead of a gas boiler to provide heat to radiators and a hot water tank. The main disadvantage is the high running cost, but, of course, no flue or local fuel supply is needed.

For domestic use, two other main types of electric heating are available: dry-core storage using high thermal capacity bricks, and wet storage. Both types are heated by off-peak supplies and require special wiring to carry the heavy loading.

Electric storage heaters are designed to take advantage of off-peak tariffs. It is usual to design systems based on storage heaters so that they are able to meet some of the heating needs by using electricity at the off-peak rate. In practice, the proportion of electricity actually consumed at the off-peak rate can vary quite widely according to the way the heating is used. Systems used for heating only part of a house generally require a greater proportion at the on-peak rate. With electric storage heaters, irrespective of storage medium, heat leakage may provide heat when not wanted, therefore their use can be less than 100% efficient.

District heating and Combined heat and power (CHP)

District heating, or the closely related community heating, is a method of providing space heating and hot water supply to a number of buildings or flats from a central source. The heat may be derived from any of the ordinary fuels, but the larger boiler plants normally associated with these schemes facilitate the use of low-grade fuels which, because of small size and low calorific value, are not suitable for individual house installations.

In CHP schemes the generation of heat in the central plant is combined with the generation of electricity, which enables heat and electricity to be produced with a lower fuel consumption than would be required to produce similar amounts of heat and electricity in separate plants.

In a thermal-electric station providing both heat and electricity, there is a problem in that the heat and electricity demands do not normally coincide. At one time it was suggested that the difficulty might be met in large stations by a combination of condensing and back pressure turbines. Another method which has been used to overcome the lack of balance is the use of heat accumulators; in these, excess heat, produced when the heat demand is low and the electricity demand high, is stored until the conditions reverse and the demand for heat exceeds that for electricity. These accumulators store heat in the form of hot water and need to have efficient insulation.

Air-conditioning and HVAC systems

Air-conditioning implies the pretreatment of ventilation air to control and modify its temperature and, sometimes, its moisture content. The air may also be filtered and cleaned of contaminants. Since air-conditioning usually involves at least one or more of these other factors, in addition to heating and cooling, it tends to be relatively expensive in terms of fuel consumption when compared with naturally ventilated buildings, using about 50% more primary energy. As people get used to having air-conditioning in their cars they may expect it at home too, and the industry is already marketing domestic air-conditioning systems to satisfy that demand. The demand is also being driven by climate change predictions. However, not only does it consume more primary energy, the space needed to accommodate air-handling ducts, especially in non-domestic buildings, can be large (Figure 9.4).

Heating, ventilating and air-conditioning (HVAC) systems can take many forms (Box 9.2).

Air-conditioning was introduced in response to the perceived need to cool modern buildings which have tended to suffer from high solar heat gains from oversized windows, or, conversely, poor natural daylighting encouraging the use of many energy-intensive lighting appliances to compensate. Increased concern over the adverse environmental impact of energy use has stimulated the design and construction of energy-efficient buildings, many of them suited to natural ventilation.

Some of the more important points to consider in preparing a specification for a HVAC system for a building are listed in Box 9.3.

Geothermal and heat pumps

It is usual to find heat pumps to be associated with geothermal low-grade heat sources.

A heat pump is a 'machine operating on a reversed heat engine cycle to produce a heating effect'. Energy from a low temperature source (eg earth, lake or river) is absorbed by the working fluid, which is mechanically compressed, resulting in a temperature increase. The high temperature energy is transferred in a 'heat exchanger'. The operation of a heat pump has been likened to that of a refrigerator, though saving and using the heat

Box 9.2 Some of the available HVAC systems

- Window ventilation and radiator perimeter heating
- Window façade ventilation and radiator perimeter heating
- Mechanical extract ventilation, window supply and radiator heating
- Mechanical supply and extract ventilation with radiator heating
- Mechanical displacement ventilation with radiator heating
- Mechanical displacement ventilation with static heating and cooling
- Ventilating chill/heat beams
- Four-pipe fan coil units with central ventilation
- Air Treatment Modules (ATM) zonal air-conditioning
- Terminal heat pump with central ventilation
- Variable air volume (VAV) air-conditioning with terminal re-heat
- VAV air-conditioning with radiator perimeter heating
- Fan-assisted terminal VAV
- Low temperature air fan-assisted terminal VAV
- Variable refrigerant flow (VRF) rate

A brief outline of the characteristics of each of the above systems is given in BRE's book *Building services*.

Figure 9.4 Air-handling ducting installed in a highly serviced building

instead of throwing it away. Most applications have achieved a coefficient of performance (COP) of around 3; that is to say, the output is something like three times as great as the input. Even installations designed for single dwellings can be economic if the conditions are right (Figure 9.5). The British Standard dealing with heat pumps is BS EN 14511.

Harrison H W & Trotman P M. *BRE Building Elements.*
Building services: Performance, diagnosis, maintenance, repair and the avoidance of defects. BR 404. Bracknell, IHS BRE Press, 2000

Box 9.3 Some considerations when specifying HVAC systems

Conflicts between HVAC and SHEV systems

There may be conflicts between a HVAC system and a smoke and heat exhaust ventilation (SHEV) system. In fact, the relationship between the two is very complex. There is a fundamental difference between a HVAC system, which is designed to put heat into a building and a SHEV system which is designed to remove it.

Legionnaires' disease

Legionnaires' disease gets its name from the American Legion Convention of 1976, when a number of legion members died from what were then unknown causes. After intensive research the cause was identified as the bacterium now called Legionella pneumophila, a strain of bacterium found in all waters, though usually at low concentrations. The bacterium is carried in aerosol emissions, largely from defective or poorly maintained air conditioning plant and domestic water systems. Infection is by inhalation of those aerosols.

Sick building syndrome (SBS)

Sick building syndrome (SBS) is a phenomenon experienced by people in certain buildings. The key symptoms are irritation of the eyes, nose, throat and skin, together with headache, lethargy, irritability and lack of concentration. The syndrome can be discriminated from other building-related problems such as physical discomfort, infections and diseases resulting from long-term cumulative hazards such as asbestos and radon. SBS is most often, but not exclusively, described in office buildings, particularly those that are air-conditioned.

Non-biological pollutants

Many non-biological pollutants can be present in buildings, although generally at low concentrations except in some industrial workplaces. There is concern that mixtures of many pollutants, each at low concentration, may have adverse effects that are not fully understood or predictable, but hard evidence is lacking. There is further information on all these items, also including pest infestation, in BRE's *Building services*.

British Standards Institution. BS EN 14511: Parts 1–4: 2004 *Air conditioners, liquid chilling packages and heat pumps with electrically driven compressors for space heating and cooling*

Figure 9.5 Heat pump installation in the Integer House on the BRE site, drawing heat from 50 m underground

Solar energy

The benefit of solar energy to our built environment is invaluable, providing us with a constant, unlimited source of heat and light. We can use the heat to warm our buildings and provide hot water, while sunlight helps to light rooms without using additional energy, and can be used to generate electricity using photovoltaics (PVs). The sun also provides an additional benefit in that it drives the wind. This gives us the potential for the use of natural ventilation to keep our buildings cool, particularly in the summer. The sun's energy can be used in buildings in three main ways:

Passive solar energy and passive solar design

Passive solar energy is ambient energy from the sun that gives us heat, light and natural ventilation. A strategy that seeks to gain the benefit of passive solar energy within buildings is called passive solar design (PSD).

A building designed along passive solar design principles seeks to exploit the benefits of solar heat, daylight and natural ventilation while minimising the unwanted results of solar energy such as excessive heat in summer or glare.

For this reason, passive solar design is also greatly concerned with shading (to prevent unwanted solar gain), with heavyweight building materials (to absorb heat and prevent rapid fluctuations of temperature in the space, or thermal mass), with window design for good daylight and

Key Point
In the UK, the beneficial effects of solar energy normally outweigh the adverse effects.

various ventilation options. PSD is also concerned with the position of the building on the site to prevent it from being overshadowed and with the internal planning of the building so that those rooms that most need ventilation, daylight and sunlight can receive them.

Active solar: solar hot water

Using specially designed mechanical systems, solar thermal systems can generate much more heat for space heating and hot water than passive solar alone. Solar collectors, at the heart of most solar thermal systems, absorb the sun's energy and provide heat for hot water, heating and other applications in residential or commercial buildings. Modern systems are highly efficient.

There are two basic types of solar heating system:
- liquid systems heat water or liquid antifreeze in a 'hydronic' collector, and
- systems based on 'air collectors'.

Both systems collect and absorb solar radiation, and transfer the solar heat directly to the interior space or to a storage system (eg hot water tanks), from which the heat is distributed. If the system cannot provide all the heat required, an auxiliary or back-up system provides the additional heat. Liquid-based systems are more often used when storage is required.

The circulation of heat can be either passive (relying on natural convection or water pressure to circulate the fluid through the collector to the point of use) or active (using pumps which increase the system's efficiency but with the additional capital cost requirements for the pump and associated controls). There are several types of hydronic solar collectors but the two most common types suitable for buildings in the UK are flat-plate collectors or evacuated tube collectors.

Flat-plate collectors are simple but effective devices, containing a dark plate within an insulated box having a glass or plastic cover. Evacuated tube collectors are more sophisticated, have higher efficiencies and are effective under a wider range of conditions, but are more expensive than flat-plate collectors.

The best locations for solar thermal systems are on a roof or wall that faces within 90° of south.

Buildings that face an easterly direction will benefit from the heat earlier in the day which can be an advantage where there are facilities to store heat. If the collector surface is in shadow for parts of the day, the output of the system decreases. The availability of solar thermal is confined to daylight hours which change seasonally. Ideally, the panel should be angled at, or near to, 45°.

Active solar: photovoltaics

Photovoltaic (PV) systems use solar cells to convert sunlight into electricity. The PV cell consists of one or two layers of a semi-conducting material, usually silicon. When light shines on the cell it creates an electric field across the layers, causing electricity to flow. The greater the light intensity, the greater the flow of electricity.

There are three basic kinds of solar cells:
- monocrystalline with a typical efficiency of 15%,
- polycrystalline with a typical efficiency of at least 13%,
- thin film which can be applied to other materials such as glass or metals and with a typical efficiency of 7%.

The cost of the materials is generally highest for the more efficient types of cell.

Building-integrated photovoltaics are used to generate electricity. Each PV cell provides a small amount of electricity so a large number form an array with a greater output. PV is usually installed in parallel with the grid, although stand-alone generation is not uncommon, particularly in isolated areas.

A key advantage of photovoltaics is that they can be integrated with the fabric of a building with relative ease. Where systems can be integrated within buildings, the structural support for PV is often available at no additional cost. In ideal circumstances, the modules replace building components such as curtain walls, roof tiles and structural glazing and vertical walls.

Framed PV modules can either be roof-mounted, free-standing or integrated into the roof or façades of a building. Recent developments have led to many different forms of solar collector, with efficiencies improving all the time.

Ventilation

Adequate ventilation is essential for the well-being and health of building occupants. Traditionally, it was achieved by natural means, ie by providing fresh air. Fresh air is needed to:

- provide sufficient oxygen for breathing,
- remove excess water vapour,
- dilute body odours,
- dilute, to acceptable levels, the concentration of carbon dioxide produced by occupants and combustion processes,
- remove or dilute other indoor pollutants.

The Building Regulation requirements for each part of the UK are:

- **England & Wales:** strictly speaking, the Building Regulations merely require a building to be adequately ventilated from the point of view of its occupants. AD F provides guidance on how to meet that requirement.
- **Scotland:** Section 3 of the Technical Handbooks, Domestic and Non-Domestic, provides for a mandatory functional standard with supporting guidance. The specifier is at liberty to use alternative means as long as they meet the standard.
- **Northern Ireland:** Technical Booklet K.

Natural ventilation

Natural ventilation is defined as ventilation driven by the natural forces of wind and temperature differences. Its use should be intentional and controlled. It should not be confused with infiltration which is the unintentional and uncontrolled entry of outdoor air, and the exit of indoor air, through cracks and gaps in the external fabric of the building. Airtightness of the building fabric has become a crucial issue, meriting the close attention of designers as well as contractors. Box 9.4 gives some important guidance on airtightness (see also the section on *Air permeability* in chapter 1).

Key Point
Natural ventilation should be intentional and controlled.
Airtightness of the building fabric has become a crucial

Air infiltration cannot be designed for, nor indeed can it be precisely built into, a building; it may therefore be considered as an unwanted addition to the running costs of the building. Moreover, infiltration is neither a reliable nor an energy-efficient substitute for properly designed ventilation.

The design of a naturally ventilated building should reflect different requirements for winter and summer occupancy. In winter, excess ventilation needs to be minimised, but background ventilation can be provided in the form of trickle ventilators in window heads to meet occupants' needs for fresh air. In summer, ventilation may need to exceed what is required solely to satisfy occupants' needs to avoid overheating. The distribution of fresh air is important within the habitable spaces, and can enhance comfort conditions and freshness.

The following basic design guidelines help to improve natural ventilation.

- Shallow plan forms are better than deep ones, in particular for single-sided ventilation.
- Shading (preferably external) minimises summer overheating.
- An airtight building envelope minimises unwanted air infiltration: *Build tight, Ventilate right!*
- Trickle ventilators provide controllable background ventilation.
- Openable windows provide controllable and draught-free ventilation but, for night cooling, they should be lockable in a secure position.
- Precise control of local ventilation should be allowed to the building occupants.
- Using low-energy lighting and IT equipment avoids unnecessary heat gains.
- Office machinery should be situated near local extract ventilation.

Windows can contribute to the balance of heating and ventilation (and therefore the control of condensation) only if the building occupants find the controls easy to reach, convenient to use, and precisely controllable so that open areas can be adjusted to best advantage in windy conditions.

Extract fans can be used as an auxiliary ventilation device to natural ventilation through

Box 9.4 Airtightness

Airtightness is now an important issue for the construction industry. Building Regulations (England & Wales) require that air leakage is to be minimised in all new buildings.

In a well-insulated building, uncontrolled air leakage can account for a high proportion of the total heat loss (up to 50% in some cases). Exposed sites and elevated positions increase air leakage.

Under the new regulations, designers are responsible for designing airtight buildings. Contractors are responsible for building an airtight building, arranging for an airtightness test to be carried out after completion, and also for any necessary rectification and re-testing. The apparatus depends on the volume of the building to be tested, with houses needing only a relatively small fan (Figure 9.6), but buildings of greater volume needing much larger apparatus (Figure 9.7)

There are essentially four main routes by which air can leak out from a building:

- joints around components (eg window-to-wall joints),
- gaps between adjacent elements (eg wall-to-roof joints),
- gaps around services passing through the external envelope,
- through permeable building materials (eg unpainted lightweight blockwork).

In order to achieve a satisfactory level of airtightness, it will be necessary for the designer to identify all relevant routes at an early stage in the design process, making sure that they can be adequately sealed to form a continuous and impermeable barrier within the envelope construction. It is then the contractor's responsibility to meet those standards.

Post-completion air leakage audits are carried out to identify air leakage paths in the envelope of a building. While the fan is pressurising the building, small hand-held smoke pencils reveal air leaking through gaps and cracks.

Further information can be obtained from BRE's *Good Building Guide 67*.

Figure 9.6 Small fan used to pressure test dwellings

Figure 9.7 BREFAN apparatus for measuring air leakage rates from an entire building

---

windows and are necessary in any case for compliance with building regulations in rooms where moisture is produced.

Spaces can be cross-ventilated to a depth of about five times their height (ie 15 m depth for a 3 m height). CIBSE Guide A and BS 5925 give simple equations to estimate single-sided and cross-ventilation flows. Computational fluid dynamics (CFD) can be used to predict natural ventilation rates in new buildings and computer design

programs are widely available. There are, however, points to watch, particularly in relation to proximity of naturally ventilated buildings and their effects on each other and the avoidance of adjacent car parking (Figure 9.8).

Trickle ventilators can provide controllable background ventilation. Minimum open areas

Jaggs M & Scivyer C. *Achieving airtightness* (3 Parts). Good Building Guide 67. Bracknell, IHS BRE Press, 2006

Chartered Institution of Building Services Engineers. *CIBSE Guide A. Environmental design.* 2006
British Standards Institution. BS 5925: 1991 *Code of practice for ventilation principles and designing for natural ventilation*

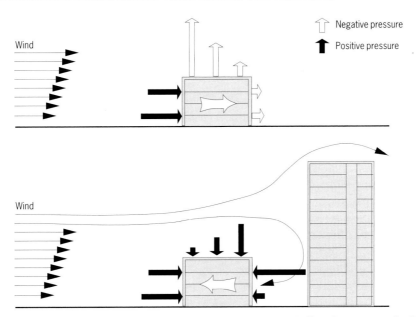

Figure 9.8 A tall building can deflect high wind speeds down to ground level, and affect the pressure distribution on a nearby low-rise building. This could change, or even reverse, the ventilation flow through the low-rise building. The actual pressures depend on wind direction and height and spacing of buildings. It is pressure differences, rather than absolute pressure, that drives flows

needed for these ventilators are 8000 mm² in habitable rooms, and 4000 mm² in kitchens, bathrooms and WCs. They may also be useful in a non-domestic situation.

Mechanical ventilation

Where natural ventilation systems prove to be inadequate, then reliance is placed on mechanical systems. In the UK domestic sector, the use of continuously operated mechanical ventilation (MV) is becoming more common. There are two main types:
- balanced supply and extract mechanical ventilation with heat recovery (MVHR),
- mechanical extract ventilation (MEV)

In both, the air is collected or distributed through a duct network. An MEV system consists essentially of the extract components of a MVHR system and is cheaper to buy and install.

MEV and MVHR installations cannot be fitted in dwellings which have open-flued combustion appliances (any appliance which draws some or all of its combustion air from the living space) since these systems might interfere with the operation of appliances and combustion products could be drawn back into the living areas.

Air-handling plant for forced air ventilation systems in non-domestic buildings is usually fairly bulky. It is best sited near to zones of maximum demand. Plant room sizes typically can be around 1 m² per 100 m² of floor area served, subject of course to a minimum smallest area for items of basic equipment, and some pro rata reduction for the very largest installations.

Passive stack ventilation

Passive stack ventilation (PSV) systems work by a combination of the natural stack effect (ie air movement that results from the difference in temperature between indoors and outdoors) and the effect of wind passing over the roof of the building. Hot air rises through its buoyancy. Natural ventilation can be effected by channelling, through a duct, this warmer indoor air, using as motive power the pressure differential created by its buoyancy.

During the heating season, and under normal UK weather conditions, the ducts will allow warm moist air from the 'wet' rooms to be vented directly outside. The airflow rate in each duct will depend on several factors, the most important, at low wind speeds, being the temperature difference between indoors and the outside air. As the temperature indoors increases, because of cooking, heating, washing or bathing, the airflow rate will also increase. Wind speed and direction also influence the rate of airflow but in a relatively complex way that depends on the interaction between factors such as the airtightness of the dwelling, the type and position of air leakage paths and the position of the duct outlet terminals.

There are two distinct types of stack ventilation systems:

- small and simple systems for dwellings described in BRE's *Information Paper* 13/94 with more guidance in *Information Paper* 6/03 (see also Figure 9.9),
- larger, more elaborate systems for non-domestic buildings.

PSV is likely to be an attractive alternative to mechanical ventilation in some domestic situations: for example, in cold houses with high moisture production where fans would run for long periods; where occupants are unwilling to run fans; or where occupants are particularly sensitive to noise.

The system should be designed to:

- avoid cross-flow ventilation between kitchen, bathroom and WC,
- prevent, as far as possible, airflow in the ducts being adversely affected by the prevailing wind speed and direction, or by sudden changes in these, and
- minimise resistance to airflow by having ducts that are as near-vertical as possible.

Although the physical principles of PSV systems are the same for both domestic and non-domestic buildings, in practice the non-domestic situation is much more complex. Solar chimneys, in which glazed elements are incorporated into the chimney structure, can enhance stack pressures (Figure 9.10). Wind pressures can also be utilised by placing the outlet in a negative pressure zone relative to the inlet. In these designs, airflows into the building are at low level and exhausted at high level; therefore, some care has to be exercised in determining the different sizes of ventilation openings for each floor

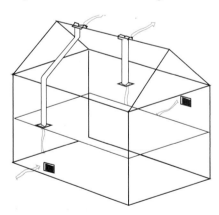

Figure 9.9 PSV layout suitable for most two-storey dwellings

Figure 9.10 Solar chimneys in a BRE office building. The sun warms the external duct to enhance buoyancy, which in turn increases air flow

Stephen RK, Parkins LM and Wooliscroft M. *Passive stack ventilation systems: design and installation.* Information Paper IP 13/94. Bracknell, IHS BRE Press, 1994

Baker P, McEvoy M & Southall R. *Improving air quality in homes with supply air windows.* Information Paper IP 6/03. Bracknell, IHS BRE Press, 2003

Chartered Institution of Building Services Engineers (CIBSE). *Natural ventilation in non-domestic buildings.* CIBSE Applications Manual AM 10. London, CIBSE, 1997

of the building if equal ventilation rates are required. CIBSE have published a relatively simple procedure to do this (see CIBSE Applications Manual AM 10).

Cold water supply and distribution

Until 1999, water was supplied in England and Wales by statutory water undertakers who were controlled by the provisions of the Water Act 1945. Installations in buildings were required to comply with relevant byelaws made under the Water Acts. These byelaws, based on the Model Water Byelaws primarily covered prevention of waste, undue consumption, and misuse and contamination of water. Byelaws have been replaced by the Water Regulations 1999 which came into force on 1 July 1999. Scotland is covered by regulations for public and private water supplies: the Water Quality (Scotland) Regulations 1990 (as amended) and the Private Water Supply Regulations 1992. Additionally, Scottish local authorities have water byelaws, the Water Authorities Enforcement Regulations, which were updated in April 2000 and are used to implement the European Drinking Water Directive.

The specific properties of drinking water supplied for public consumption varies widely in degree of hardness, and the amount of chemical treatment which is necessary for adequate purification. Hard water contains more calcium and magnesium than soft water which is often shown by build-up of 'lime' deposits within pipes and on the surfaces of sanitary ware.

Conservation of water

> **Key Point**
> Wholesome water is a scarce and precious commodity.

Most of the population, including the construction industry, need radically to revise present attitudes to water consumption (Figure 9.11).

Ways of conserving water now include measures to:
- reduce the demand for water,
- improve the efficiency of water using appliances,

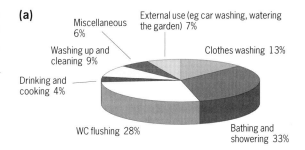

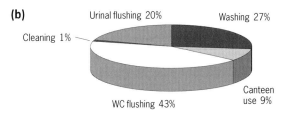

Figure 9.11 Typical use of water in (a) domestic buildings in southern England, and (b) office buildings in the UK

- reduce the loss and waste of water,
- harvesting of rainwater to increase usage of greywater.

One possible means of encouraging the reduction of overall consumption is to install water meters (an option available to existing customers of water authorities but compulsory for new properties) which measure the volume of water consumed: it is presumed that knowledge by users of rates of water usage should help to reduce profligacy and so conserve supplies. There are, however, many social, political and financial implications. Metering of supplies to non-domestic buildings is usual.

Increasingly, automatic leak detectors are becoming available in the UK. These devices, which are fitted into the incoming mains, close when a leak is detected and so are able to prevent wastage of

The Stationery Office
Water Supply (Water Fittings) Regulations 1999. Statutory Instrument 1999 No 1148. 1999
The Water Supply (Water Quality) Regulations 2001
The Water Supply (Water Quality) (Scotland) Regulations 2000
The Water Supply (Water Quality) (Amendment) Regulations (Northern Ireland) 2003

Scottish Water Authorities
Water Byelaws 2000 (Scotland)

water and damage to property. Some operate by sensing a high flowrate and others use conductivity detectors to activate valves.

Policy on purity

In the last years of the 20th century, the policy was still to supply all water from the public mains of a quality suitable for drinking, even though only a small proportion of this was used for drinking and cooking. Depleted rainfall figures and the consequent reduction of river flows, and reservoir and aquifer levels (most years of the last decade of the twentieth century have been officially classed as drought years) has led to water shortages in many areas of the UK. This policy will not be sustainable indefinitely. Greater use of greywater (or water defined as wastewater not containing faecal matter or urine, and which is not of drinkable quality) may become preferred or even obligatory for applications such as flushing WCs, car washing and garden watering. However, most black water, defined as wastewater which contains faecal matter or urine, will still need to be treated off-site for the foreseeable future. Waste water is defined in the later section on *Above ground drainage*.

Water which is stored, even though it might have originally been of drinking quality, may deteriorate. For many building types other than housing, in which large quantities of water are used for sanitary and washing purposes, supplies may be prone to interruption, making it necessary to store large quantities of water which then becomes unfit for drinking. In those circumstances, drinking water is normally supplied separately, directly from the mains. It is arguably the case now that interruptions in the public supply are so infrequent that all outlets can be supplied off the mains with no storage.

Rainwater

One of the most popular techniques for saving water is to harvest rainwater for tasks where wholesome water has normally been used hitherto, though large storage capacities are required. Potential uses include WC flushing, garden watering and car washing.

Delivery rates

When preparing performance specifications for the selection of fittings, the flowrates shown in Table 9.1 may prove helpful. BS 6700 gives flowrates for other installations. These flowrates are easily checked on site by timing the filling of graduated containers.

Other criteria are as listed below.
- Stored water pressures to be not greater than 300 kPa (3 Bar) at point of delivery. The temperature of cold water in storage tanks should be maintained below 20 °C, and if possible below 15 °C, to prevent the growth of the legionella bacteria.
- Mains pressure not to exceed 600 kPa (6 Bar). If supply is in excess of this value, a pressure-reducing valve should be installed.
- Cold water storage capacity to be in proportion to size of household; for a household of four people, say 450 litres.
- The system to be sterilised before use (eg chlorine dioxide of 50 ppm).
- Inlet valves to cisterns to be of a type appropriate to the pressure.
- Cold water pipes at high levels wherever possible to be run beneath hot water pipes, or the hot pipes to be insulated.
- Cold water pipes may also be insulated to help avoid the risk of condensation.

Table 9.1 Minimum rates of flow at point of delivery	
Appliance	Minimum rates at mains pressure (litres/sec)
Sink, 15 mm pipe dia	0.20
Sink, 22 mm pipe dia	0.30
Wash handbasin, 15 mm pipe dia	0.15
WC flushing cistern	0.10
Bidet	0.15

British Standards Institution. BS 6700: 1987 *Specification for design, installation, testing and maintenance of services supplying water for domestic use within buildings and their curtilages*

Current but partly replaced by BS EN 806-2:2005 *Specifications for installations inside buildings conveying water for human consumption. Part 2: Design*

● Mainly in relation to future maintenance work, cold water tanks and system pipework to be installed so that they are not obstructed by, and do not obstruct, other equipment and pipework.

Hot water supply and distribution

In houses, domestic hot water (DHW) is usually provided by the same boiler that meets space heating requirements. Summer operation, supplying hot water only, will be met by either the boiler or an immersion heater. In non-domestic buildings, DHW is usually taken from a separate boiler to improve summer efficiency.

Unvented systems

Unvented DHW systems have been permitted since 1986. One typical system heated indirectly by means of a primary coil and secondary circuit is shown in Figure 9.12.

Expansion vessels are required to take up changes in water volume; cold water filling points with pressure gauges are also needed. Normally components are bought as a package with all the safety devices in place. There are regulations on the safety aspects, not only on the risk of explosion but also on backsiphonage into the mains supply. The advantages of unvented systems are that there is no risk of cold water freezing in the roof space, no airlocks, and, with combi-boilers, hot water is available at mains pressure. A potential disadvantage is the risk of explosion where the design or installation does not conform to requirements, but this risk is now considered to be low.

Storage capacity

Although 120 litres capacity is usual for small dwellings, larger dwellings with more than one bathroom will require larger storage. BS 6700 recommends a minimum capacity of 100 litres for a solid fuel installation, 200 litres for an electrical off-peak, or 45 litres per occupant, whichever of these produces the largest figure, unless the design of the system justifies a smaller capacity. A five-person house would therefore need 225 litres. High performance cylinders with rapid heating coils are available which reduce the time taken to reheat a cold cylinder, and may also reduce boiler cycling. Modern practice is to pump both primary and secondary circuits with priority for hot water if required. Recovery times are more rapid than for gravity fed systems.

Energy conservation

In most homes, the water heater is the second largest user of energy in the home after space heating. However, in highly insulated homes, the contrary may be true. By reducing the amount of hot water used, some of the energy needed to heat it can be saved. In the UK, domestic water heating consumes about 75 PJ each year. Also, a reduction in the delivered water temperature of 5 °C will produce an energy saving of about 10%.

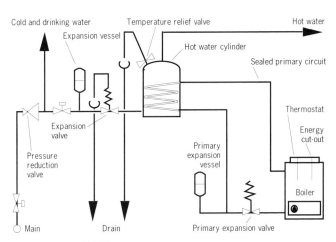

Figure 9.12 An indirectly heated unvented DHW system

Energy is saved by insulating against heat losses from the hot water system (not always done in practice) by:
● reducing water temperatures,
● using controls to improve appliance efficiencies, and
● taking short showers instead of baths.

Insulation of the hot water storage cylinder is extremely cost-effective and controls are well worthwhile, but measures which employ heat exchangers and extra plumbing may not be economic. Changes to building regulations have been introduced to improve tank and pipe insulation. A further economy measure to consider is heat recovery from wastewater, but equipment is not usually installed for this purpose unless the volumes of waste hot water are considerable.

Key Point

In most homes, the water heater is the second greatest user of energy after space heating.

Performance criteria

When preparing performance specifications for the design of a hot water system, the following guidelines may prove useful.
● The system ought to be capable of supplying water at a temperature adjustable between 50 °C and 70 °C, and at a flowrate of not less than 0.45 l/s for a period of up to 15 minutes in any one hour. However, temperatures above 40 °C can cause scalding so delivery temperatures should be limited (see BRE's *Information Paper IP 14/03*).
● Minimum flowrates at each of the individual outlets should be not less than those given above for cold water.
● Thermostats should be subject to a tolerance of ±2 °C in both switch-on and switch-off modes.
● The predetermined temperature, ±3 °C, at any outlet, should be reached not later than 10 seconds after flow commences.

BRE. *Preventing hot water scalding in bathrooms: using TMVs.* Information Paper IP 14/03. 2003

● Expansion loops and resilient packing should be used in fixing the pipes to the building carcass.

Combi-boilers

Both the domestic hot water storage cylinder and associated cold water storage tank can be eliminated by using the combination boiler (the so-called combi-boiler). The boiler incorporates both central heating and domestic water heating direct from the mains via separate heating coils, the central heating coil being shut off automatically when domestic hot water is required. Combi-boilers will always give a supply of (nearly) mains pressure hot water while storage cylinder systems can run cold when hot water demand is high. Combi-boilers are particularly well suited to smaller dwellings, where hot water demand is limited. For larger dwellings, storage combis can be fitted.

Sanitary fittings

The numbers and disposition of sanitary fittings will depend on particular circumstances, but remember that in practically all buildings, possibly including houses, sanitary facilities should provide for both privacy and unassisted use by a person who may use a wheelchair. Further information can be found in chapter 12.

Around 1 in 8 dwellings have more than one bath or shower, and this provision is expected to grow with the current tendency to install en-suite bathrooms. With this growth of provision of course may come a concomitant potential growth in the frequency of bathing and therefore in the total consumption of water, which needs to be addressed by appliance designers and specifiers and users alike.

The requirements for discharge rates from appliances should be a primary consideration in the design and subsequent checking of waste systems. Typical expected maximum discharge rates for the UK are given in relation to each appliance in the following sections and the concept of discharge units and their combination with respect to drainage systems is dealt with in the section on *Above ground drainage* later in this chapter. The sizes of outlets, traps and pipework should be such that the discharges from sanitary appliances are not restricted. Pipes serving more than one appliance should be sized taking account of simultaneous discharge by using the concept of a discharge unit value. This technique of a discharge unit value is not described here but more information is available in BS EN 12056.

From the point of view of health and safety any shower thermostat should fail safe. Metal baths, sinks and pipework must have provision for, or must be capable of, being provided with an effective and approved means of connecting to an earth continuity conductor.

Most appliances, unless subjected to impact damage are inherently durable once installed. BRE investigations put the expected lives of most sanitary fittings at around 30 years.

Baths

A common rectangular bath size is 1700 mm × 700 mm with coordinating heights in multiples of 50 mm and other plan sizes increasing in multiples of 100 mm. Corner baths and baths incorporating space for a shower have become popular in recent years, but their dimensions vary considerably and so does the space needed to use them. They also use much more water than a conventional bath. Materials include acrylic, sheet steel and cast iron. European Standards, in addition to the British Standards, are applicable. They include, for example, BS 4305-1/EN 198, BS EN 232 and BS EN 263. A clear floor space of at least 1100 mm × 700 mm is needed in order to use the bath, with the longer dimension adjacent to one side of the bath (Figure 9.13). With an 80-litre capacity and 40 mm diameter branch pipe, the maximum discharge rate will be around 1.3 l/s.

British Standards Institution
BS EN 12056: 2000 *Gravity drainage systems inside buildings.* Parts 1–5
BS 4305-1: 1989 and EN 198: 1987 *Baths for domestic purposes made of acrylic material. Specification for finished baths*
BS EN 232: 2003 *Baths. Connecting dimensions*
BS EN 263: 2002 *Crosslinked cast acrylic sheets for baths and shower trays for domestic purposes*

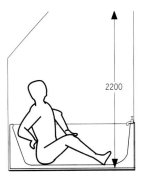

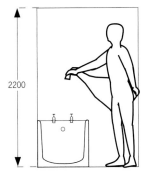

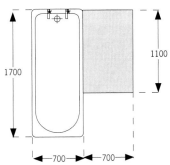

Figure 9.13 Space needed to use a bath

Showers

Plan sizes of 800 mm × 800 mm and 900 mm × 900 mm are the most common. A tray height of 150 mm is common but level access trays are available. If enclosed on one or two sides, an activity area is needed on plan of 400 mm × 900 mm (or the width of tray used) adjacent to one open side of the shower.

There are many different materials and designs for shower surrounds; they include plastics curtains, tiling of the wall surfaces, glazed screens and folding doors. These sizes apply to all types of shower: an unenclosed but drained corner of the bathroom, as more commonly found abroad; basic floor trays, either set into or standing on the floor plane, and which can be fitted with various types of screening; and the packaged shower unit supplied as an integral floor tray and enclosing cabinet. Showers are covered by the various parts of BS 6340-1. Materials from which the trays are commonly made include acrylics to BS 6340-5, porcelain enamelled cast iron to BS 6340-6, vitreous enamelled sheet steel to BS 6340-7, and glazed ceramics to BS 6340-8. European Standards, in addition to the British Standards, are applicable. They include BS EN 251, BS EN 263 and BS EN 329.

Flow rates from single-head showers are small so that the 40 mm discharge pipes usually fitted do not require venting. However, difficulties may arise in achieving a self-cleaning velocity and adequate provision should be made for cleaning the outlet. A shower unit with more than one shower head may produce considerably greater flowrates than through a single outlet. For an electric shower of 7–8 kW, maximum discharge rate will be around 0.07 l/s. A low pressure or low volume shower will have a maximum discharge rate of around 0.15 l/s, but a high pressure shower will have a maximum discharge rate of 0.35 l/s (without a plug 0.4 l/s, with a plug 1.3 l/s). The average amount of water used for a conventional shower is about 30 litres (a bath requires about 80 litres).

British Standards Institution
BS 6340: 1983 Shower units:
    Part 1: Guide on choice of shower units and their components for use in private dwellings
    Part 5: Specification for prefabricated shower trays made from acrylic material
    Part 6: Specification for prefabricated shower trays made from porcelain enamelled cast iron
    Part 7: Specification for prefabricated shower trays made from vitreous enamelled sheet steel
    Part 8: Specification for prefabricated shower trays made from glazed ceramic
BS EN 251: 2003 *Shower trays. Connecting dimensions*
BS EN 263: 2002 *Crosslinked cast acrylic sheets for baths and shower trays for domestic purposes*
BS EN 329: 1997 *Sanitary tapware. Waste fittings for shower trays. General technical specifications*

Wash hand basins

Wash hand basins are now 600 mm × 400 mm on plan, sometimes larger. Front rim height is 800 mm for pedestal sets. Ideally, the basin needs to be fixed at a lower height for washing face and hair than for hand-rinsing. The 800 mm height is really a compromise for family dwellings. Where the basin is more likely to be used by adults, a height of 900 mm may be acceptable. BS 8300 requires a rim height of 720–740 mm for use by wheelchair users. Wash hand basins, both pedestal and wall-hung basins, are covered by BS 5506-3. European Standards, additional to British Standards, are applicable for products, eg BS EN 32, BS EN 37 and BS EN 111. An area of at least 1000 mm × 700 mm is needed in order to be able to use the basin. With a 6.1 litre capacity and 32 mm diameter branch pipe, the maximum discharge rate will be around 0.6 l/s.

Sinks

A single sink in a dwelling should be large enough to take an oven shelf, the largest object commonly washed there, and therefore be at least 500 mm × 350 mm × 175–200 mm deep. If the depth is, say, 250 mm or more, it will also be suitable for washing clothes and household linen. The type of double sink now most usually found in the UK has two equal bowls about 400 mm × 400 mm, but a more useful combination might be one large bowl (say 500 mm × 350 mm and 175 mm deep) big enough for cleaning oven trays and, if necessary, hand washing clothes and household linen, and a smaller rinsing bowl alongside. With a 23-litre capacity and 40 mm diameter branch pipe, the maximum discharge rate will be around 1.3 l/s.

British Standards Institution
BS 8300: 2001 *Design of buildings and their approaches to meet the needs of disabled people. Code of practice*
BS 5506-3: 1977 *Specification for wash basins. Wash basins (one or three tap holes). Materials, quality, design and construction*
BS EN 32: 1999 *Wall-hung wash basins. Connecting dimensions*
BS EN 37: 1999 *Pedestal WC pans with independent water supply. Connecting dimensions*
BS EN 111: 2003 *Wall-hung hand rinse basins. Connecting dimensions*

WCs

Three types of WC appliances may be available (Figure 9.14):

- siphonic,
- wash down,
- wash out.

The wash-out pan leaves the excreta largely exposed in the shallow bowl and away from the trap until the flush operates to wash it out. This type is fairly rare, and was formerly used to a considerable extent in health care buildings. The wash down is the most common type encountered, with the excreta largely submerged in the front part of the trap until the flush removes it. The siphonic type is similar in some respects to the wash down, but in this case the siphon uses atmospheric pressure to force the evacuation of the bowl when the vacuum is created in the discharge pipe. The siphon may be of the single-trap or double-trap kind. Under the Water Regulations 1999, valve flushing has been permitted from January 2001. Low-level WC suites are around 800 mm wide × 700 mm deep with a 50 mm seal. High-level suites may be slightly less deep. Rim height is normally around 400 mm, but may be lower for use by children in schools. An activity space of 800 mm wide × 600 mm deep is needed, and to full ceiling height. European Standards apply; for example, BS EN 37.

WCs are now required to have a maximum flush volume of 6 litres and urinals with flushing valves are now permitted in some circumstances.

WCs that are currently available in the UK should be BS EN 997 Class 2 WCs, not Class 1. Class 2

WCs will have passed the stringent performance requirements of BS EN 997 that include the UK Regulators Specification for WC suites.

Typical discharge rates are, for:

- wash down with high- or low-level cistern and 7.5 litre capacity: maximum discharge rate of 1.8 l/s,
- wash down with close-coupled cistern and 7.5 litre capacity: maximum discharge rate of 1.2 l/s.

(BS EN 12056-2 suggests for a 6-litre cistern 1.2–1.7 l/s, for a 7.5-litre cistern 1.4–1.8 l/s, and for a 9-litre cistern 1.6–2.0 l/s).

Dual-flush cisterns are expected to become mandatory. Smaller capacity cisterns (eg as low as 4 litres) are expected to become more common in the future.

Macerators enable WCs to be connected to internal drains using comparatively small diameter pipes: down to 22 mm for short runs, but 32 mm for longer runs. They are particularly useful where access to foul drains by normal means is restricted or perhaps impossible, and they may take as little as 3 litres of water to flush. Heavy duty models are available which can pump the slurry discharge up to

British Standards Institution

BS EN 37: 1999 *Pedestal WC pans with independent water supply. Connecting dimensions*
BS EN 997: 2003 *WC pans and WC suites with integral trap*
BS EN 12056-2: 2000 *Gravity drainage systems inside buildings. Sanitary pipework, layout and calculation*

(a)
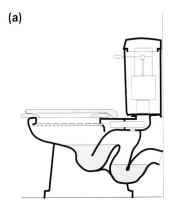

(b)
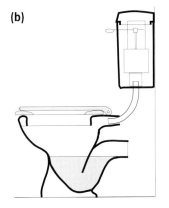

(c)

Figure 9.14 Types of WC appliance: **(a)** siphonic, **(b)** wash down and **(c)** wash out

two storeys high and for considerable horizontal distances. Power for all models can be taken from normal 240 V AC supplies. Different designs of macerator have different capabilities for dealing with bowl contents, and users can overload them with fibrous material such as disposable nappies leading to blockages. Systems using large radius bends tend to be more trouble-free than those with sharp bends. Discharge rates vary according to model: generally about 0.4–1.5 l/s.

Composting toilets

Of the toilets that use no water for flushing, the most common type in the UK is the composting toilet. Usually, in its domestic form, it is electrically powered, heating the waste material to promote the composting action. Size is its main problem; the smallest domestic model is about twice the size of a conventional WC suite. Large models (greater than 15 m^3 capacity) usually do not require external energy to start and sustain the process as the aerobic decomposition is sufficiently exothermic to be self-sustaining. Large composting toilets may be more environmentally acceptable as they consume only a small volume of water, require no drainage pipework and produce compost that can be used in the garden. However, the questions of adequate hand-cleansing facilities, if there is no available water supply, and the safety of children using toilets with open chutes, must be considered.

Bidets

Appliance dimensions are 700 mm × 400 mm on plan. An activity space of 800 mm × 600 mm on plan is needed to use this appliance. Most bidets can be filled from taps fitted to the rim, but some are fitted with a spray to which special regulations apply; that is to say, hot and cold supplies must be drawn from tanks and not direct from the mains, and there must be no tees serving other appliances. British and European Standards are applicable, eg BS EN 35. With a 23-litre capacity and 40 mm diameter branch, the maximum discharge rate will be around 0.9 l/s.

Washing machines

From the point of view of economy in the use of energy, depending on circumstances, it is perhaps better to use machines that draw water from the main domestic hot supply, rather than heating it in the machine; this requires a permanent connection to the DHW supply. On the other hand, overnight off-peak heating can be used on cold-filled machines. Most machines are around 600 mm × 600 mm on plan, and need an activity space of 1000 mm × 1100 mm. With a 4.5 kg capacity, the maximum discharge rate will be around 0.6 l/s. (BS EN 12056-2 suggests for a capacity of up to 6 kg, 0.6 l/s, and for up to 12 kg, 1.2 l/s.)

Dishwashers

Front-loading machines are convenient to load and can stand permanently in a position under a worktop. According to the old British Standard, now replaced by BS EN 1116, a plan area of 600 mm × 600 mm should accommodate most machines. Activity space requirements are similar to those for a washing machine. With a 12–14 place setting capacity, the maximum discharge rate will be around 0.25 l/s.

Above ground drainage

The requirements for drainage systems are generally split between above and below ground drainage systems. In this book, above ground drainage systems are included in this chapter, and below ground systems, most of which occur outside the building footprint, are included in chapter 10, *External works*.

Waste water is normally divided into two categories, reflecting the treatment that is necessary before re-use or discharge:

- greywater is water not containing faecal matter or urine, and
- blackwater is water containing faecal matter or urine.

British Standards Institution
BS EN 35: 2000 *Specification for bidets*
BS EN 12056-2: 2000 *Gravity drainage systems inside buildings. Sanitary pipework, layout and calculation*
BS EN 1116: 2004 *Kitchen furniture. Co-ordinating sizes for kitchen furniture and kitchen appliances*

Rainwater collection and disposal is covered in chapter 8 *Roofs* so it is not dealt with here. However, some rainwater pipes may be accommodated internally, and will require provision for checking for both capacity and leaks since the consequences of overflow internally are much more serious than for pipes sited externally.

Characteristics of systems

Most drainage systems rely on gravity to maintain the water flows within them. However, some fully or partially pumped systems are also available. Examples of gravity discharged appliances include: WCs, basins, sinks, baths and showers. Examples of pumped discharged appliances include: washing machines, dishwashers, small-bore pumped WCs (often known as macerating WCs), basement appliance drainage pumps and sump pumps. Some of these appliances are included in BS EN 12050 and BS EN 12056-4.

Most drainage piping is now installed internally rather than externally, and there is currently greater provision for anti-siphonage devices than in the past, but there are still many different drainage layouts which are possible, including vertical and horizontal main drains within the building. Where street level drainage is at a higher level than the lowest part of the drainage system in a building, a storage system will have to be constructed from which the sewage is pumped to the main sewer.

To prevent odours escaping from the drainage system, water-sealed traps are used at each appliance. Large fluctuations in pressure in the pipework system can, under certain conditions, destroy the water seals; for this reason, the positive and negative pressures in the system should be contained within the limit of 38 mm water gauge in order that at least 25 mm of water seal is retained within the traps.

There is an alternative to water-sealed traps, a self-sealing waste valve made from a polypropylene casing surrounding an elastomeric flat tube, which may be used in a variety of situations (BRE's *Information Paper IP 5/05*).

Venting of stacks is usually by an open pipe the top of which is protected from nesting birds taken at least 900 mm above any opening into the building which is within 3 m of it. For internal stacks this means penetrating the roof covering. An alternative is to use an air admittance valve (AAV) fitted to terminate the stack above the flood level of the highest appliance. The AAV opens to allow air into the system when pressure in the system becomes negative (Figure 9.15).

The wastewater load produced by the various appliances determines the required airflow needed to ventilate the sanitary pipework system. There are various ways of determining these values, but the currently accepted method for the UK is the use of discharge units (DUs) as specified in the European code of practice for above ground drainage systems BS EN 12056.

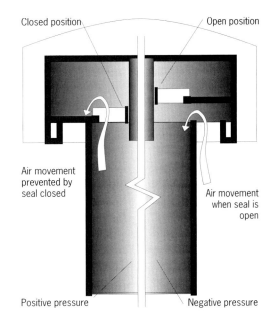

Closed position　　　　　　　　Open position

Air movement prevented by seal closed

Air movement when seal is open

Positive pressure　　　　　　　Negative pressure

Figure 9.15 An air admittance valve

British Standards Institution
BS EN 12050: 2001 *Wastewater lifting plants for buildings and sites. Principles of construction and testing*
BS EN 12056-4: 2000 *Gravity drainage systems inside buildings. Wastewater lifting plants. Layout and calculation*

White MK, Griggs JC & Sutcliffe S. *Self-sealing waste valves for domestic use: an assessment.* Information Paper IP 5/05. Bracknell, IHS BRE Press, 1999

The appropriate diameter and gradient of a drain, a discharge pipe or a stack are determined using the sum of all the discharge units and flows that each carries. Appliances generally have discharge units related to their peak flowrates (BS EN 12056-2). From this standard, the system that is applicable to the UK is System III.

The use of floor gullies in the UK is not widespread, and they tend to be used less frequently and in different manners from those in continental Europe. Waterless urinals have no DU.

The total wastewater load is calculated and compared with the maximum loads determined for the different sizes of drainage stack and associated ventilation systems. Minimum requirements are set out in the relevant accompanying documents for the Building Regulations for each part of the UK:

- **England and Wales:** AD H,
- **Scotland:** Section 3 of the Technical Handbooks, Domestic and Non-Domestic,
- **Northern Ireland:** Technical Booklet N.

Materials and durability

Materials currently in common use for pipework in buildings are: cast iron, copper, galvanised mild steel, plastics, but lead can still be found in older properties. Less frequently used materials include: stainless steel, borosilicate glass.

In some pipework systems, more than one material may be used, but using dissimilar metals in the same system should take account of the effects of electrolytic corrosion. The list below shows the effect of combining metals: an earlier-named metal will be attacked by a later-named one; and the closer the metals are in the list, the smaller the severity of attack:

- zinc,
- iron,
- lead,
- brass,
- copper, and
- stainless steel.

Lead, copper and iron systems have proved to be durable, often with service lives in excess of 50 years if properly maintained. These systems are frequently replaced because of the high cost of adaptation rather than because of material failure. Neoprene, or a similar non-metallic component, should be used for making connections in mixed metal systems.

Relevant British and European Standards include those mentioned above, and BS EN 274 and BS EN 12763.

Pipes for soil and waste systems are manufactured in a number of plastics materials including PVC-U, MUPVC, high density polyethylene (HDPE), polypropylene (PP) and ABS. All are light in weight, easy to handle and highly resistant to corrosion. Attention should be paid, though, to expansion in plastics drainage systems as the coefficients of expansion of these materials are much higher than for metals. Horizontal runs of plastics pipe require more support than metal pipes. Plastics material exposed to direct sunlight may require protection to resist ultraviolet degradation. It is advisable to seek guidance from manufacturers of any materials other than PVC-U or MUPVC. Also, some solvents and organic compounds damage plastics materials.

Embrittlement of plastics piping systems may be hastened by excessive UV exposure which may cause failure within 20 years, or make adaptation of the system difficult after perhaps 10 years. In particular, polypropylene pipes are not suitable for use externally, or even internally when exposed to sunlight since photo-oxidation results in the pipe becoming brittle and the pipe inevitably fails. PVC-U pipes are not affected in the same way.

ABS, high and low-density polyethylene, polypropylene and MUPVC are sometimes suitable for use in high temperature conditions but manufacturers should be consulted on the choice of appropriate materials and grades.

British Standards Institution. BS EN 12056-2: 2000 *Gravity drainage systems inside buildings. Sanitary pipework, layout and calculation*

British Standards Institution
BS EN 274: 2002 *Waste fittings for sanitary appliances.* (Parts 1–3)
BS EN 12763: 2000 *Fibre-cement pipes and fittings for discharge systems for buildings. Dimensions and technical terms of delivery*

PVC-U is the most commonly used plastics material for domestic-scale discharge pipes but should not be used in applications where large volumes of water above 80 °C are discharged; for instance, in dwellings where wash-boil programme washing machines or water-heating appliances without thermostatic control are used.

The durability of cast iron pipework to BS 416-1 depends on its protective coating; each item of these systems should be examined carefully before installation to ensure the protective coating is undamaged. The introduction of flexible joints has considerable advantages in labour costs over the previous hemp and lead joints. Cast iron pipes may be fixed satisfactorily by means of the lugs on the pipe sockets; by cast iron, malleable iron or steel holderbats for building in, nailing or screwing to the structure; or by purpose-made straps or hangers.

Copper tubes to BS 2871-1 are available in long lengths, which reduces the number of joints. Copper plumbing assemblies are suitable for MMC or simple prefabrication techniques. Jointing is normally by compression or capillary fittings. Copper U bend traps can be problematical, particularly when household bleach from sinks has been passed through them. Copper tubing which cannot be bent or welded is also available to other specifications, and therefore it may prove unsuitable for sanitary plumbing installations.

Where the building is sited on contaminated land, the pipe materials will require particular care in specification. For further information, see BRE's Report *Performance of building materials in contaminated land.*

Provision for maintenance

Suitable access should be provided to all pipework so that it can be tested and maintained effectively during its lifetime. Access covers, plugs and caps

British Standards Institution
BS 416-1: 1990 Discharge and ventilating pipes and fittings, sand-cast or spun in cast iron. Specification for spigot and socket systems
BS 2871-1: 1971 *Specification for copper and copper alloys. Tubes. Copper tubes for water, gas and sanitation*
Paul V. *Performance of building materials in contaminated land.* BR 255. Bracknell, IHS BRE Press, 1994

should be sited so as to facilitate the insertion of testing apparatus and the use of equipment for cleaning parts of the installation and for removing blockages. Their installation should not be impeded by elements of the structure or by other services.

Access points should not be located where their use can lead to nuisance or danger if spillage occurs. It is preferable that they are above the spill-over level of the pipework likely to be affected by a blockage or are extended to suitable positions at the face of the duct or casing, or at floor level. On discharge stacks, access should be provided to the stack at intervals of not less than three storeys.

All access positions and inspection covers should be readily accessible, and have adequate room around them for using cleaning rods. Cleaning access points additional to those at pipe junctions should also be made in long waste branches. Consideration should also be given, during the design of a pipework system, to access for replacing pipe lengths and fittings which may become damaged.

Fire protection

> **Key Point**
> The safety of people in case of fire is an over-riding consideration for the designer.

Nearly 1000 people die every year as a result of fires in buildings. About 60% of the deaths are due to inhaling smoke and toxic products, and it is necessary, therefore, that particular attention is paid to reducing the spread of smoke in fire as well as to providing means of escape. Some aspects of smoke control involving pressurisation and depressurisation require powered systems, and so come within the definition of building services.

Passive fire protection measures are those features of the fabric (eg cavity barriers) that are incorporated into building design to ensure an acceptable level of safety. These features have been dealt with in chapters 5–8 on floors, walls and roofs, respectively. Measures which are brought into action in the event of a fire (such as fire and smoke detectors, sprinklers and smoke extraction systems) are referred to as active fire protection and are discussed in this chapter. It is widely thought that active and passive systems of fire protection in some way compete. In practice, each system should complement other fire protection systems, working together to enhance the safety of a whole building and its occupants. It is generally thought that fire protection is a simple procedure which does not require the continued application of specialist engineering expertise. This is not so. For an overall examination of the main issues relating to fire protection see BRE's report *Fire safety in buildings*.

Means of escape

Provision is made for means of escape in all kinds and sizes of buildings in the relevant accompanying documents for the Building Regulations for each part of the UK:
- **England and Wales:** AD B,
- **Scotland:** Section 2 of the Technical Handbooks, Domestic and Non-Domestic,
- **Northern Ireland:** Technical Booklet E.

The main means of protection include:
- automatic fire detection and alarm,
- automatic fire-extinguishing systems (eg sprinklers),
- fire-fighting devices,
- structural fire barriers (eg between and within dwellings and common escape routes),
- self-closing fire doors,
- alternative exits and escape routes,
- restriction of travel distance,
- provision of smoke control.

Only some of the above come within the definition of building services and are dealt with in this chapter. The remainder are primarily related to layout, height above or below ground, compartmentation and final exits to safety. Fire detection is covered in the next section.

Malhotra HL. *Fire safety in buildings.* BR 96. Bracknell, IHS BRE Press, 1987

Dry risers, hosereels and handheld extinguishers

Dry risers are for use by the fire brigade after its arrival on site. The risers are connected to the pumps on fire appliances, and serve outlets positioned on each floor. Hosereels and hand-held extinguishers are for use by building occupants in the early stages of a fire when the outbreak is small and there is little personal risk.

Sprinklers

Sprinkler systems are generally recognised as one of the first means of dealing with fires in buildings like hostels and commercial premises. There are two main requirements of sprinkler systems:
- to detect and control a fire by preventing its vertical and horizontal spread, enabling fire-fighters to deal with it as a small outbreak,
- to cool surfaces at the perimeter of a fire to prevent or delay their ignition.

The main function of sprinklers is to contain fire and to reduce the frequency of large fires. They are usually effective, and have played a major role in the protection of property against fire. However, it has increasingly been recognised that they also may have a role in life safety.

Sprinkler systems may be wet or, rarely, dry. That is to say, either the systems are permanently charged, ready at all times to discharge water into the seat of the fire, or they are dry. A dry system can be used where the water might freeze if allowed to stand in the pipes. The effectiveness of sprinklers in this case relies on an extremely rapid filling of the system once the alarm is triggered. A relatively recent development is the early suppression fast response (ESFR) sprinkler. Water leaking from sprinkler heads is comparatively rare, though the risk might need to be assessed for building contents of high value.

One of the main potential problems with sprinklers when they operate is that the supply system must have sufficient capacity to allow full functioning of all the heads that open. This provision is covered in the basic design of the system. The performance of sprinkler systems is covered in BS 5306-2.

Other protection features and systems

Where the building is non-domestic, systems which may need to be considered include:

- pressurisation of spaces,
- depressurisation of spaces,
- gas extinguishing systems,
- smoke and heat exhaust ventilators (SHEVs).

Pressure differential systems are designed to protect against smoke leakage through small gaps in passive protection of an escape route (Figure 9.16). Air is pumped into an escape route, usually a stairwell or corridor to maintain pressure sufficient to oppose the buoyancy of hot gases on the fire floor, the stack effect due to the building's own warmth, and wind pressure effects. BS 5588-4 recommends a pressure level of between 50 Pa and (approximately) 100 Pa as represented by the maximum 100 N force required for a person to open a door.

In the process of depressurisation, gases are removed from fire compartments to maintain the desired pressure drops across, and air speeds at, escape doors. The method is not often used in the UK, though it is included in BS 5588-4. Fire-rated fans and extract ducts are usually essential. Depressurisation works best when fire compartments are relatively well sealed.

Gas-extinguishing systems are normally used where people are not present, and where the stored contents are valuable, such as computer rooms or strongrooms. Halon has been used extensively in the past to fight fires in these enclosed situations, but the use of halons is being phased out following international agreement. Carbon dioxide extinguishing systems also exist; the principle, basically, is to starve the fire of oxygen.

The principles of smoke and heat exhaust ventilation (SHEV) are simple. Hot buoyant gases from a fire rise to form a layer at sufficient height that allows a layer of cooler, clear air to remain at lower levels for sufficient time for occupants to be safely evacuated. This clear lower layer can be maintained by exhausting smoke from the high level either by natural or mechanical means (Figure 9.17). European Standards for equipment for use in SHEV systems have also been developed (BS EN 12101).

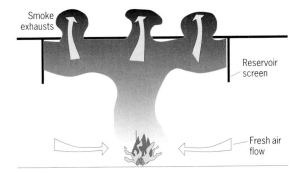

Figure 9.17 Smoke ventilation

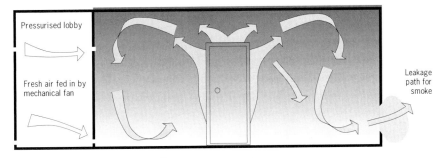

Figure 9.16 Pressurisation to force hot gases away from an escape route

British Standards Institution
BS 5306-2: 1990 *Fire extinguishing installations and equipment on premises. Specification for sprinkler systems*
BS 5588-4: 1998 *Fire precautions in the design, construction and use of buildings. Code of practice for smoke control in protected escape routes using pressure differentials*

British Standards Institution. BS 12101: *Smoke and heat control systems*
    Part 2: 2003 *Specification for natural smoke and heat exhaust ventilators*
    Part 3: 2002 *Specification for powered smoke and heat exhaust ventilators*

Utilities and service supplies

Gas and oil

The main legislation relating to the safety of gas installations is the Gas Safety (Installation and Use) Regulations 1998. Many of the Approved Documents to the Building Regulations (England & Wales; for example B, E, F, G and J) are in part concerned with some of the wider aspects of gas safety. For those areas outside the range of mains distribution, there is the possibility of using propane or butane, stored either in small portable pressure cylinders, or distributed by tanker vehicles for transfer to large fixed storage tanks.

Gas can be supplied at different pressures according to location and type of buildings served. It may be necessary to provide space for pressure-reducing plant near the buildings to be served.

Propane cylinders (propane is explosive and toxic) are normally stored outside the building, whereas butane (which is explosive but not toxic), is normally used in portable appliances and can be stored indoors. Large domestic heating installations are fuelled from a fixed tank to which deliveries are made by road tanker.

Gas pipes must not be run in wall cavities, and where they pass through cavity walls they must be sleeved (BS 6891). They must have electrical equipotential bonding.

Tanks for the storage of oil for central heating use are normally sited outside the building alongside a hard-standing area accessible to road tankers. Supply pipes to the appliance are buried.

Refuse

Waste collection authorities in the UK were required to produce recycling plans under the Environmental Protection Act 1990. Most of the authorities in England and Wales submitted to the Government the recycling options they planned to develop; nearly all of these have indicated they would introduce 'bring' schemes, and over one-third of them that they would introduce kerbside collection recycling schemes.

That proportion is expected to increase significantly in the near future. A change in the means of storage of waste at the household level may be required in order to implement efficient recycling.

Many local authorities use plastics wheeled bins, examples of which include:
- 120 and 240 litre plastics wheeled bins for both mixed waste (no segregation of recyclables) and segregated wastes (Figure 9.18),
- 40- and 50-litre plastics boxes (with or without lids) for the storage and collection of dry recyclables such as paper, cans and plastics bottles,
- subdivided 120- and 240-litre wheeled bins with compartments used to store different types of wastes, for example recyclables and residual wastes (non-recyclables),
- separate plastics bags for residual wastes (usually black bags) and recyclables (usually green, blue, or clear bags for different categories of recyclables).

AD H of the Building Regulations (England & Wales) gives guidance about the minimum volume

Figure 9.18 A large bin designed for mechanical handling

HMSO. *Gas Safety (Installation and Use) Regulations 1998.* Statutory Instrument 1998 No. 2451. London, The Stationery Office

British Standards Institution. BS 6891: 2005 *Installation of low pressure gas pipework of up to 35 mm (R1 1/4) in domestic premises (2nd family gas). Specification*
HMSO. *Environmental Protection Act 1990* (c. 43). London, The Stationery Office

of a waste container, 0.12 m³, based on the waste produced by a typical household of 0.09 m³ per dwelling and assumes weekly collections.

EcoHomes criteria (see chapter 1) call for a set of four containers for household waste, space for them, and access for removal to a collection point.

- At least four containers should be provided which are distinctly identified as being for different purposes (eg different colours) and have a collective capacity of at least 240 litres (0.24 m³) per household, ie an average of 60 litres per bin.
- The bins should be in a suitable hard-standing area, within 3 m of an external door of a single-family house and within 10 m of an external door of other dwellings.
- To accommodate the possibility of additional bins being required in the future, the bin storage area must normally be at least 2 m² without interfering with pedestrian or vehicle access.
- Bins should be protected from wind on at least two sides by a wall, fence or suitable hedge at least as high as the tallest bin.
- Access for a car or refuse collection vehicle to within 20 m of the location of the bins.

Electricity

The new AD P on *Electrical safety* came into force in England and Wales on 1 January 2005. Essential features of the requirements are that the installation shall be carried out by a competent and registered person and that a certificate to confirm conformance be issued. A copy of this certificate may also need to be transmitted to the building control body.

Single phase 230–240 V, 60, 80 or 100 A, 50 Hz, earthed neutral installation to BS 7671 is the usual type of supply provided in individual dwellings. These installations all come within the definition of low voltage installations (extra-low voltage is up to 50 V AC or 120 V DC, low voltage up to 1000 V AC or 1500 V DC between conductors). Electricity meters should be accessible from the outsides of houses in weatherproof lockable cupboards.

For certain purposes of BS 7671, system types are referred to as TN, TN-C, TN-S, TN-C-S, TT, and IT, depending on the relationship of the source and of exposed conductive parts of the installation to earth; before any work on a system begins, the type must

be correctly identified. Further information is given in the Standard BS 7671.

It is the maximum demand for electrical energy, the connected load, that determines the minimum overall capacity of the system, and the necessary sizes of all distribution systems and other appliances. Whether or not the system should take into account the non-simultaneous use of appliances (the diversity factor) depends on the importance of having a non-interrupted supply, clearly crucial in some building types such as hospitals.

The installation should not prejudice the supply in any way. In other words, no failure within the building curtilage should cause failure to the mains, and within the installation it is usual for its various parts to be separately protected so that complete breakdown does not occur. Many existing domestic installations are protected by fuses, though an increasing number by residual current devices (RCDs) miniature circuit breakers (MCBs) or residual current breakers with overcurrent protection (RCBOs).

The consumer, or distribution, unit (the box that contains the terminals for the individual circuits and the fuses or circuit breakers that protect them) should provide for at least one ring circuit (in a modern system) per 100 m² of floor area. In an average two-storey dwelling this usually means:

- two separate power circuits, each serving half of both floors or the whole of one floor,
- two separate lighting circuits, again each serving half of both floors or the whole of one floor,
- electric cooker, immersion water heater, and any electric space heating, normally have dedicated circuits.

An ample number of socket outlets should be provided (Table 9.2). These figures should be treated as absolute minima, and can, bearing in mind the proliferation of electrical appliances in modern homes, with advantage be doubled. To minimise the need for adaptors, twin or triple socket outlets are

British Standards Institution. BS 7671: 2001 *Requirements for electrical installations. IEE Wiring Regulations.* 16th edition

| Table 9.2 Minimum provision of sockets for different rooms in a dwelling ||
Room	No. of sockets
Working area of kitchen	4
Utility room	2
Dining area	2
Living area	4
First (or only) double bedroom	3
Other double bedrooms	2
Single bedrooms	2
Hall and landing	1
Store, workshop or garage	1

recommended in locations where more than one appliance might be connected.

Undoubtedly one of the main risks from electrical supplies is electric shock from faulty installations or appliances. Protective devices designed to prevent or minimise shock should function when any leakage to earth is detected; current is cut almost instantaneously, usually sufficiently quickly to avoid death, although not necessarily injury. Provisions against electrocution are covered in some detail in BS 7671.

Shock (or fire) risk could result from drilling and putting nails or screws through cables. This happens sometimes because electrical contractors try to economise by running cables diagonally instead of in vertical or horizontal runs that can be more readily traced from the positions of electrical fittings and appliances. This can be prevented by strict specification on the part of the designer. Awareness by electricians, and by other contractors in buildings, of the effects of their work methods and competence, and the work of the other skilled trades on site, is probably a contributory factor in reducing the risks of electrocution.

All electric cables give off heat in use, depending on the load they carry. The heat emitted by any cable operating within its design loading is normally safely dissipated. Overheating due to electrical load alone should not occur if installations comply with BS 7671. Sometimes, however, the normal heat

---

British Standards Institution. BS 7671: 2001
Requirements for electrical installations. IEE Wiring Regulations. 16th edition

dissipation from a cable is impeded by thermal insulation or because the temperature of its environment is raised by other heat sources such as proximity to boilers. Consequently, overheating of the cable may damage its insulation. Cables in power circuits which may be loaded to full capacity are more at risk than those in lighting circuits. If these conditions are not taken into account, there can be a risk of short circuiting or fire. All cables should be de-rated where the ambient temperature exceeds 30 °C. Again, this needs the attention of the specifier.

Electric lighting

> **Key Point**
> Balance the need for lighting, providing a well-lit space for occupants and their tasks against energy and maintenance costs as well as capital cost.

The specification of electric lighting involves reviewing the need for lighting in each space in the building against the cost of lighting (see Box 9.5).

Regulations
A number of regulations cover lighting in buildings but, in general, they do not stipulate a particular level of lighting.
The Workplace Regulations 1992 require:
● neither insufficient nor excessive light,
● lack of glare,
● well lit stairs,
● extra local lighting where needed,
● clear and safe positions for light switches,
● natural lighting (see chapter 5),
● emergency lighting where lighting failure would result in substantial increase in risk.

Guidance on safe lighting in workplaces is given in the HSE publication *Lighting at work*. This guidance is not compulsory but is a way of showing compliance with the law.

In workplaces with computer screens, the Display Screen Equipment Regulations usually apply. They require that lighting be adequate and appropriate, and cause no disturbing glare or reflections on the screens.

Box 9.5 Reviewing the need for lighting against cost

The needs may be:

Task needs
- Are particular illuminances required for the tasks to be performed?
- Do the tasks involve identification of colours, requiring good colour rendering of lighting?
- Is glare, for example bright lighting in the field of view, or reflections in computer screens, likely to be an issue for the types of task that are carried out?

Occupant needs
- Will the lighting be flexible enough to meet the different requirements of different people, or the same person doing different tasks?
- Is emergency lighting provided so people can escape if necessary?
- Does the lighting meet special visual needs, eg of the elderly or partially sighted?

Space needs
- Is the distribution of light in the space comfortable?
- Does the lighting fit in with the architecture and function of the space, for example by highlighting areas of importance or visual interest?
- Does the lighting complement the daylight in the space, with an appropriate colour appearance and distribution?
- Is the lighting safe for hazardous areas?

Against the needs is balanced the cost of lighting, eg:

Capital costs
- Lamps and luminaires.
- Lighting controls.
- Installation and wiring; a flexible lighting control system can sometimes reduce the cost of wiring.

Maintenance costs
- Replacement of lamps. Some lamps have much longer life than others. This can have environmental implications, especially if lamps contain mercury or other chemicals; disposal cost as well as new lamp cost is important.
- Cleaning of luminaires, the frequency of which depends on type of fitting.
- Inaccessible lighting locations will drastically increase maintenance costs.

Energy costs
- These are usually significantly greater than capital costs over system life.
- They depend on lamp efficacy, luminaire efficiency, effective lighting controls.

Salary-related costs
- A bad lighting system may have an impact on productivity, either directly through poor task performance or indirectly through lowered morale and increased stress.

The 1985 Housing Act stipulates that a dwelling in England and Wales should have adequate provision for lighting. Guidance on the application of this requirement is contained in the DCLG Housing Fitness Standard. However, this guidance is advisory; local housing authorities are asked to:

'have regard to it when applying the standard but must form their opinion in the light of all relevant circumstances'.

Health and Safety Executive (HSE)
The Workplace (Health, Safety and Welfare) Regulations 1992 . Statutory Instrument 1992 No.3004. London, The Stationery Office, 1992
Lighting at work. HSG 38. London, The Stationery Office, 64 pp, 1998
HMSO. *Health and Safety (Display Screen Equipment) Regulations 1992. Statutory Instrument 1992 No. 2792.* London, The Stationery Office, 1992

Section 8 of the Annex of the Housing Fitness Standard deals with lighting, and recommends at least one 'lighting outlet' per room providing enough light to enable normal domestic activities to be carried out.

There are proposals to replace the Housing Fitness Standard by a Housing Health and Safety Rating System.

Since the 1995 edition of the England & Wales Building Regulations, AD L has included requirements for energy-efficient lighting. In the 2006 edition, these requirements have been revised and extended. Four ADs give guidance on ways of complying with the requirements.

The Building Regulations cover the following:
- new dwellings and those where there has been a material change of use (eg a hotel or office becoming a dwelling),
- buildings other than dwellings, both new buildings and in replacement work, where the

new lighting covers more than 100 m² of floor area.

Further details of the recommendations in the ADs are given below. Similar regulations, although the detailed recommendations are slightly different, apply in:
- **Scotland:** Section 6 of the Technical Handbooks, Domestic and Non-Domestic and
- **Northern Ireland:** Technical Booklet F.

Standards
For workplaces, guidance on lighting provision is given in BS EN 12464-1. This standard is not mandatory but is normative; buildings constructed under public procurement procedures should follow its recommendations. In the UK, the CIBSE *Code for interior lighting* is widely used; its recommendations are intended to follow those in BS EN 12464-1.

CIBSE also publishes a range of lighting guides for different types of buildings; these give much useful additional information. Other documents give helpful guidance for offices, hospitals and schools (see chapter 13, *References and further reading*).

Illuminance recommendations
BS EN 12464-1 and the CIBSE *Code for interior lighting* give recommended illuminances for a variety of tasks in a wide range of building types. Both documents, however, give provision for the recommended illuminances to be varied, for example when:
- accurate visual work is critical, or errors are costly to rectify,
- the task itself is unusually easy or difficult to see (maybe because it has small details or low contrast),
- the task is undertaken for an unusually short or long time,
- people doing the work have poor vision.

British Standards Institution. BS EN 12464-1: 2002 *Light and lighting. Lighting of work places. Indoor work places*
Chartered Institution of Building Services Engineers (CIBSE). *Code for interior lighting.* London, CIBSE, 2004

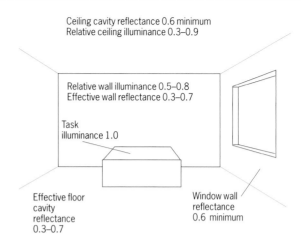

Figure 9.19 Recommended illuminance ratios. Modified from the CIBSE *Code for interior lighting*

For a surface to appear bright, not only does light have to fall on it, but it also has to have a reasonably high reflectance. Thus, a space will appear bright if the walls and ceiling have light-coloured finishes and some direct lighting, eg from wall washers or uplights. Consequently, the CIBSE Code recommends ranges of surface reflectance and illuminance ratios. It suggests walls should reflect between 30% and 70% of the light falling on them and a ceiling should have a minimum reflectance of 60%. It recommends a wall illuminance of 0.5–0.8 times the task illuminance, and a ceiling illuminance of between 0.3 and 0.9 times the task illuminance (Figure 9.19). With lower reflectances and illuminance ratios than these, the space will look gloomy, even if the horizontal illuminance is high.

Lamps and control gear
A range of lamp types is available, including tungsten, tungsten halogen, fluorescent and various types of discharge lamps. Lamp output is given in lumens; if lamps have a high lumen output fewer of them may be needed to light an interior, but the light may be more difficult to control, and glare and non-uniformity could result.

The power a lamp consumes (Watts) depends on the type of lamp and also on the ballast or control gear. Fluorescent and discharge lamps require a starter and ballast to control the current through the lamp. The least efficient ballasts are now banned by the Energy Efficiency (Ballasts for Fluorescent Lighting) Regulations 2001. The more efficient electronic ballasts operate the lamp at high frequency, reducing lamp power consumption. Typically, the total circuit power of a high frequency circuit is around 25% less than a lamp with an electromagnetic ballast giving a similar light output. High frequency electronic ballasts are also lighter in weight than their electromagnetic equivalents, are silent in operation and automatically switch the lamp off at the end of its life, eliminating flashing. They may, however, have a shorter life themselves. The high frequency operation eliminates flicker and has been shown to reduce the reported incidence of headaches and eyestrain in some individuals (Wilkins et al 1989).

Tungsten lamps do not require a ballast. However, low voltage tungsten halogen lamps will need a transformer.

Different lamps of similar lumen output may have different power consumptions. For example, a 15 W compact fluorescent lamp may have similar output to a 60 W tungsten lamp. The energy efficiency of a lamp is normally given by its luminous efficacy, defined as the lamp output (lumens) divided by the power in Watts. In addition, the circuit luminous efficacy includes the power consumed by the control gear.

AD L gives recommendations in terms of luminous efficacy. For lighting in dwellings, a way to comply would be to provide some lighting fittings or luminaires that only take lamps with a circuit luminous efficacy greater than 40 lm/W (eg fluorescent tubes and compact fluorescent lamps; Figure 9.20). ADs L1A and L1B give guidance on how many of these fittings might reasonably be installed in a dwelling, based on its floor area.

For general lighting in office, industrial and storage spaces in buildings other than dwellings, the

DCLG. *The Energy Efficiency (Ballasts for fluorescent lighting) Regulations 2001*. Statutory Instrument 2001 No. 3316. London, The Stationery Office, 2001

Wilkins AJ, Nimmo-Smith I, Slater AI & Bedocs L. Fluorescent lighting, headaches and eyestrain. *Lighting Research & Technology* 1989: **21**(1): 11–18

Figure 9.20 A pendant fitting that can only take compact fluorescent lamps. Instead of the traditional bayonet fitting it has a pin based lampholder that contains the ballast and starter (Courtesy of Lampholder 2000)

AD recommendations include the efficiency of the luminaire or light fitting. These are described in the next section on *Luminaires*. For other spaces in non-domestic buildings, lighting with an average lamp and ballast efficacy of not less than 50 lumens per circuit watt would comply.

Display lighting in non-dwellings is also covered if it is fixed to the building. A way of complying with AD L (or Section 6 in Scotland and Booklet F in Northern Ireland) would be to show that the installed display lighting has an initial efficacy of not less than 15 lumens per circuit watt. This includes tungsten halogen as well as most discharge/fluorescent lamps.

Though not covered by the Building Regulations, lamp colour is also important. The spectrum of a light source will affect its colour appearance and colour rendering. Colour appearance describes whether the light from the lamps looks 'warm' or 'cool'. If illuminances are below 300 lux, 'warm' or intermediate lamps are best; interiors with 'cold' colour lamps can look gloomy at these low light levels. In daylit interiors, intermediate colour lamps will blend best with the natural light. Replacement lamps should match the existing ones in colour.

Colour rendering is also important. This is the ability of a light source to show surface colours as they should be. Lamps with poor colour rendering will distort some colours, and make a room look less bright and colourful. Daylight has colour rendering index (CRI) of around 100, while a single colour light source like low pressure sodium has a CRI of 0. For most interior lighting, a CRI of 80 or more is preferred. Most triphosphor lamps have a CRI in this range. Most older halophosphate types have lower CRIs and will give poorer colour rendering; they are also slightly less efficient than the triphosphor tubes.

Luminaires

The efficiency of a luminaire is quantified by the Light Output Ratio (LOR) (Figure 9.21). This is the ratio of the light emitted by the luminaire to the light output of the lamps in it. In practice, the LORs of different ranges of luminaires can vary considerably, even for luminaires that look similar. The LOR depends on the quality of the materials used as well as the basic design of the luminaire. Most luminaires for general office lighting would be expected to have an LOR of 0.6 or more.

The 2006 Building Regulations (England & Wales) AD L2 includes a consideration of the luminaire for office, storage and industrial spaces. A way to comply would be to provide lighting with a luminaire efficacy averaged over the whole building of not less than 45 luminaire-lumens/circuit watt. The luminaire efficacy depends on the light output ratio of the luminaire, the luminous efficacy of the lamp and the efficiency of the control gear.

When choosing a luminaire for a regular array, the manufacturer's recommended Spacing to Height Ratio (SHR) should be noted. This is the ratio of the maximum spacing of the luminaires, divided by their height above the working plane. If luminaires are spaced further apart than this, there may be dark areas between fittings.

The Utilisation Factor (UF) of a luminaire is the proportion of the light emitted by the lamps that reaches a particular plane (eg the horizontal working plane) either directly or by reflection. It takes into account the properties of the room, its shape and surface reflectances, as well as the luminaire characteristics. A high utilisation factor indicates that much of the light reaches the horizontal plane, but may also mean that there is little reaching the walls or ceiling.

For luminaires in particular, the specifier should ensure that the equipment actually installed meets the specified performance standards. A substitute luminaire may look similar to a specified one, but have markedly worse performance.

The choice of luminaire also affects glare. In a conventional ceiling-mounted lighting installation, glare can occur if the luminaire gives out a lot of sideways light, which people may view directly or by reflection in a shiny surface like a computer display screen. The CIBSE *Code for interior lighting* and BS EN 12464-1 give limiting values for glare rating for a wide range of tasks. The glare rating may be calculated from the manufacturer's luminaire data.

In most workplaces with computer screens, control of glare is a requirement of the *Display Screen Equipment Regulations*; *CIBSE Lighting*

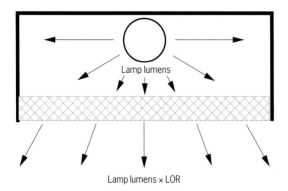

Figure 9.21 Definition of light output ratio (LOR)

Chartered Institution of Building Services Engineers (CIBSE).
Code for interior lighting. London, CIBSE, 2004
Lighting Guide: The visual environment for display screen use. LG03. London, CIBSE, 1996
British Standards Institution. BS EN 12464-1: 2002 *Light and lighting. Lighting of work places. Indoor work places*
HMSO. *Health and Safety (Display Screen Equipment) Regulations 1992. Statutory Instrument 1992 No. 2792.* London, The Stationery Office, 1992

Guide 3 gives guidance on this. In 2001, CIBSE SLL issued a major addendum to this guide. The new addendum withdrew the widely used category system for rating downlighting luminaires. For each category, there was a recommended cut-off angle (from the vertical) for light from the luminaire, to reduce glare and screen reflections. There had been problems when specifiers ignored the broader range of guidance given in CIBSE's *Lighting Guide 3* and just installed a particular category of luminaire everywhere. Lighting installations designed in this way can appear 'cave-like' or gloomy, caused by the sharp cut-off angles of the category system luminaires.

Sharp cut-off luminaires are not needed everywhere. In small spaces, such as cellular offices, the geometry between a near-vertical screen and the luminaires is such that reflections in the screen are very unlikely, however intensive the screen use. Current practice in display screen types has also altered. Most tasks involve dark letters on a white background, whereas the luminance limits in the category system were set for dark screens with light coloured letters, for which reflections are more noticeable. Also, most modern screens are treated to reduce reflections.

As well as abolishing the category system, the SLL addendum relaxes the luminance limits for spaces where treated screens, and those with dark letters on a white background, are in use; and introduces additional recommendations for minimum wall and ceiling illuminances.

Special luminaires are required for hostile and hazardous environments. These may include damp or dusty environments, extreme temperatures, or where there are corrosive chemicals or a risk of fire or explosion. The CIBSE guide *Lighting for hostile and hazardous environments* gives details.

Lighting controls
Appropriate lighting controls form an essential part of any lighting system. Controls allow the building occupants to take charge of their environment. They can also give significant energy savings, up to 30–40% or more in some types of building. Modern types of control can help the building manager rearrange the internal spaces, avoiding costly

wiring. And controls can change the lighting at preset times (scene setting) giving changes of mood in, for example, restaurants and public spaces (Figure 9.22).

A wide range of control types is available and they are listed in Box 9.6. BRE's *Digest 498* gives guidance on selecting lighting controls depending

Figure 9.22 Boots the Chemist have fitted refurbished stores with controls that ensure all display lighting and illuminated signs are switched off when the store is closed. The main lighting is also reduced via the time control

---

Box 9.6 Available control types

- New forms of manual control (including infra-red switching) and switching using the telephone or a computer.
- Occupancy sensing: especially valuable for infrequently used spaces.
- Photoelectric control: switching or dimming the lamps in response to daylight.
- Time switching, eg switching off the display lighting in a shop outside opening hours (see Figure 9.22).
- Lighting energy management systems, where the building manager can control the forms of lighting operation throughout a large building, identifying faults and monitoring usage.

Chartered Institution of Building Services Engineers (CIBSE). *Lighting for hostile and hazardous environments.* Application Guide. 1983
Littlefair P. *Selecting lighting controls.* Digest 498. Bracknell, IHS BRE Press, 2006

on the type of space, and whether it is daylit or continuously occupied.

AD L2 has a requirement for lighting controls in non-dwellings. The AD presents flexible guidance which depends on the type of building. A way to comply is to install local switching, with time switching and photoelectric switching where appropriate. Local switches can include dimmers, pull cords and infra-red controls, as well as the simple rocker switch.

Alternatively, the AD suggests that the requirement can be met by following the guidance in BRE's *Digest 498*.

The AD also gives guidance on control of display lighting. According to AD L, requirements would be met if display lighting were on separate circuits that could be switched off at times when people are not inspecting the displays.

AD L2 includes guidance on commissioning of building services systems including lighting (see also CIBSE *Commissioning Code L*). A way of complying with AD L would be to provide a report by a competent person showing that the lighting had been properly inspected and commissioned. Commissioning of lighting controls is particularly important. If the system does not perform as it should, energy consumption can be excessive. Alternatively, the system can be over-zealous in switching off and dimming lighting, resulting in occupant complaints and eventually system disconnection. Frequent switching can also reduce lamp life.

Emergency lighting
Emergency lighting may be required for two reasons. Escape lighting allows occupants to leave the building or a hazardous area in case of emergency or failure of the main lighting. Standby lighting enables work to continue in the event of an emergency power failure. An example could be in a factory where potentially dangerous equipment has to be shut down, or in a hospital operating theatre where an operation needs to be completed. The CIBSE *Code for interior lighting* explains that

standby lighting can be treated as a special form of conventional lighting, and that the level of standby lighting can range from 50–100% of the normal illuminances.

Most published guidance and legislation covers escape lighting. The Building Regulations (England & Wales) AD B (Fire safety) defines the types of building and areas within buildings that require emergency lighting. BS 5266-1 gives detailed guidance. Local fire authorities may impose additional standards.

BS 5266-1 gives guidance on:
● route marking, including exit signs and location of luminaires,
● minimum illuminances,
● speed of operation following power failure,
● glare from luminaires,
● illumination of fire equipment,
● lifts and escalators,
● special areas and high risk areas.

Designers should consider the use of low-level mounted powered wayfinding emergency lighting as this is particularly useful in smoky conditions or when used by people with poor vision. For more details see BRE's *Information Paper IP 9/97*.

Further guidance is given in CIBSE's publication *Emergency lighting*.

Aerogenerators
Increasing interest is being shown in the possibility of generating electricity from small aerogenerators or wind generators attached to individual buildings; indeed recently the Scottish Parliament has examined the possibility of providing automatic planning consent for such schemes. At present,

Chartered Institution of Building Services Engineers (CIBSE)
Code for interior lighting. London, CIBSE, 2004
Emergency lighting. Technical Memorandum TM12. 1986
British Standards Institution. BS 5266-1: 2005 *Emergency lighting. Code of practice for the emergency lighting of premises*
Webber GMB, Wright MS & Cook GK. *Emergency lighting and wayfinding for visually impaired people*. Information Paper IP 9/97. Bracknell, IHS BRE Press 1997

Chartered Institution of Building Services Engineers (CIBSE). *Commissioning Code L: Lighting*. 2003

power output is still small in relation to the average consumption of even a single dwelling, although a highly insulated dwelling may prove more worthwhile. Some aerogenerators have outputs as low as 1 kW, and are mainly used for battery charging, or specialised applications such as running heat pumps (Figure 9.23).

Wind does not blow consistently, even in the most exposed locations of the British Isles. Consequently, either the electricity generated by this method has to be used immediately, such as for space or domestic hot water heating, or there has to be a system which stores it for supply at some future time when, say, the aerogenerator is not working. This is seen as a significant drawback, at least for small-scale plant. Where the electricity is being used immediately for heating water, all that is required is that the heating elements have a receptive capacity greater than the maximum output of the aerogenerator; but electricity generated for storage requires an expensive battery system. For most of the land area of the UK, the total electricity available from each aerogenerator, if all wind energy is tapped, will be less than 1600 kWh per m^2 of rotor area per annum (see Rayment (1976) and Figure 9.24). Of course, aerogenerators at exposed coasts will generate more power.

Figure 9.23 This aerogenerator supplies only a small amount of power, a fraction of the total requirements for a dwelling

Where the generator is to be sited directly on a building, it will be necessary to check expected noise and vibration when the rotor is running at full speed.

TV, IT and communications

Telephone lines in rural areas, and in some urban areas too, will often be found strung overhead from distribution poles to terminals at eaves level on dwellings. In urban areas, such wiring is normally underground.

Provision of wired services in non-domestic buildings depends very much on the circumstances of use, and will be a matter for individual decision. For all buildings to which the public have access, and in which conferences, meetings and entertainment will be accommodated, consideration should be given to installing hearing enhancement systems for people with hearing impairment. There is further information about this in chapter 12. For domestic telephones, consideration should be given to installing permanently wired standard line jack sockets in at least the entrance hall, living room and main bedroom.

There is an increasing demand for sophisticated technologies for more efficient handling of telephone, TV, CCTV, computer systems, remote monitoring safety (low voltage) and security devices. A convenient term has been coined: the so-called 'smart home'. An introduction to smart home technology can be found in *Smart homes, a briefing guide for housing associations*.

Television rod and dish aerials now abound, mostly individual installations, but some blocks of flats have communal aerials. Their design and installation is the responsibility of specialist firms. The rapid increase in the number of different systems and wiring needs leads to an increase in the space which needs to be allocated to house them.

Rayment R. Energy from the wind. *Building Services Engineer* 1976: **44**(3)
Bromley K, Perry M, Webb G & Ross K. *Smart homes, a briefing guide for housing associations*. Bracknell, IHS BRE, 2003. Available at:
www.bre.co.uk/pdf/smarthomesbriefing.pdf

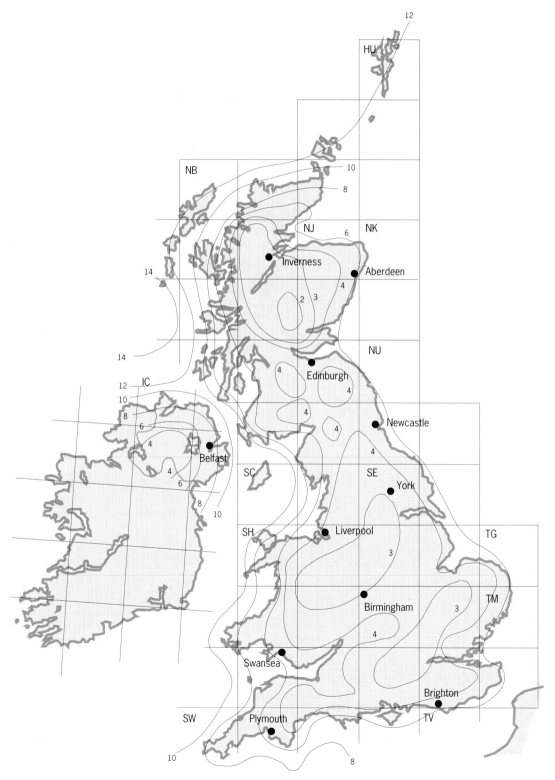

Figure 9.24 Wind energy in the UK at an effective height of 10 m over an open site in GJ/m² per annum (1 GJ = 278 kWh)

Although not specified in purely domestic buildings, platform floors may be required in highly serviced domestic-scale buildings, for example those used for offices. A platform floor, complete with its supporting deck, is required to provide the full range of performance and loading characteristics, in addition to special requirements relating to demountability and accessibility . These floors normally provide a system of loadbearing fixed or removable floor panels, supported by adjustable pedestals or jacks placed at the corners of the panels, to provide an underfloor void for the housing and distribution of services.

Electrical interference in buildings has been identified as one of the main causes of malfunction in electronic building services equipment. The EU have issued the Electromagnetic Compatibility Directive, which requires installations not to cause interference with other installations. Electrical disturbances on power cables can couple into signal and data cables to cause interference to computer networks, building management systems (BMSs), and other systems used in building services. To avoid interference, power and signal cables must be segregated, with a minimum separation distance maintained between them; if they cannot be separated, ideally they should be screened (see BS 6701 and BRE's *Digest 424*).

People using display screen equipment have the protection of the Health and Safety (Display Screen Equipment) Regulations. Employers have duties to assess workstations where display screens are used:
- to identify and reduce risks,
- to observe minimum requirements for the correct ergonomic use of the equipment,
- to offer suitable breaks in using the equipment,
- to provide suitable training and spectacles, if required.

EU. *The Electromagnetic Compatibility (EMC) Directive 89/336/EEC*
British Standards Institution. BS 6701: 2004 *Telecommunications equipment and telecommunications cabling. Specification for installation, operation and maintenance*
BRE. *Installing BMS to meet electromagnetic compatibility requirements*. Digest 424. Bracknell, IHS BRE Press, 1997

Fire detection and fire alarm

> **Key Point**
> Every occupied building must have a means of alerting people in the event of fire.

The requirements for this means of alerting people should be determined by a fire risk assessment carried out by the person responsible for the building, or may be specified by the authority responsible for enforcing fire safety legislation in the building.

Apart from small premises, where warning can be given adequately by word of mouth, hand-operated bells or similar, some form of electrical fire alarm system is generally required.

Manually operated systems with manual (break-glass) call points and sounders distributed throughout the building may be adequate where occupants do not sleep on the premises. However, automatic fire detection will normally be necessary to supplement the manual call points where:
- people may sleep on the premises or may have limited mobility,
- the level of occupancy is such that a fire may jeopardise the means of escape before the occupants can be warned,
- fire protection measures such as fire doors, smoke and heat control systems or fire-extinguishing systems have to be operated automatically.

All new dwellings are required to have a means of fire detection, consisting of at least one mains-operated smoke alarm on each storey, covering the circulation spaces. Where two or more smoke alarms are installed, they should be interconnected, so that any alarm signal is given by all of them.

Automatic fire detection may also be required by the owner or occupier of the building or their insurance company to help mitigate the risk of financial losses due to fire.

Further guidance on the need for fire detection and fire alarm systems can be found in the relevant parts of BS 5588 for the type of building, the

HMSO. *Health and Safety (Display Screen Equipment) Regulations 1992. Statutory Instrument 1992 No. 2792*. London, The Stationery Office

national building regulations, DCLG's series of *Fire safety — risk assessment* guides and the website: www.RedBookLive.com.

Where fire detection and/or fire alarm systems are required these should be designed and installed to BS 5839-1, for buildings in general, and to BS 5839-6, for dwellings.

The warning of fire may also be given by means of a voice alarm system instead of simple sounders (see BS 5839-8). The ability to give clear and informative voice messages may be particularly useful for large or complicated buildings (eg where phased evacuation procedures are required), and where the occupants are untrained and unfamiliar with the building. Consideration should also be given to providing adequate warning to people with hearing difficulties who may be alone and unable to hear the audible alarms (eg by flashing beacons or vibrating devices or pagers).

For the fire alarm system to have the maximum benefit, the fire service should be summoned as soon as possible in the event of an alarm. The fire service should be called via the 999 public emergency call system and for many buildings this is sufficient. However, an additional automatic means of transmitting the alarm signal to an alarm-receiving centre (ARC) should be considered where the complexity of the building or low level of occupancy may result in a significant delay.

Fire detectors are available that work on a number of different detection principles. The relevant parts of BS 5839 give recommendations on the selection, siting and spacing of fire detectors to give adequate fire detection performance with a minimum of false alarms. All equipment making up the system should meet the appropriate standards (see the list of regulations and standards in chapter 13) and should be certified by a third party such as The Loss Prevention Certification Board (LPCB; see www.RedBookLive.com).

Even with the best equipment, fire detection and alarm systems are only effective if correctly designed, installed, commissioned and maintained by a company with demonstrated competence and a proven track record, eg a company certified to LPS 1014.

Security systems

The choice of security system will depend on the level of security required. The appropriate British Standards are BS EN 50131-1 and PD 6662. These standards grade security into four levels based on the anticipated risk and measures intruders are likely to take to avoid detection; Grade 1 offers the lowest level of security and Grade 4 the highest.

There are many kinds of intruder detection devices available, eg:
- movement detectors based on passive infra red (PIR), microwave or less common ultrasonic Doppler technology,
- protective switches,
- acoustic detectors,
- volumetric capacitive detectors,
- pressure mats,
- beam interruption detectors,
- glass-break detectors,
- vibration sensors,
- CCTV image detection devices.

There are also detectors that comprise combinations of the above. The most popular of these being the combined PIR and microwave Doppler detector, known as 'dual detectors'. They often incorporate

---

British Standards Institution
BS 5839 *Fire detection and alarm systems for buildings*
    Part 1: 2002 *Code of practice for system design, installation, commissioning and maintenance*
    Part 6: 2004 *Code of practice for the design, installation and maintenance of fire detection and fire alarm systems in dwellings*
    Part 8: 1998 *Code of practice for the design, installation, commissioning, and maintenance of voice alarm systems*
BS EN 50131-1: 1997 *Alarm systems. Intrusion systems. General requirements*
PD 6662: 2004 *Scheme for the application of European Standards for intruder and hold-up alarm systems*

Loss Prevention Certification Board (LPCB)
Requirements for certificated fire detection and fire alarm firms. LPS 1014. Available from www.RedBookLive.com

---

British Standards Institution. BS 5588 *Fire precautions in the design, construction and use of buildings* (Various Parts)

DCLG. *Fire safety — risk assessment* (Series of titles, various dates). London, The Stationery Office

sophisticated electronic signal processing, providing good detection performance with a high immunity to unwanted alarms.

The selection of an appropriate detection device should be based on the perceived threat and most likely method of intrusion coupled with an assessment of the environment in which the detector is to be installed. For example, it would be unwise to install an acoustic-type glass-break detector in an environment subject to high levels of ambient noise when the intruder alarm system is in operation.

BS EN 50131 also prescribes levels of environmental classification enabling products to be selected for their suitability to the anticipated service environment. Four levels of environmental classification are provided, ranging from Class I (a benign, conditioned internal room) to Class IV (an outdoor location exposed to the elements).

At the time of writing, standards for the component parts of the intruder alarm system within the BS EN 50131 series are in preparation. In future, each component of the system will be labelled with the applicable security grade and environmental classification. It is recommended that where possible, products be used that have been evaluated for compliance with the criteria for these standards and approved by an independent certification body such as LPCB.

Intruder alarm systems requiring Police response will in addition need to comply with DD243 and the ACPO policy. Alarm systems meeting these requirements incorporate means to confirm an alarm condition has been caused by an intrusion and not simply a false alarm. This is usually achieved by requiring an alarm trigger from two separate detection devices of dissimilar technologies to have occurred within a defined time window before a 'Confirmed Alarm' signal is sent to an ARC. Alternatively, confirmation that an intruder is present may be by audible or visual verification.

LPCB maintains a list of approved products and installers known as the *RedBook* (see www.RedBookLive.com). The list is circulated worldwide and is available free of charge. Also, the National Inspectorate for Security (NSI) systems and the SSAIB list approved installers for intruder alarms, access control systems and CCTV systems.

Regulations and standards

Building Regulations
All construction, including external works, must be carried out in accordance with:
● Health and Safety at Work etc Act 1974, and
● The Construction (Design and Management) Regulations 1994.

These place an onus on designers to eliminate or reduce hazards where possible at design stage, and to pass on information on residual risks, so that they can be adequately managed during construction.

Some of the more important aspects of building services which are specifically covered in the Building Regulations for each part of the UK are as follows. The footnotes list other references.

Conservation of fuel and power
● **England and Wales:** AD L1 and L2,
● **Scotland:** Section 6 of the Technical Handbooks, Domestic and Non-Domestic,
● **Northern Ireland:** Technical Booklets F, L and P.

Drainage
● **England and Wales:** AD H,
● **Scotland:** Section 3 of the Technical Handbooks, Domestic and Non-Domestic,
● **Northern Ireland:** Technical Booklet N.

Ventilation
● **England and Wales:** AD F,
● **Scotland:** Section 3 of the Technical Handbooks, Domestic and Non-Domestic,
● **Northern Ireland:** Technical Booklet K.

Water
● **England and Wales:** The Water Supply (Water Quality) Regulations 2001.
● **Scotland:** The Water Supply (Water Quality) (Scotland) Regulations 2000, the Private Water Supply Regulations 2000 and the Water Byelaws 2000 (Scotland).

HSE. *The Health and Safety at Work etc. Act 1974*
DCLG. *The Construction (Design and Management) Regulations 1994.* Statutory Instrument No. 3140

- **Northern Ireland:** The Water Supply (Water Quality) (Amendment) Regulations (Northern Ireland) 2003

Also comply with The Water Supply (Water Fittings) Regulations 1999.

Gas
Gas Safety (Installation and Use) Regulations 1998.

Electricity
- **England and Wales:** AD P,
- **Scotland:** Section 4 of the Technical Handbooks, Domestic and Non-Domestic,
- **Northern Ireland:** Technical Booklet E.

Waste management
The Waste Management Licencing Regulations 1994.

Fire protection
- **England and Wales:** AD B,
- **Scotland:** Section 2 of the Technical Handbooks, Domestic and Non-Domestic,
- **Northern Ireland:** Technical Booklets E.

Standards
Some of the more important aspects of building services covered in the national standards are listed below. Other standards for reference will be found in the footnotes on the relevant page of this chapter.

Conservation of fuel and power
- BS 8207: 1985.

Drainage
- BS EN 12056-2: 1995.

The Stationery Office
Water Supply (Water Fittings) Regulations 1999. Statutory Instrument 1999 No 1148. 1999
The Water Supply (Water Quality) Regulations 2001
The Water Supply (Water Quality) (Scotland) Regulations 2001
The Water Supply (Water Quality) (Amendment) Regulations (Northern Ireland) 2003
Private Water Supply Regulations 2000
Gas Safety (Installation and Use) Regulations 1998.
The Waste Management Licencing Regulations 1994.

Scottish Water Authorities
Water Byelaws 2000 (Scotland)

Ventilation
- BS 5925: 1991.

Water
- BS 6700: 1987,
- BS EN 806-2:2005,
- Code of practice CP 342-2: 1974,
- BS EN 12056-2:1995.

Electricity
- BS 7671: 1992.

Fire protection
- BS 5306-2: 1990,
- BS 12101-6,
- BS 5839 Part 1: 2002, Part 6: 2004, Part 8: 1998.

British Standards Institution
BS 8207: 1985 *Code of practice for energy efficiency in buildings*
BS EN 12056-2: 2000 *Gravity drainage systems inside buildings. Sanitary pipework, layout and calculation*
BS 5925: 1991 *Code of practice for ventilation principles and designing for natural ventilation*
BS 6700: 1997 *Specification for design, installation, testing and maintenance of services supplying water for domestic use within buildings and their curtilages*
BS EN 806-2: 2005 *Specifications for installations inside buildings conveying water for human consumption. Design*
CP 342-2: 1974 *Code of practice for centralized hot water supply. Buildings other than individual dwellings*
BS EN 12056-2: 2000 *Gravity drainage systems inside buildings. Sanitary pipework, layout and calculation*
BS 7671: 2001 *Requirements for electrical installations. IEE Wiring Regulations. 16th edition*
BS 5306-2: 1990 *Fire extinguishing installations and equipment on premises. Specification for sprinkler systems*
BS EN 12101-6: 2005 *Smoke and heat control systems. Specification for pressure differential systems. Kits*
BS 5839 *Fire detection and fire alarm systems for buildings*
    Part 1: 2002 *Code of practice for system design, installation, commissioning and maintenance*
    Part 6: 2004 *Code of practice for the design, installation and maintenance of fire detection and fire alarm systems in dwellings*
    Part 8: 1998 *Code of practice for the design, installation, commissioning, and maintenance of voice alarm systems*

10 External works

This chapter covers freestanding and retaining walls, fencing and security, as well as external lighting, paving and landscaping. It also deals in outline with the supply of water, and other utilities, together with wastewater and land drainage.

At the same time as foundations and site layout are considered in the design process, initial consideration of other external and ground works is required. It is sometimes forgotten that drains and external works must be matched to ground conditions in the same way as foundations, but site investigations and subsequent provision of facilities are sometimes inadequate in this respect. Issues such as weak, loose or unstable soil will affect the stability of drains and pedestrian and vehicle routes, and local unstable areas can affect retaining and free-standing perimeter walls and other walls too.

All construction must be carried out in accordance with the Health and Safety at Work Act (1974) and the Construction (Design and Management) Regulations (1994). These place an onus on designers to eliminate or reduce hazards where possible at design stage, and to pass on information on residual risks, such that they can be adequately managed during construction and occupation. These apply to all construction work, including external works.

Slope stabilization and retaining walls

Types of embankments

Where building on sloping land cannot be avoided there are several techniques for stabilising the slope, including regrading, anchors, retaining walls, grouting, deliberate use of vegetative root systems and improvement of drainage. Soil stabilisation is, however, a specialist subject, and design should normally be undertaken by a suitably experienced geotechnical engineer.

The most obvious, and often the most effective, way to improve the stability of a slope is to regrade it using cut and fill. Reducing the overall height of the slope tends to be the most feasible solution where the suspect ground or prospective failure is near the surface. Conversely, adding material to the toe of a slope is most effective for alleviating deeper seated prospective failures. The choice between cut or fill is normally constrained by the need to maintain either the crest or the toe at a specific position. Where this is not the case, a combination of

Health & Safety Executive (HSE)
The Construction (Design and Management) (CDM) Regulations 1994. Available as a pdf from www.hse.gov.uk/pubns
HMSO. *Health and Safety at Work Act 1974.* London, The Stationery Office, 1974

techniques can normally be adopted so that material does not need to be imported or removed from site. Cut slopes can normally be left a little steeper than filled slopes where the unconsolidated material has a shallow natural angle of repose. Root systems of suitable planting can also be used to stabilize slopes.

Soil can be retained by:

- sheet piling,
- piles,
- gabions (stone-filled baskets),
- cribwork (stacked box-like structures containing granular soil), or
- burying reinforcement in the earth.

Grouting can help to arrest movement in slips where there are significant voids or zones of weak ground into which the grout can penetrate. Grout can also improve shear strength in clay soil, but its use is limited by the relatively low permeability of this soil. Inexpert grouting can also block natural drainage pathways and buried pipes.

Types of retaining walls

Small retaining walls are used to support slopes steeper than those at which they would be naturally stable. Typical uses are to provide flat terraces in a sloping site for gardens or access paths. Small is defined here as a height typically up to 2 m, or occasionally 3 m. Any retaining walls larger than 2 m (reinforced) or 3 m (unreinforced) should be designed by a competent engineer. Retaining walls are normally specifically designed for each separate application. They fall into the following categories:

- gravity walls,
- reinforced walls on spread foundations,
- embedded walls, which may be anchored, propped or cantilevered,
- gabions and cribs.

It is also sometimes possible to combine different features of these types.

Gravity walls

Gravity walls are the oldest type of design. They rely on dead weight to resist applied soil forces. They are made from bonded stone, brick or concrete where, normally, no tension is permitted in the wall,

or, alternatively, they may be designed to permit small tensile forces in the mortar. Walls higher than 2 m are likely to be reinforced.

Reinforced walls on spread foundations

By using spread foundations and reinforcement, the weight of some of the retained soil contributes to the stability of walls. Typically, such walls are made of reinforced concrete so the vertical stem can be relatively thin. Buttressing may also be added to either the inside, or sometimes the outside, face.

Embedded walls

Embedded walls depend mainly on mobilising passive earth pressure on the buried section of the wall. They may be constructed of contiguous or secant piles (see the section on *Piles* in chapter 4, *Foundations and basements*) or steel sheet piles. Cantilevered walls rely on the pile stiffness to limit serviceability movements. Anchored or propped walls use additional structural elements to limit movements. Props are normally temporary (eg where there are faces situated opposite each other during excavation) but anchors are normally permanent. There is more information in the *Piling Handbook*.

Gabions and cribs

Gabions are increasingly used in the landscaping of building sites, and occasionally for loadbearing purposes. Most kinds of durable stone can be used. Mesh size of the cage is normally related directly to stone infill size; durability depends on wire material and its protection from corrosion. Marine and industrial locations may need extra precautions. Shaped precast concrete units (cribs) stacked one above another and interlocked also allow pockets of earth in which plants can be encouraged to grow.

Retaining wall design

There are three main modes of failure of small retaining walls (Figure 10.1; see also BS 8002 and CIRIA's *Design of embedded retaining walls*):

Arcelor. *Piling handbook.* 8th edition. Luxembourg, Arcelor RPS. 2005

British Standards Institution. BS 8002: 1994 *Code of practice for earth retaining structures*

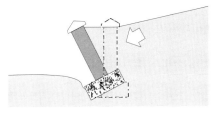

Overturning failure

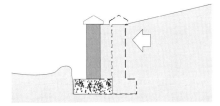

Sliding failure

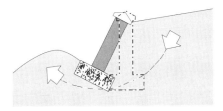

Bearing failure

Figure 10.1 Typical modes of failure in small retaining walls

- overturning about the toe,
- sliding,
- bearing failure.

The wall may also crack, usually by horizontal shear fracture in the brickwork.

The factor of safety against overturning is calculated by estimating the limiting values of the forces acting on the wall and taking moments about the most critical point, which would normally be the toe. The factor of safety, which is the ratio of the resisting moment to the overturning moment, must be at least 2.

The vertical thrust on the base of the wall should not exceed the allowable bearing pressure for the soil. This is based on ultimate or settlement bearing capacity calculations, with due allowance for inclination and eccentricity of the loading, and a

suitable factor of safety. For granular soils the factor of safety should be 2, and for cohesive soils it should be 3.

Small retaining walls are unlikely to be designed initially to remain stationary against lateral movement. This lateral movement can be limited by placing loose granular backfill between the wall and the retained soil, or by using proprietary compressible material between the wall and retained soil.

On clay soil the desiccation caused by large trees can be reversed by the construction of a wall, and associated cutting of roots and introduction of a pathway for moisture. Construction is therefore likely to be followed by swelling, and the soil may open and close adjacent to the wall over a period of years. There is a risk that backfill material will fall down the cracks when the soil shrinks away, thereby inducing a ratcheting action that will prevent the soil from re-occupying its original space. The provision of compressible material may help to prevent this.

For small walls, calculations are not normally necessary and standard wall sections can be used unless any of the following apply.

- The wall is higher than 3 m above the top of the foundations.
- There is supporting backfill on which banked-up soil, stored materials or buildings are to be placed close to the wall.
- The wall is higher than 2 m above the top of the foundations supporting backfill on which vehicles or other heavy items will stand or pass close by.
- Retaining soil where a slope adjacent to the wall is steeper than 1 in 10.
- The wall is supporting a fence of any type other than a simple guard rail.
- The wall retains very wet earth, peat or water (eg a pond).
- The wall is not constructed of bricks or blocks, or is of dry masonry construction.
- The wall is in an area of mining subsidence or other unstable ground.
- The water table lies within 0.5 m of the underside of the wall foundations.

CIRIA. *Design of embedded retaining walls.* Report C580. London, CIRIA, 2001

If any of the above apply, the advice of a chartered civil or structural engineer or similarly qualified person should be sought.

The choice of concrete mix for foundations is important. A wide range of mixes can be used if the soil conditions are generally dry. If, however, the soil is consistently wet or damp a sulfate-resisting mix is required. Similarly, if the soil is contaminated by industrial waste an appropriate mix for the conditions will be required. Where contaminants are present in the soil, degradation of the cementitious material by chemical attack may lead to the formation of leachable salts and in turn to increased porosity of the concrete, thereby exposing it to further attack; on the other hand, different chemical reactions may lead to expansion of the concrete, leading to deformation, cracking, and eventual disintegration (see BRE's Report, *Performance of building materials in contaminated land*). Precast units should be checked for suitability with the manufacturer (see also chapter 2).

Avoiding ground collapse

> **Key Point**
> To avoid ground collapse, excavation of the footings needs to be undertaken with care.

Faces higher than 1.2 m should be sloped back to a safe angle (45° is normally recommended). Exposed faces in cohesive soils should not be left exposed longer than necessary as the risk of instability increases with time. Particular care should also be taken with setting out of wall footings. With the exception of piered walls, the bases of retaining walls should be constructed centrally on their foundations. BRE's *Good Building Guide 27* provides guidance on locating wall footings and structural reinforcement of small retaining walls. For the connection between the wall and its concrete footings the reinforcing bar should have a bend to optimise anchorage (Figure 10.2).

Paul V. *Performance of building materials in contaminated land*. BR 255. Bracknell, IHS BRE Press, 1994
BRE. *Building brickwork or blockwork retaining walls*. Good Building Guide 27. Bracknell, IHS BRE Press, 1996

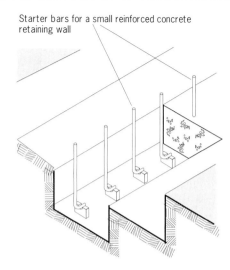

Starter bars for a small reinforced concrete retaining wall

Figure 10.2 Starter bars for a small reinforced concrete retaining wall

Resisting water penetration (Figure 10.3)

The retaining side of the wall should be backfilled with lightly compacted non-cohesive material. Clay and organic soils should be avoided. A drainage layer of free-draining material such as coarse aggregate, clean gravel or crushed stone should be incorporated next to the wall. If the backfill is fine-grained it should be separated from the drainage material by a geotextile filter. The drainage layer should discharge through weep holes in the wall. Weep holes should be at least 75 mm diameter, but sometimes can be 50 mm where the risk of blockage can be minimised by the use of a geotechnical

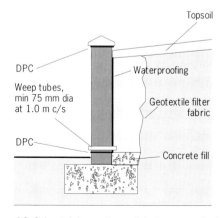

Topsoil

DPC

Weep tubes, min 75 mm dia at 1.0 m c/s

DPC

Waterproofing

Geotextile filter fabric

Concrete fill

Figure 10.3 A retaining wall needs to be properly detailed to resist water penetration

membrane, and such holes need to be spaced at intervals not exceeding 1 m horizontally. Weep holes or tubes should be incorporated in the base of the wall, either below or within the lower-level dpc. Concrete should be placed behind the wall below the weep holes.

Below ground drainage

Design aspects

Below ground drainage and sewer systems are currently specified within BS EN 752. The wastewater loadings determined for the above-ground system are used in the design of the below-ground systems. The minimum requirements are set out in the Building Regulations for each part of the UK:

- **England & Wales:** AD H.
- **Scotland:** Section 3 of the Technical Handbooks, Domestic and Non-domestic: Wastewater drainage.
- **Northern Ireland:** Technical Booklet N.

Wherever possible, drainage systems designed under the recent policy imperatives of Sustainable Drainage Systems (SUDS) should be powered by gravity alone. In some circumstances, however, gravity drainage systems may not be appropriate or would be uneconomic. The usual alternatives are vacuum or pressure systems. Situations where non-gravity systems will need to be considered include:

- high water tables,
- contaminated land,
- mountainous countryside,
- settlements separated by rivers,
- supermarkets,
- buildings above tunnels, and
- sites with negligible or back (negative) falls.

Key Point
Wherever possible, drainage systems should be powered by gravity alone.

British Standards Institution. BS EN 752: *Drain and sewer systems outside buildings* (7 Parts)

The appropriate Standards for alternative drainage systems include BS EN 1671 and BS EN 1091. In some situations where redevelopment is taking place, existing drains or sewers may need to be repaired before their re-use can be envisaged. There are various rehabilitation techniques available, including those given in BS EN 13380.

If water-conserving appliances are to be used in buildings there could be some impact on the drainage systems. BRE *Information Paper IP1/04* sets out the principles of designing drain and sewer systems for low-water-use dwellings.

Drains and private sewers

The aim should be to design a system that minimizes:

- excavation of ground,
- length of drain runs,
- number of inspection chambers or manholes,
- number of sewer connections,
- risk of blockages.

The design of the drainage system must ensure that waterborne waste is carried away efficiently with minimal risk of blockage or leakage of effluent into the ground. There are two main aspects of design:

- hydraulic aspects,
- the selection of suitable pipes and procedures for burying the pipes to ensure that they are adequately protected and do not cause subsidence of nearby foundations or are surcharged by foundations.

Drain pipes fall into two categories (Table 10.1):

- flexible pipes, which are made from materials such as plastics,

---

British Standards Institution
BS EN 1671: 1997 *Pressure sewerage systems outside buildings*
BS EN 1091: 1997 *Vacuum sewerage systems outside buildings*
BS EN 13380: 2001 *General requirements for components used for renovation and repair of drain and sewer systems outside buildings*

Lauchlan C, Griggs J & Escarameia M. *Drainage design for buildings with low water use.* Information Paper IP1/04. Bracknell, IHS BRE Press, 2004

Table 10.1 Pipe characteristics				
Material	Appearance	Nominal outside diameters (mm)	Lengths (m)	Notes
PVC-u	Orangey brown	110, 160	3, 6, 9	Not for high temperature effluent as they become brittle at low temperatures. Suitable for most domestic drainage or surface water drainage.
Concrete	Grey	150 and upwards	0.15–1.8 or more	Only use cement mortar joints where some leakage is acceptable. Flexible joints are available.
Fired clay	Terracotta	100–800	0.3–2.5 or more	Preferred diameters given in BS 65. Three strength classes: standard, extra and super.
Ductile iron	Grey-black	150 and upwards	Typically 4	Care required during handling. Useful at shallow depths under roads.
Spun iron	Black or red	75 upwards	Various	Replaced traditional cast iron pipes.

- rigid pipes, which are made from concrete, vitrified clay or spun iron.

The behaviour of the two types of pipe in the ground is significantly different: while rigid pipes have an inherent strength, flexible pipes will deform under the application of loads and require support from the surrounding backfill material to prevent excessive deformation of the pipe. A flexible system can, however, be made with rigid pipes and flexible couplings. Most modern rigid pipes are plain-ended and joined using elastomer-sealed couplings.

The hydraulic design of drains and sewers is now covered in BS EN 752-4, and is not dealt with in detail here except to record that there are a number of parameters which need to be considered, including:
- flowrates (range and peak),
- velocity of flow (including self-cleansing velocity),
- viscosity of fluid,
- pipe diameter,
- pipe roughness,
- head-loss coefficients caused by manholes and bends,
- pipe lengths.

British Standards Institution. BS EN 752-4: 1998 Drain and sewer systems outside buildings. Hydraulic design and environmental considerations

The static load on a pipe depends on the diameter of the pipe, the depth, the trench width and traffic, foundation or other applied loading. The loading is also greatly influenced by the backfilling around it and support under it. Flexible pipes depend on appropriate backfilling to function without excessive distortion. The choice is often between a high-strength pipe on a weak bedding or a weaker pipe on a stronger bedding.

Drains should be laid in straight runs between access points at an approximately constant gradient. Research has shown that formerly widely-held fears concerning the use of very steep gradients were unfounded, hence many systems which have been installed in hilly areas during recent years will have steeper drain runs and fewer back-drop manholes. The design of these manholes has also tended to change, with 45° backdrop slopes instead of vertical ones. A shallow bowl system is frequently used in such situations, employing a hemispherical bowl of around 0.6 m diameter, at or near the surface, with a central outlet which connects with the main drain using a suitable length of pipe.

Pipe sizes and gradients are dealt with in the various documents which exemplify the national building regulations. For example, in England & Wales AD H requires that wastewater drains carrying blackwater (foul water) are to be laid at not less than 1 in 40 for diameters of 75 mm or 100 mm for peak flowrates of less than 1 litre per second, and 1 in 80 where greater peak flows are anticipated. Where a minimum of 5 WCs are connected, a minimum gradient of 1 in 150 is recommended.

Trench positional and excavation limits, together with bedding and backfilling requirements are given in AD H. Bedding and backfilling specifications depend on the depth of the drain and the strength and size of the pipes (Box 10.1).

Most, but not all, pipes are now supplied with flexible joints. In addition to allowing rapid assembly, these joints allow a limited amount of movement to take place between adjacent lengths of pipe and still remain watertight. The watertightness of the drain is therefore less likely to be impaired by ground movements caused, for example, by seasonal shrinkage and swelling of clay soils, mining subsidence, or settlement of nearby foundations.

Where drains connect to manholes or structures, the connecting length of pipe should be kept as short as possible (ideally, not exceeding two pipe diameters). Where large differential movements are anticipated, for example when passing under or through walls, a short length of pipe with flexible joints at each end is used to connect to the projecting pipe; this allows the short length to flex and reduces the potential for breakage or distortion.

A drain is permitted to pass under a building provided it is surrounded by at least 100 mm of granular or other flexible filling, although rocker pipes may be required at both ends and concrete encasement to the underside of the floor slab is required if the crown is within 300 mm of the underside of the slab. Where a drain passes through a wall or foundations, either an opening of at least 50 mm all around or a short length of pipe built into the walls is required, with joints a maximum of 150 mm from the wall and connected to short lengths fitted with flexible joints. The entry will need to be packed to make it gastight (Figure 10.4).

Beneath foundations a pipe trench within 1 m of loadbearing wall foundations is required to be filled with concrete to the level of the underside of foundations.

So far as standards are concerned, drainage systems outside the building footprint are covered in

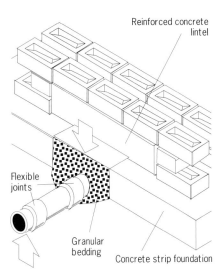

Figure 10.4 Drain detail passing through a wall

BS EN 752 which includes the following aspects of design:
- generalities and definitions,
- performance requirements,
- planning,
- hydraulic design and environmental considerations,
- rehabilitation,
- pumping installation,
- maintenance and operations,
- construction and testing of drains and sewers,
- pressure sewerage systems outside buildings.

The standards listed in the footnote below will need to be consulted. Specific plastics materials for drainage are covered in the standards listed on the next page.

Workmanship is covered in BS 8000-14.

British Standards Institution
BS EN 752: *Drain and sewer systems outside buildings:*
    Part 1: 1996 *Generalities and definitions*
    Part 2: 1997 *Performance requirements*
    Part 3: 1997 *Planning.*
    Part 6: 1998 *Pumping installations*
BS 8000-14: 1989 *Workmanship on building sites. Code of practice for below ground drainage*

Box 10.1 Bedding and backfilling for rigid and flexible pipes. Diagrams reproduced from AD H under the terms of the Click-Use Licence

Rigid pipes

It is possible to lay rigid pipes without granular bedding in a trench which has been excavated by hand to a high standard of workmanship, but most rigid pipes laid in machine-excavated trenches will need to be laid on a compacted granular bed with material conforming to BS EN 1610, Annex B, Table B15. (The permitted size of the aggregate depends on pipe diameter, for example, 5–10 mm for 100 mm diameter pipes). The trench is then overlaid with selected fill free from large stones and vegetable matter.

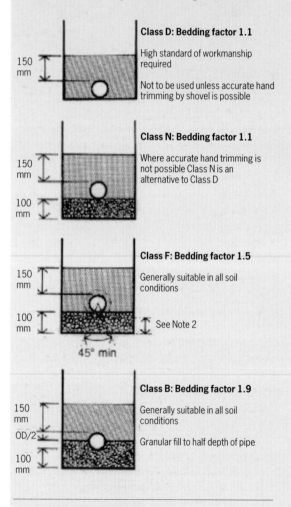

Class D: Bedding factor 1.1

150 mm

High standard of workmanship required

Not to be used unless accurate hand trimming by shovel is possible

Class N: Bedding factor 1.1

150 mm

100 mm

Where accurate hand trimming is not possible Class N is an alternative to Class D

Class F: Bedding factor 1.5

150 mm

100 mm

Generally suitable in all soil conditions

See Note 2

45° min

Class B: Bedding factor 1.9

150 mm

OD/2

100 mm

Generally suitable in all soil conditions

Granular fill to half depth of pipe

British Standards Institution. BS EN 1610: 1998
Construction and testing of drains and sewers

Flexible pipes

Flexible pipes need to be surrounded entirely by granular material. The diagrams from AD H illustrate two alternatives.

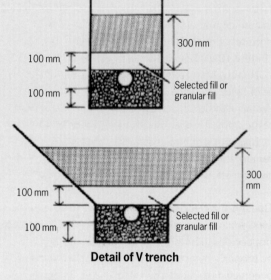

300 mm

100 mm

100 mm

Selected fill or granular fill

300 mm

100 mm

100 mm

Selected fill or granular fill

Detail of V trench

Key:

Selected fill should be free from stones larger than 40 mm, lumps of clay over 100 mm, timber, frozen material, vegetable matter.

Granular material should conform to BS 882: 1983 Table 4 or BS 8301: 1985 Appendix D. Compaction fraction > 0.3 for Class N > 0.2 for Class F and B.

Selected fill or granular fill should be free from stones larger than 40 mm.

Notes:
1. Provision may be required to prevent ground water flow in trenches with Class N, F or B type bedding.
2. Where there are sockets these should be not less than 50 mm above the floor of the trench.

Where pipes are laid under or in close proximity to buildings, they will need to be surrounded with a minimum of 100 mm concrete. Where pipes are to be laid at shallow depths they need to be completely surrounded by granular fill, covered first with a mat of compressible material and subsequently by a concrete slab.

Inspection chambers and manholes

Precast concrete sectional manholes are now commonly specified for deep drains. They can be used in hilly areas where backdrops may be employed. The backdrop may take place either outside (Figure 10.5) or inside (Figure 10.6) the manhole.

Inspection chambers in plastics are mainly used in connection with plastics systems, and are normally limited to inverts not exceeding 2 m (Figure 10.7). BS 7158 gives more information.

Health and safety legislation has led to the avoidance of maintenance engineers entering drainage systems, and therefore manholes are much less commonly specified now.

Inspection chamber covers of cast iron are remarkably durable provided they are not damaged by impacts, but steel covers are prone to corrosion.

Cesspools, cesspits and settlement tanks

Cesspools and cesspits are normally required to be impervious to liquids both from outside and inside, be properly covered and ventilated, and not to become a nuisance to occupants of buildings. Details of construction are given for England &

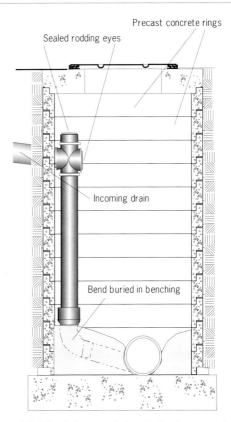

Figure 10.6 Backdrop manhole with the drop contained within the manhole

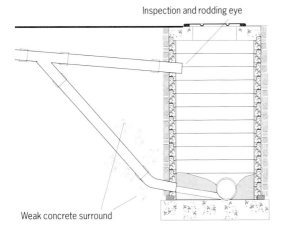

Figure 10.5 One design of a backdrop manhole with the backdrop outside the manhole

British Standards Institution
BS 7158: 2001 *Plastics inspection chambers for drains and sewers. Specification*

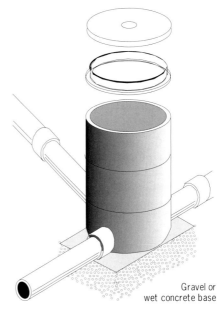

Figure 10.7 A typical plastics manhole or inspection chamber

Wales in AD H and for Scotland in Section 3 of the Technical Handbooks, Domestic and Non-Domestic. Cesspools are not permitted in Scotland. BS 6297 is increasingly viewed as out of date. It will eventually be replaced by BS EN 12566 when all the parts have been published. Maintenance frequency depends on varying factors but can require annual removal of solids. Suitable access for maintenance should be provided, which may include hardstanding or roads for vehicles within a specified distance of the tank.

Wastewater (blackwater) treatment plants

Septic tanks are small, on-site sewage treatment tanks that are used where sites are remote from public sewers. The principle of operation is solids settlement by gravity. Blackwater flows into one side, the solids are partially broken down by anaerobic bacteria and the partially treated water is drained off to ground for final treatment. Discharge consent will be required from the appropriate Environmental Agency.

The design criteria are given in BS EN 12566-1. They include:

- structural, hydrostatic and backfilling loads,
- watertightness under test,
- hydraulic efficiency under test,
- access,
- durability,
- installation instructions,
- operating and maintenance instructions.

As well as BS 12566 and BS 6297, see also British Water's publication *Flows and loads* and CIRIA's publications *Septic tanks and small sewage treatment works* and *On-site sewage dispersal options* for more information.

Most prefabricated chambers are made from reinforced concrete, glass-fibre-reinforced resin or polyethylene.

British Standards Institution
BS 6297: 1983 *Code of practice for design and installation of small sewage treatment works and cesspools*
BS EN 12566: *Small wastewater treatment systems for up to 50 PT*

Soakaways and wastewater (greywater) treatment plants

> **Key Point**
> An increasingly common approach in the conservation of water is to use semi-treated greywater for, eg flushing toilets and watering gardens.

Similar techniques to septic tanks are used for settling any solids, while biological treatment is used to ensure the water is suitable for discharge consent. One common method is by filtering the semi-treated water through a reed bed.

If the house drains are not to be connected to mains foul and surface water sewers, the infiltration options of soakaways and on-site wastewater treatment plants will need to be considered. Box 10.2 lists sources of further information.

Reed beds

There are two basic types of reed beds, one of which has the surface permanently flooded with the wastewater, and the other in which the level of the water is below the surface of the bed. It is the latter kind that is of most interest in the UK climate. The method of construction of the bed takes one of two general forms:

- horizontal, or
- vertical.

Horizontal type: the flow is from a discharge pipe and distributor trench filled with gravel through a 0.6 m deep bed planted with reeds, to a discharge at the base at the far end of the bed.

British Standards Institution
BS EN 12566-1: 2000 *Small wastewater treatment systems for up to 50 PT*
BS 6297: 1983 *Code of practice for design and installation of small sewage treatment works and cesspools*
British Water. *Flows and loads 2: Sizing criteria, treatment capacity for small wastewater treatment systems (package plant).* British Water, London, 2004 revised 2005. 6 pp
CIRIA
Septic tanks and small sewage treatment works. TN 146. 1993
On-site sewage dispersal options. SP 144/L2. 1998

Box 10.2 Sources of information on soakaways and greywater treatment plants

Rainwater
BRE Good Building Guide 38. *Disposing of rainwater.* 2000

Surface water soakaways
BRE Digest 365. *Soakaway design.* 1991

Foul water (blackwater) soakaways
BS EN 752: *Drain and sewer systems outside buildings* (7 Parts)

Small wastewater treatment plants
BS 6297: 1983 *Code of practice for design and installation of small sewage treatment works and cesspools*
until it is superseded by:
BS EN 12566-1: 2000 *Small wastewater treatment systems for up to 50 PT. Prefabricated septic tanks*

Reed beds for wastewater treatment
BRE Good Building Guide 42. *Reed beds.* (Griggs J & Grant N) 2000
    Part 1: Application and specification
    Part 2: Design, construction and maintenance
Grant N & Griggs J. *Reed beds for the treatment of domestic wastewater.* BR 420. Bracknell, IHS BRE Bookshop. 2001.

Drain infestation by rodents
BRE Information Paper IP 6/90 *Rats in drains.* (Hall J & Griggs J) 1990

Vertical type: the water flow is fed intermittently downwards from a perforated discharge pipe at the level of the reeds through several layers of sand and gravel of various sizes. More than one bed is needed in order to allow resting and recovery of the sand layer, the choking of which is one of the commonest modes of failure of this type of bed.

Reed beds need annual maintenance, including cutting of the reeds. They also need to be fenced to keep out grazing animals.

Surface water drainage, soakaways and flood storage
Surface water drainage
Design rules for surface water drainage are only applicable where the drainage is separate from foul water drainage. Design procedures for urban water drainage (sustainable drainage systems or SUDS) are regularly reviewed which then leads to changes in building regulations. Further guidance is given in CIRIA publications C521, C522 and C523.

Surface drains remove water from the vicinity of buildings and surrounding hard surfaces to appropriate disposal points. The major difference from foul water drainage is in the provision of access, including intercepting traps and trapped gullies. The latter two are needed only where the system eventually connects to a foul sewer.

Open drains, swales and ditches are cheap and easy to install, and can carry high levels of discharge for a short time, but require regular maintenance. The outfall of the ditch may require careful detailing to ensure that discharge and any entrained material do not cause problems. Shallow gravel-filled trenches are often used to intercept run-off from a slope. They need protection by a filter to prevent fines blocking them and it may be necessary to line the base to prevent erosion. A perforated pipe can increase capacity.

Deep rubble or gravel-filled drains (French drains) can be used to stabilise small ground slides by lowering groundwater levels. Underdrains may be necessary where soils do not drain quickly. They consist of a perforated drain pipe laid in a trench, covered with gravel and sand.

Land drains typically are of coarse-fired clay, concrete or plastics with open joints allowing water to flow in or out. They silt up in time and are vulnerable to blockage by roots.

More guidance can be found in BS 1196, BS 4962 and BS 5911.

Soakaways
Soakaways are pits or trenches used to collect storm water from buildings and paved areas and allow it to

CIRIA
Martin P et al. *Sustainable urban drainage systems: design manual for Scotland and Northern Ireland.* CIRIA Report C521. 2000
Martin P et al. *Sustainable urban drainage systems: design manual for England and Wales.* CIRIA Report C522. 2000
Martin P et al. *Sustainable urban drainage systems: best practice manual for England, Scotland, Wales and Northern Ireland.* CIRIA Report C523. 2001

percolate gradually into the ground without causing flooding. They are increasingly used to limit the impact of new developments on existing sewer systems. They can also be used for the accommodation of treated wastewater (subject to Local Authority permission).

Soakaways must have an adequate capacity to store immediate storm water run-off and must discharge the water sufficiently quickly to provide the necessary capacity before the next storm. The time taken for discharge depends on the soakaway shape and size, and the infiltration characteristics of the surrounding soil. Generally, clay soils will have relatively poor characteristics in comparison with granular soils.

Traditionally, soakaways are compact, square or circular pits suitable to store run-off from areas of less than 100 m². They can also take the form of trenches which follow convenient contours. These have a larger surface area for a given stored volume and therefore a better rate of discharge. For larger areas, soakaways can be specified consisting of precast rings or trenches. They should not be deeper than 3–4 m. They should be at least 5 m from any building, and avoid positions where the ground beneath foundations could be adversely affected.

Perforated precast concrete ring unit soakaways are installed within a square pit, and the granular backfill must be separated from the surrounding soil with a geotextile membrane to prevent migration of fines into the soakaway and consequent settlement of the surrounding soil. The top surface of the granular fill should also be covered with a geotextile membrane.

Site data should be examined to assess whether any factors are likely to affect percolation or stability of the area (eg variable soil conditions, filled land, variations in groundwater levels). Infiltration characteristics of soil where soakaways are to be located are determined by excavating a trial

pit, filling it with water and timing how long it takes to empty. The trial pit should be of sufficient size to represent a section of the proposed soakaway (typically 0.3–1 m wide and 1–3 m long). It should be filled several times in swift succession to ensure the moisture content of the surrounding soil is typical of the site once occupied.

A simple catch pit or silt trap to protect soakaways from silting up can be formed within a small shallow inspection chamber with the inlet and outlet at approximately the same level (Figure 10.8).

For areas not exceeding 100 m² the percolation trial pit should be 1–1.5 m below the invert level of the drain discharging to the soakaway, or typically 1.5–2.5 m below ground level. If, for some reason a pit of deeper than 3 m is required, it will probably not be possible to carry out a full depth test. In these circumstances, the infiltration rate should be calculated on the basis of the time taken for the water to fall from 75% to 25% of the full depth. In this case, the calculation will also be amended. Where soakaways are proposed on chalk or other soils subject to instability, the advice of a geotechnical engineer should always be sought.

The steps involved in calculating the rate of percolation and associated frequency of soakaways are given in BRE's *Digest 365*. The steps are:

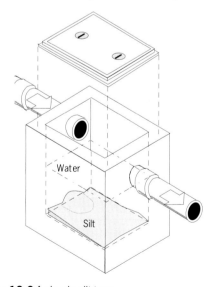

Figure 10.8 A simple silt trap

British Standards Institution
BS 1196: 1989 *Specification for clayware field drain pipes and junctions*
BS 4962: 1989 *Specification for plastics pipes and fittings for use as subsoil field drains*
BS 5911: *Concrete pipes and ancillary concrete products* (Various Parts)

BRE. *Soakaway design.* Digest 365. 1991

- excavate trial pit(s),
- select form of soakaway (ie pit or trench),
- calculate design rainfall and frequency of occurrence,
- calculate inflow and outflow from the soakaway, and therefore the storage volume required.

Scottish practice requires a slightly different procedure. A 300 mm square hole, 300 mm deep is excavated at the bottom of at least two trial pits, not less than 5 m apart. The holes are filled with at least 300 mm depth of water and allowed to drain overnight. On refilling the next day, the time taken for the drop from 75% full to 25% full is measured. The test is repeated three times. The percolation value is obtained by dividing the time in seconds by 150 to give the time taken for the water level to drop by 1 mm. The area of the soakaway is calculated from:

$$A = P \times V_p \times 0.25$$

where:

A = area of the subsurface drainage trench (m^2),
P = number of persons served by the treated wastewater,
V_p = percolation value obtained (s/mm).

Flood storage

For large-scale developments with extensive hard surfacing the run-off may exceed the capacity of existing surface water drains. Flood storage reservoirs may be provided, and may be suitable for amenity use provided adequate safety precautions are taken to protect the public, especially children and the disabled. Design of flood storage reservoirs is a specialist area and more information can be found in CIRIA's *Design of flood storage reservoirs*.

Hall MJ, Hockin DL, Ellis JB. *Design of flood storage reservoirs.* CIRIA Report B014. London, CIRIA, 1993

Freestanding walls

> **Key Point**
> Freestanding wall collapses are one of the more common forms of serious and fatal accidents associated with the immediate environment of buildings.

Freestanding walls are exposed to the weather and to wind loadings on both faces. Where even a small part of a wall becomes unstable it can lead to progressive collapse of the whole wall. Such collapses are one of the more common forms of serious and fatal accidents associated with the immediate environment of buildings.

These walls are normally formed in brickwork or blockwork, and generally will be reinforced where they are over 2.5 m. In sheltered areas, where wind speeds are low, they may be unreinforced up to 3.25 m in height, whereas in areas with high wind speeds lower heights may require reinforcement. As a general rule, walls of over 2.5 m in height will require design by a structural engineer or other suitably qualified person.

The structural design is primarily based on wind loading with the design based on published wind speed data for the UK. Regional values will require adjustment for the local topography (size and frequency of features which will act as windbreaks). For more details, see the section on *Wind* in chapter 3.

A rule of thumb approach to design is included in BRE's *Good Building Guide 14*, based on four wind exposure zones (Figure 10.9). Within each zone, proposed wall locations are categorised as either sheltered or exposed. Sheltered locations are typically in urban areas, but may also occur where there is considerable local interruption of wind flow. Exposed locations are typically rural areas, or other areas where there is a clear view over open country. If in doubt, either seek specialist advice or design as if the wall is exposed.

BRE. *Building simple plan brick or blockwork freestanding walls.* Good Building Guide 14. Bracknell, IHS BRE Press, 1994

Rules of thumb for wall thickness and height are given for each zone in Table 10.2. Walls in the numbered wind speed zones shown in Figure 10.9 should not exceed the heights given in Table 10.2.

Wall height is measured from the lowest ground level to the top of the capping or coping. For ground slopes of between 1 in 10 and 1 in 20 wall heights should be reduced by 15%. Formal design procedures should be adopted for sites with steeper slopes.

Note: The rules of thumb listed in Table 10.2 are not applicable in the following circumstances:

● where there is a risk of vehicle impact,
● where there may be pressure of a large number of people (eg on public right of way near a stadium),
● where there is excessive vibration (eg from heavy traffic),
● where there is higher than normal wind loading (eg close to high-rise buildings, on the crest of a hill, near mountains),
● where there is atypical loading (eg supporting a heavy gate or door or where the difference in ground level either side exceeds twice the wall thickness),
● where there is excessive ground movement (eg as a result of settlement or shrinkage and swelling of clay or peat soil).

Both English garden wall bond and Flemish garden wall bond are often used, but half-brick walls can be strengthened by staggering the construction on plan or by using piers.

As for retaining walls, the foundation should be designed to ensure the thrust of wind loading and dead weight passes through the central third of the foundation. For walls not exceeding 2.5 m in height a foundation depth of 0.5 m is normally adequate in good ground. For higher walls in cohesive soils the foundation depth should be 0.75 m minimum. Generally, it should be possible to found a freestanding wall at half the depth recommended for low-rise buildings. However, where walls are to be founded on highly shrinkable soils or close to large trees (or where large trees have been removed) foundation depths should be based on the recommendations given in chapter 4, *Foundations and basements*.

Foundation width is normally governed by the need to resist overturning, although allowable bearing pressure may be relevant in poor ground. Rules of thumb for width are given in BRE's *Good Building Guide 14*. Wherever possible, avoid construction on made ground unless properly compacted; remove soft spots and backfill using lean mix concrete.

Where the wall is intended to provide security, ensure it does not provide a platform to access roofs, garages or sheds. Security walls need to be at least 3 m high to deter trespassers. Use of glass, spikes or wire on top of the wall has implications for property owners' liability to trespassers.

As mentioned above, simple retaining or freestanding walls can normally be designed in accordance with rules of thumb (see BRE's *Good Building Guides 14, 19* and *27*.

Table 10.2 Rules of thumb for wall thickness and maximum height above ground		
Wall thickness	Wall height limit (mm)	
	Sheltered	Exposed
Half brick	725	525
One brick	1925	1450
One-and-a-half-brick	2500	2400
Half brick	650	450
One brick	1750	1300
One-and-a-half-brick	2500	2175
Half brick	575	400
One brick	1600	1175
One-and-a-half-brick	2500	2000
Half brick	525	375
One brick	1450	1075
One-and-a-half-brick	2450	1825

BRE
Building simple plan brick or blockwork freestanding walls. Good Building Guide 14. 1994
Building reinforced, diaphragm and wide plan freestanding walls. Good Building Guide 19. 1994
Building brickwork or blockwork retaining walls. Good Building Guide 27. 1996

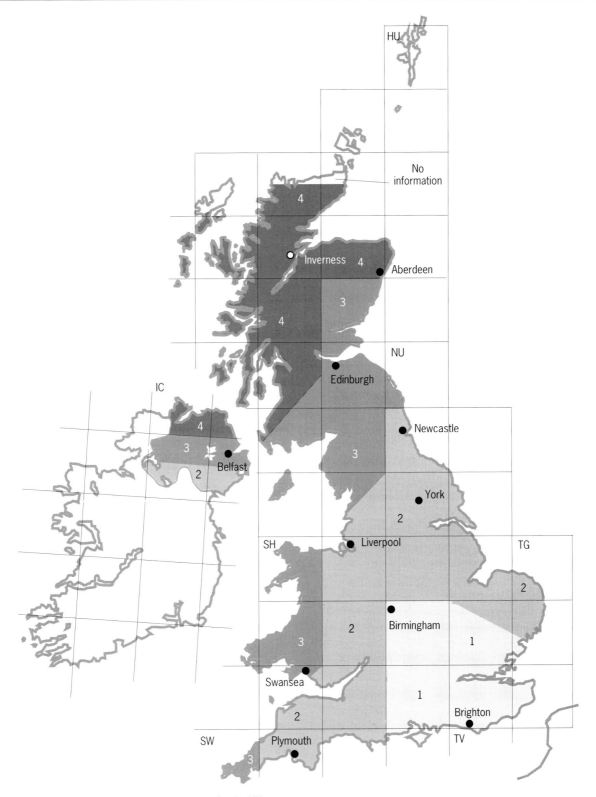

Figure 10.9 Simplified wind speed zones for the UK

Detailing, materials and workmanship for freestanding garden walls

High- and low-level dpcs are recommended in all walls that are not constructed from frost-resistant materials. BRE's *Good Building Guides 14* and *19* provide guidance on the selection of dpcs. High-level dpc detailing is more complex in staggered walls and flexible dpcs may be difficult to incorporate unobtrusively due to requirements for lapping and bedding on both surfaces.

The minimum solution for low-level dpcs is to build up with dpc Type 1 or 2 clay bricks (or equivalent) to one course above the weep holes. It is more practical to form the whole construction from foundations to approximately 200 mm above finished ground level with dpc Type 1 or 2 bricks. Flexible materials or slate should not be used for dpcs at low level.

Frost-resistant cappings and copings to BS 4729 or other copings to BS 5642-2 should be used. They should incorporate an overhang and drip if the wall is not in a very sheltered location, or is not going to be built of frost-resistant bricks. BRE's *Good Building Guide 17* contains guidance on copings. Mortars should normally be class (i). Where there is a risk of children playing or vandalism, an interlocking capping should be provided. Some complex wall designs may make staining of the wall surface by concentrated water run-off almost inevitable.

Other design issues to consider include:
- specification of appropriate bricks and blocks,
- the correct positioning of piers and movement

British Standards Institution
BS 4729: 2005 *Clay and calcium silicate bricks of special shapes and sizes. Recommendations*
BS 5642-2: 1983 *Sills and copings. Specification for copings of precast concrete, cast stone, clayware, slate and natural stone*

Brick Development Association
A range of design guides and file notes, eg:
Design of brick diaphragm walls
Design of free-standing walls
BDA guide to successful brickwork
The design of brickwork retaining walls
External walls: design for wind loads
See www.brick.org.uk/publications for a full list of publications

joints (which should be continuous through copings), and
- protection from vehicle impact where appropriate.

Good workmanship practices include:
- lay bricks frog up,
- finish joints with a bucket handle profile,
- rake out mortar joints to 10–12 mm before rendering (except blockwork),
- specify render and mortar mix to suit exposure conditions,
- lifts should not to exceed 1.5 m per day,
- protect new masonry from frost, wind and rain,
- allow 28 days for mortar to develop full strength before backfilling.

Various design guides on freestanding and retaining walls are published by the Brick Development Association (see footnote on previous page for examples). Among the topics covered are:
- the use of reclaimed bricks,
- detailed structural guidance.

Fencing

Main characteristics and selection criteria
The main types of fencing are described in BS 1722, while guard rails and fences are covered in BS 7818. The main types in various materials include those listed in Box 10.3.

Heights quoted in BS 1722 are 0.9 m, 1.2 m, 1.4 m, 1.8 m and 2.15 m. The lowest standard heights are used for the fronts or sides of domestic dwellings. Where the property adjoins public land, railways, or commercial/industrial property, higher fences will typically be provided. If required for security 340 mm cranked extension arms can be provided on top of certain types of fence. Guard rails are normally around 1 m, but horse riders require 1.8 m. For sites up to slopes of around 20°, and depending on design and circumstances, fences are

British Standards Institution
BS 1722: *Fences* (Various Parts)
BS 7818: 1995 S*pecification for pedestrian restraint systems in metal*

Box 10.3 Main types of fencing and materials

Wire
- Chain link
- Rectangular wire mesh and hexagonal wire netting
- Strained wire with timber, concrete or steel posts
- Anti-intruder fences in chain link and welded mesh

Wood
- Woven wood and lap boarded
- Cleft chestnut pale
- Close boarded
- Wooden palisade
- Wooden post and rail

Steel
- Mild steel continuous bar
- Steel palisade
- Open mesh steel panel

normally stepped; above this, it is usual to follow the ground surface.

Selection of fences depends on the materials to be used (particularly for posts), the degree of security required, the excavation for posts, the type of infill required and the life expectancy required. The effects of strong winds where fences offer a virtually solid barrier must be considered, and calculations may be required to ensure stability.

Wire fencing depends for its strength on the frequency and size of posts since it is largely transparent to wind forces. Wood fences are not transparent to wind, and they become increasingly liable to damage in strong winds.

Steel panels and guard rails are required to deflect not more than 8 mm under a load of 2.5 kN for normal fencing, and not more than 10 mm under 3.5 kN loading for security fencing.

The durability of wire mesh is highly dependent on its location; urban or polluted locations and marine environments are more aggressive. Galvanised steel fencing should be protected by a minimum of 350 g/m^2 on both sides to give a 30-year life, but a minimum of 275 g/m^2 should be provided in marine environments to give a 10-year life. Preservative-coated wood posts should also last about 10 years, although some may rot at ground level. More guidance on deterioration of fencing is given in BRE's *Foundations, basements and external works*.

Types of fencing
Steel wire fencing

Straining posts and struts are normally used at all ends, corners and significant changes in level, but also in straight runs. Intermediate posts normally occur at around 3 m centres. The lowest 300 mm may be buried below ground for increased security or as protection against burrowing animals.

Chain link fencing

Increased durability is provided by chain link instead of wire netting. The mesh size is normally 40 or 50 mm. Wire diameter can be 2.5 mm (medium), 3 mm (heavy) or 3.55 mm (extra heavy). Gates are normally steel-framed, infilled with chain link.

Wire netting

This type is normally used in agricultural applications, and there are various sizes of mesh depending on the animal to be enclosed or excluded. Straining posts are normally provided at each end and at 150 m centres on long runs. Intermediate posts normally occur every 3.5 m.

Strained wire fencing

This type is supported on posts of rolled or hollow section steel, timber or reinforced concrete. The distances between intermediate posts are often too great to entirely eliminate sagging in the wires (similar spacing to steel wire fencing above). Droppers of wood or steel are therefore introduced to act as spacers. The wires are maintained in tension for the life of the fence, and BS 1722 specifies a tension not less than 1600 N four days after tensioning.

Harrison H W & Trotman P M. *BRE Building Elements. Foundations, basements and external works: Performance, diagnosis, maintenance, repair and the avoidance of defects.* BR 440. Bracknell, IHS BRE Press, 2002

British Standards Institution. BS 1722: *Fences* (Various Parts)

Anti-intruder fencing

The type of fence described in BS 1722-10 is mounted on substantial concrete posts, with main cross-sections of 150 mm × 150 mm and intermediate ones of 125 mm × 120 mm. Steel rectangular or circular hollow section posts can also be used. Zinc or plastics-coated chain link or welded mesh consists of 3 mm diameter wires, and the tops of the posts carry cranked or straight extension arms with up to 3 strands of barbed or razor wire.

Woven wood and lap-boarded fencing

This type of fence gives a good standard of privacy, but is often not very durable. It is used almost exclusively in domestic applications. Heights range from 0.6 m to 1.8 m. Posts may be square section timber or plain or slotted, or slotted reinforced concrete. Slats for woven fencing are normally 75 mm × 5 mm thick, held in place by rectangular sawn battens. Lapped boards may be square, waney-edged or feather-edged, and are slightly more robust than woven panels.

Cleft chestnut pale

This traditional type of fence is mostly used in domestic applications. Wiring encircles the pale during manufacture, giving a robust construction. Heights vary from 0.9 m to 1.8 m. Wiring is at heights varying from 450 mm to 750 mm, depending on the total height of the fence. Pale spacing is at 50, 75 or 100 mm centres, depending on the level of security required. Intermediate posts are normally placed between 2.25 m and 3 m for concrete, and slightly closer for wood. Protection of the wire is a key element of durability.

Close-boarded fencing

The wood is usually either treated softwood or oak left untreated. Details are shown in Figure 10.10.

Wooden palisade

This is a decorative fence, normally used in domestic situations. Details are shown in Figure 10.11.

British Standards Institution. BS 1722-10: 1999 *Fences. Specification for anti-intruder fences in chain link and welded mesh*

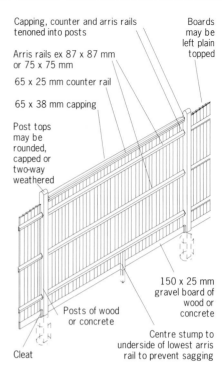

Capping, counter and arris rails tenoned into posts
Arris rails ex 87 x 87 mm or 75 x 75 mm
65 x 25 mm counter rail
65 x 38 mm capping
Post tops may be rounded, capped or two-way weathered
Posts of wood or concrete
Cleat
Boards may be left plain topped
150 x 25 mm gravel board of wood or concrete
Centre stump to underside of lowest arris rail to prevent sagging

Figure 10.10 Close-boarded fencing

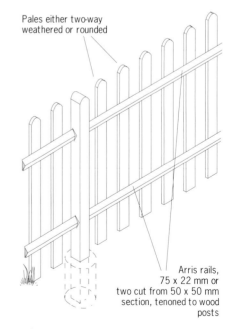

Pales either two-way weathered or rounded
Arris rails, 75 x 22 mm or two cut from 50 x 50 mm section, tenoned to wood posts

Figure 10.11 Wooden palisade fencing

Wooden post and rail

This type of fence is not fully stock-proof, but is often used in agricultural situations. Heights are normally around 1 m. Posts can be sawn rectangular, cleft or left in the round. Main posts are normally 75 mm × 150 mm at 2.85 m centres (where the rails are mortised) or 1.8 m centres (where the rails are butt-jointed). Thinner intermediate posts are usually around 38 mm × 87 mm. Rails are normally rectangular, and two, three or four may be provided depending on the circumstances. Rails are normally scarf-jointed, but wider posts give the option of butt-jointing.

Continuous bar fencing and hurdles

These are normally used for farming purposes. BS 1722, refers to light, medium and heavy duty.

Steel palisade

Palisade fences offer higher security than wire fences. They are available in heights up to 2.4 m and above. Steel posts are normally set at 2.75 m centres. Pales for normal use are 2.5 mm thick, and for security uses 3 mm thick. They are usually shaped profiles with a variety of tops. High security fencing pales may be embedded 350 mm into the ground or 150 mm into concrete.

Open mesh steel panel

For ordinary use this fencing is up to 2.4 m high, and for security fences it is up to 3.0 m high. All are available with extension arms. The mesh is constructed from wires up to 5 mm thick, spaced at 12.5–75 mm centres and welded into panels. Durability depends on protection given to the wire.

Sheeted steel

Fences sheeted with corrugated or profiled, shaped galvanised steel will occasionally need to be specified where additional privacy is required. Galvanising is post-manufacture so should eliminate cut-end corrosion. The conditions of service are different from the same materials used in roofs so they will not necessarily behave in a similar fashion.

Public utilities

This section deals with water, electricity and gas installations outside the building, but within the boundaries of the site. Installations within the building are described in chapter 9, *Building services.*

Cables and pipes already on the site should be colour-coded if they are reasonably recent and for new work. The colour codes are given in Box 10.4.

Most of the cabling work will be done by specialist contractors, but adequate reinstatement of trenches and pavings above cables will be part of normal design. The main problems to be aware of in terms of general designers' responsibilities will be:

● inadequate consolidation of fill in trenches,
● loose or sagging brackets supporting overhead supplies,
● non-registration of CCTV cameras in accordance with the requirements of the Data Protection Act 2000,
● buried cables and pipes that are unmarked or unidentified.

Box 10.4 Colour coding for cabling	
● Low-voltage electricity cable	**Black**
● High-voltage electricity cable	**Red**
● Water pipe	**Blue,** but older pipes will be black
● Cable TV	**Green**
● Telecommunications cable	**Grey**
● Gas pipe	**Yellow**

Water supply

Wholesome water is supplied in England and Wales by statutory water undertakers controlled by the provisions of the Water Supply Regulations (see

The Stationery Office
Water Supply (Water Fittings) Regulations 1999. Statutory Instrument 1999 No 1148. 1999
The Water Supply (Water Quality) Regulations 2001
The Water Supply (Water Quality) (Scotland) Regulations 2000
The Water Supply (Water Quality) (Amendment) Regulations (Northern Ireland) 2003
Private Water Supply Regulations 2000
Scottish Water Authorities
Water Byelaws 2000 (Scotland)

references in the foonote). Installations in buildings are required to comply with these regulations, and they apply also to taps on the exterior of buildings and to stand-pipes. Reference should also be made to BS EN 805.

Mains now tend to be buried at a depth of a minimum of 750 mm, but up to 1350 mm where the risk of freezing is greatest. The durability of water mains, particularly those laid in contaminated land, will depend on the type of soil, soil moisture, chemical and physical–chemical factors, and the material of which the pipes are composed (see BRE's report *Performance of building materials in contaminated land*.

The commonest materials for older supply mains included copper, lead and cast iron, but most mains now are installed in plastics. These are largely inert once buried, but are vulnerable to damage before installation and in contaminated ground or in the presence of some micro-organisms (eg bacteria and fungi). For example, PVC-u can lose its structural integrity when exposed to organic contaminants, while polyethylene and polybutylene can be damaged by petroleum products. Polyethylene has poor corrosion resistance to oxidising acids and can become cracked and crazed when attacked by certain fungi. High molecular weight polyethylene is resistant to microbial degradation.

Light sandy soils and chalk are not generally aggressive, while acid peaty soils, and ground containing cinders or builders' rubble can be aggressive to steel, copper and aluminium. Heavy anaerobic clays may be corrosive to ferrous metals, but aluminium, lead and copper are resistant. Saline environments corrode aluminium and galvanised steel, but have little impact on copper or lead.

There is increasing interest in the harvesting of rainwater for domestic use, and the publications from the Water Regulations Advisory Scheme (WRAS) listed in the footnote below are relevant. Further information is also available in BSRIA and CIRIA publications.

Electricity, cable TV and telecommunications

Small installations will still sometimes be served by overhead cables, either single- or three-phase, carried on treated wooden poles. The last few metres are often carried down the pole and buried underground adjacent to the entry point to the building. Plastics have largely superseded bitumen wrapping for sheathing underground mains.

Single-phase 230/240 volt, 60, 80 or 100 amp, 50 Hz earthed neutral supplies are the normal supply to single dwellings. These qualify as low voltage.

Cables are laid in sand or other suitable bedding before backfilling. Cables are normally black for low-voltage installations and red for high-voltage. Highly coloured marker tapes should be applied over the bedding surround of the cable before backfilling to warn during future excavation.

Cables are normally buried at least 0.5 m deep, but existing cables may be much shallower. There are about 250 accidents a year involving site workers accidentally touching live buried cables.

Cables are normally laid within public highways limits, terminating in a buried access trap at the boundary. CCTV may also need to run below ground for security reasons. Telephone lines in rural areas may still be run on overhead supplies. Treated wooden poles are inspected at least every 10 years, and have a typical service life of around 60 years.

Telecommunications cables are normally grey and TV cables green. Many telecommunications copper cables are 'jelly filled' (ie grease-filled and sheathed, often in polyethylene).

Gas

Gas may be supplied at different pressures according to location and type of building served, and it may be necessary to provide pressure-reducing plants near the buildings to be serviced. Yellow plastics piping is almost universal. Pipe diameters for small installations could be as small as 25 mm, but for larger domestic or commercial installations pipe diameters will be greater.

British Standards Institution. BS EN 805: 2000 *Water supply. Requirements for systems and components outside buildings*
Paul V. *Performance of building materials in contaminated land.* BR 255. Bracknell, IHS BRE Press, 1994

Water Regulations Advisory Scheme (WRAS)
Reclaimed water systems. Information about installing, modifying or maintaining reclaimed water systems. 1999
Reclaimed water systems. Marking and identification of pipework for reclaimed (greywater) systems. 1999

For areas outside main gas distribution networks, small pressure cylinders of propane or butane or large fixed storage tanks can be provided.

There are extensive requirements for the safe use of gas in the Gas Safety Regulations and in the accompanying documents of the national building regulations.

Exterior lighting and security

Lighting

Exterior lighting covers various types:

- security lighting, to deter crime, or to enable intruders to be visible to CCTV or security staff,
- amenity lighting (eg in gardens or car parks) which will normally be left on while the space is being used,
- access lighting to allow safe access to and between buildings on a site,
- decorative lighting such as building floodlighting or festive lighting.

CIBSE's *Lighting guide: The outdoor environment*, gives general guidance on exterior lighting, BS 8220-1 gives recommendations on security lighting of dwellings and prEN 12464-2 gives recommendations for outdoor areas of non-domestic buildings including circulation areas and car parks.

Exterior lighting schemes will normally use one of the types of lamp listed in Box 10.5.

The 2006 Building Regulations (England & Wales) AD L only covers exterior lighting to dwellings. Exterior lighting includes lighting in porches, but not lighting in garages and carports. It does not include lighting that is not attached to the

HMSO. *Gas Safety (Installation and Use) Regulations 1998. Statutory Instrument 1998 No. 2451.* London, The Stationery Office

Chartered Institution of Building Services Engineers (CIBSE). *Lighting guide: The outdoor environment.* LG 06. London, CIBSE, 1992

British Standards Institution
BS 8220-1: 2000 *Guide for security of buildings against crime. Dwellings*
prEN 12464-2: *Light and lighting. Lighting of work places. Outdoor work places*

> **Key Point**
> Safety and security issues are paramount for pedestrian areas, and particularly for flights of steps.

dwelling. So lighting at the end of a garden, or at the garden gate, would not come under the requirements of AD L, although it would need to comply with AD P.

Exterior lighting would comply with the requirement if it was a maximum of 150 W and automatically went out when there was enough daylight, and when not required at night (eg if it had a presence detector that turned the light on for a limited time). Even a tungsten or tungsten halogen lamp would comply if it had this form of control.

Alternatively, external luminaires would also comply if they could only be used with lamps with better than 40 lm/W circuit efficacy (fluorescent and discharge lamps). So a dedicated compact fluorescent exterior luminaire would be satisfactory even under manual control. The aim is to avoid inefficient tungsten lighting being left on continuously.

Safety and security issues are paramount for pedestrian areas, and particularly for flights of steps. Bollard fittings are available for paths, and recessed

Box 10.5 Lamps suitable for exterior lighting

Tungsten halogen*
Warm light and good colour rendering

Tubular or compact fluorescent
Light varying from warm to cool with moderate colour rendering]

Low pressure sodium
Warm light with very poor colour rendering

High pressure sodium
Warm light with average colour rendering

High pressure mercury
Cool light with average colour rendering

Metal halide
Intermediate or cool light with average to good colour rendering

* *Note:* Shorter life and higher running costs than the other lamps listed.

luminaires can be set into the side walls of flights of steps.

Floodlighting of buildings must be set in consideration of both general levels of artificial light in the area and the reflection factor of the building fabric. Good modelling of architectural features will require selective siting of appropriate luminaires. Consideration should also be given to fittings which do not impair the appearance of the building during daylight.

Irrespective of the actual average values achieved for artificial lighting, it is important to achieve even coverage, except for floodlighting of features, where uniform coverage can look bland. The recommended average lighting levels for different use categories of roads and footpaths are given in Table 10.3, but actual operating regimes will inevitably be a compromise between necessary hours of use and economy.

Roads providing access to residential, business or industrial areas are normally lit from 30 minutes after sunset to 30 minutes before sunrise. Lighting columns are commonly placed at 30 m intervals where crime rates are low. Columns should not be nearer than 0.8 m from the carriageway for roads

Table 10.3 Categories for illumination of roads, footpaths and verges	
Use	Lux requirement
High (eg with high crime rates)	Average 10 Minimum 5
Moderate	Average 6 Minimum 2.5
Minor (eg residential)	Average 3.5 Minimum 1

British Standards Institution. BS 5489: 2003 *Code of practice for the design of road lighting*
    Part 1: *Lighting of roads and public amenity areas*
    Part 2: *Lighting of tunnels*
CIBSE. *Lighting the environment: a guide to good urban lighting.* ILEO1. London, CIBSE, 1995. 40 pp
Institute of Lighting Engineers (ILE). *Guidance notes for the reduction of obtrusive light.* Rugby, ILE, 2005. Available as a pdf from www.ile.org.uk

designed for traffic speeds of up to 50 km/h. Mounting heights are normally between 4 and 6 m. Lamps should not overhang the road, and normally are placed on the outside of bends.

For car parks the primary consideration is security. Large open car parks are often treated as roads and lit using the road lighting standard BS 5489. Overall illuminance levels should not exceed those recommended for street lighting in the neighbourhood. Further guidance is given in CIBSE's *Lighting the environment.*

Unwanted spill light should be minimised to avoid disturbance to nearby householders, and to reduce sky glow. Luminaires should have a high downward light output ratio with little or no upward light. Where fittings are close to the site boundary, cut-off luminaires should be chosen to reduce spill light reaching windows or adjacent properties (see Institute of Lighting Engineers' *Guidance notes for the reduction of obtrusive light* for more information).

High-efficiency light sources for exterior lighting are available at relatively low cost to both the building owner and the environment. Security floodlighting in car parks using high power tungsten halogen lamps is particularly poor in terms of high glare and deep shadows. The controlling of lamps by passive infrared sensors can also mean that this type of lamp uses more energy than a more efficient type running continuously.

Lighting fitments should be chosen and located so that each lamp is easily accessible for cleaning and re-lamping, or longer life lamps should be chosen. Annual cleaning of lamps is common.

Security

Exterior lighting enhances security. If an area needs to be protected by CCTV, the exterior lighting levels must be adequate for the cameras. Unobtrusive night surveillance can be achieved with infrared CCTV and dark-filtered tungsten halogen illumination sources.

Luminaires should be positioned to cast a pool of light around front doors, windows, etc. They (and the cables) should be out of reach of intruders. Both shadows and nuisance to neighbours should be avoided.

Table 10.4 Illumination recommended by CIBSE for external crime risk areas	
Area	Lux requirement
Pedestrian	
Low crime risk	20
High crime risk	100
Car parking	
Low crime risk	20
High crime risk	50

Intruder alert systems using a narrow beam of energy (infrared or microwave) transmit to a receiver which triggers an alarm. Since the beams travel in straight lines, three or more paired receivers and transmitters will be required to cover a large area. More pairs will be needed for irregular areas. LED infrared beams cover a narrower area than microwave beams. CIBSE lighting recommendations for crime risk areas are summarised in Table 10.4.

See also the section on *Soft landscaping* later in this chapter for recommendations on plants likely to deter intruders.

Pavings and vehicle access

Hard landscaping provides a means of safe and dry access to the building for both people and vehicles. The specialist requirements for public roads and highways are not covered here, but both planning and adoption requirements will need to be met where applicable.

Vehicle access

Vehicle access ramps to single-family dwellings should not be steeper than 1 in 6, and transition points should be eased to stop vehicles grounding. Ramps to basement car parking should not exceed 1 in 10, or 1 in 7 for short lengths with a transition length at the top and bottom of the ramp. Ramps should have a textured or ribbed surface. Ramps falling towards the basement should incorporate drainage channels.

Footpaths and pavings

Footpaths and other access routes should be designed for ease of use, especially by disabled people. For desirable widths of access routes, see chapter 12.

Pavings are exposed to rain, frost and sunlight, and paving materials need therefore to be robust and frost-resistant if they are to have adequate durability. Pavings can be categorised as flexible or rigid structurally, and they are also either permeable or impermeable to water.

Preparation and structure

All pavings will require adequate hardcore and bases. Base materials will generally consist of inert aggregate materials, including hardcore, hoggin and cement or lime-bound aggregates, and may include site soils. They must be protected from water ingress both from below and above for all but the lightest of loadings. Soil excavation, removal of soft spots, preparation, weed killing and compaction are critical to durability of the paving.

Pavings should be specified for the maximum load anticipated; it is common for infrequent but foreseeable loads such as fire engines to be ignored in design, and failures can occur very quickly as a consequence.

In order to retain the integrity of the bases it will be necessary to provide kerbs or edgings. Different heights, lengths and profiles are available, most often in concrete. Kerbs in natural stone are also available. Support should be provided at horizontal or inclined features, such as steps. Slope creep on shrinkable clays can be difficult to distinguish from clay shrinkage, which may be exacerbated by desiccation as the paving prevents rainwater from wetting soils previously exposed.

Where paving is provided adjacent to an existing building, care must be taken not to bridge the dpc. If necessary, the paving can be stopped short of the building and a gravel layer provided instead.

A key criterion for all paved surfaces is adequate surface water drainage. Cross-falls or crowning are essential to shed water and prevent puddling (Figure 10.12). Manufacturers of pavings generally provide drainage channels and covers with guidance on how to calculate drainage requirements for different

Figure 10.12 Coursed stone setts laid to camber

surfaces. Pea shingle in proximity to drainage slots should be of an adequate size to avoid displaced shingle blocking the drainage slots.

Finishes

Flags or paving stones in concrete or stone with various finishes are available. The traditional bedding material for these is to lay the flags on lime mortar spots on a bed of gravel topped with sand, and with dry mix brushed into the joints. Flags range from 600 × 450 mm to 300 × 300 mm, and are either 50 or 63 mm thick.

Both brick and concrete pavers generally of 100 × 200 mm and rectangular in shape are available in thicknesses of not less than 60 mm. Access roads and car parks are normally formed in 80 mm thick pavers. Interlocking concrete pavers are suitable for light traffic at low speeds only and for hardstandings. Various other shapes and sizes are available, such as cobbles or setts. They are normally laid on consolidated sand, on a concrete base for vehicular areas or where the bearing ratio of the soil is poor. Drainage should be provided for all areas other than small ones.

In-situ concrete is also used for paving (and particularly for paths). A tamped finish will

normally be required. Sleeved stainless steel bars at movement joints will prevent differential settling.

Coated macadam wearing courses comprise crushed rock or slag bound with bitumen. Hot rolled asphalt bases and wearing courses can also be provided. Top dressings of sprayed tar, rock and gravel provide a final course to some pavings.

Since pavings are likely to be at least occasionally wet, and may also be icy or covered in moss growth, slipperiness in use must be considered. Slipperiness should be tested in accordance with BS 8204-2. Values below 20 indicate dangerous slipperiness, while values above 40 are satisfactory. Polishing over time may occur but may be improved by pressure-hosing plant and fungal growth or coating with epoxy resins into which carborundam or crushed flint is sprinkled. Ramps will require laying to a minimum fall to avoid water accumulating and freezing. A nominal fall of at least 2% should be provided. For gradients of pedestrian ramps, bearing in mind ease of access for all building users, especially disabled people, see Table 12.1.

Compaction of finished surfaces should be adequate to ensure that deviations do not exceed 15 mm under a 3 m straight edge. Deviations in sub-bases below concrete blocks should not normally exceed +/- 20 mm, and in other circumstances +/- 30 mm. Curing of in-situ concrete is important. Covers should be applied in hot weather (and watering may be needed). Concrete should not be laid in frosty conditions.

Specific paving materials are dealt with in the following standards: BS 7263, BS 6717, BS EN 1343, BS 7533, BS 4987 and BS 3690.

Soft landscaping

Selection criteria

Soft landscaping provides a setting for buildings, and may also have a security aspect, as well as environmental benefits. Grassed areas may also provide a leisure facility. All will require at least periodic maintenance to retain their functionality and appearance.

British Standards Institution
BS 8204-2: 2003 *Screeds, bases and in situ floorings. Concrete wearing surfaces. Code of practice*

Plant material should be selected to suit the setting and also the site conditions. Landscape architects are the best qualified to advise. The aim is generally to provide foliage of a range of sizes and shapes, with a variety of colours through the seasons. There are nearly 2000 plants widely available for use in landscaping, and these are now commonly specified in accordance with *The National Plant Specification* or *Trees and shrubs for landscape planting*.

Semi-mature trees can be transplanted, but may be difficult to establish, although native species and acers also transplant and re-establish fairly easily. More typically, extra-heavy standards will be used (4–5 m tall with a trunk circumference of 12–14 cm). A range of trees down to light standards can be specified.

Staking and protection

Trees should be staked to allow some movement and two posts with the tree tied to each are now often used (Figure 10.13). Protection of the trunk should last the required time of staking. There may also need to be protection against grazing animals.

British Standards Institution
BS 7263-1: 2001 *Precast concrete flags, kerbs, channels, edgings and quadrants. Precast, unreinforced concrete paving flags and complementary fittings. Requirements and test methods*
BS 6717: 2001 *Precast, unreinforced concrete paving blocks. Requirements and test methods*
BS EN 1343: 2001 *Kerbs of natural stone for external paving. Requirements and test methods*
BS 7533: *Pavements constructed with clay, natural stone or concrete pavers* (Various parts)
BS 4987: *Coated macadam (asphalt concrete) for roads and other paved areas* (Various parts)
BS 3690: *Bitumens for building and civil engineering* (3 Parts)
Horticultural Trades Association. *The national plant specifications*. Reading, Berkshire, Horticultural Trades Association, 2002
Joint Council for Landscape Industries. *Trees and shrubs for landscape planting*. London, The Landscape Institute, 1989

Figure 10.13 A newly transplanted tree supported by twin stakes and a flexible rubber tie

Plants as security enhancements

Some plant species, being spiny or thorny, can act as effective security barriers in conjunction with security fencing. Suitable shrubs include those listed in Box 10.6.

Plants for ground cover

Ground cover planting is used to cover large areas of open ground and reduce ongoing maintenance. Typical ground cover species are listed in Box 10.7. See also The Horticultural Trades Association's *The national plant specification*.

Box 10.6 Spiny or thorny shrubs suitable as a security barrier

- Hawthorn
- Blackthorn
- Berberis
- Chaenomeles
- Holly
- Pyracantha
- Sea buckthorn
- Gorse
- Roses

Box 10.7 Typical ground cover species

- *Berberis wilsoniae*
- *Ceanothus repens*
- *Cotoneaster horizontalis*
- *Euonymus fortunei*
- *Hebe* spp
- *Hedera* spp
- *Hypericum calycinum*
- *Lonicera pileata*
- *Pachysandra* spp
- *Potentilla* spp
- *Rosa* spp
- *Rubus calycinoides*
- *Sarcococca* spp
- *Vibernum* spp
- *Vinca* spp

Grassed areas

Turf provides an instant grassed surface, but it is essential that a suitable topsoil base is prepared properly and consolidated. Turf should be laid within 18 hours of delivery in summer and 24 hours during spring or autumn. Following top-dressing, brushing and rolling, it will require copious watering for several weeks. Seeding is less expensive, but requires the seeded areas to be kept free of traffic during the first growing season. Seed mixes are normally selected on the basis of the primary use of the area. Where grassed areas are used to provide hardstanding or access for fire engines, perforated pavings (eg of concrete) can be laid on a consolidated base. Plastics mesh and grid reinforcement can be used to allow occasional hardstanding on grass areas.

Water consumption

Effective landscaping can reduce the amount of water required for irrigation and watering of plants. Currently, external use of water for gardening is between 2–3% of total water consumption for properties. Drought-tolerant planting and automatic watering systems can both reduce water wastage. It is also critical to ensure that herbicides and pesticides do not contaminate local water courses.

Regulations and standards

Building Regulations

All construction, including external works, must be carried out in accordance with:

- Health and Safety at Work etc Act 1974, and
- The Construction (Design and Management) Regulations 1994.

These place an onus on designers to eliminate or reduce hazards where possible at design stage, and to pass on information on residual risks, such that they can be adequately managed during construction.

Some of the more important aspects of external works which are specifically covered in the Building Regulations for each part of the UK are as follows. Other references can be found in the footnotes.

Drainage

- **England and Wales:** AD H,
- **Scotland:** Section 3 of the Technical Handbooks, Domestic and Non-Domestic,
- **Northern Ireland:** Technical Booklet N.

Also comply with The Groundwater Regulations 1998.

Water

- **England and Wales:** The Water Supply (Water Quality) Regulations 2001.
- **Scotland:** The Water Supply (Water Quality) (Scotland) Regulations 2000, the Private Water Supply Regulations 2000 and the Water Byelaws 2000 (Scotland).
- **Northern Ireland:** The Water Supply (Water Quality) (Amendment) Regulations (Northern Ireland) 2003

HMSO

Health and Safety at Work etc Act 1974. London, The Stationery Office, 1974

The Construction (Design and Management) Regulations 1994. Statutory Instrument 1994 No 3140. London, The Stationery Office, 1994

The Groundwater Regulations 1998. Statutory Instrument 1998 No 2746. London, The Stationery Office

Also comply with The Water Supply (Water Fittings) Regulations 1999.

Gas
Gas Safety (Installation and Use) Regulations 1998.

Electricity
- **England and Wales:** AD P,
- **Scotland:** Section 4 of the Technical Handbooks, Domestic and Non-Domestic,
- **Northern Ireland:** Technical Booklet E.

Waste management
The Waste Management Licencing Regulations 1994.

This legislation applies to geotechnical treatment of contaminated soils. For more information see chapter 2 *Site investigation and preparation.*

The Stationery Office
Water Supply (Water Fittings) Regulations 1999. Statutory Instrument 1999 No 1148. 1999
The Water Supply (Water Quality) Regulations 2001
The Water Supply (Water Quality) (Scotland) Regulations 2000
The Water Supply (Water Quality) (Amendment) Regulations (Northern Ireland) 2003
Private Water Supply Regulations 2000
Gas Safety (Installation and Use) Regulations 1998. Statutory Instrument 1998 No. 2451
The Waste Management Licencing Regulations 1994. Statutory Instrument 1994 No 1056

Scottish Water Authorities
Water Byelaws 2000 (Scotland)

Standards
Some of the more important aspects of external works specifically covered in the national standards are as follows. Other references will be found in the footnotes on the relevant page.

Drainage
- BS EN 752 Parts 1–3, 6,
- BS EN 1610: 1998,
- BS EN 1671:1997,
- BS EN 1091:1997,
- BS 8000-14: 1989,
- BS EN 12566: 2000.

Water
- BS EN 805: 2000,
- BS 6700: 1987.

Electricity
- BS 7671: 1992.

Fencing
- BS 1722,
- BS 7818: 1995.

British Standards Institution
BS EN 752: *Drain and sewer systems outside buildings:*
    Part 1: 1996 *Generalities and definitions*
    Part 2: 1997 *Performance requirements*
    Part 3: 1997 *Planning.*
    Part 6: 1998 *Pumping installations*
BS EN 1610: 1998 *Construction and testing of drains and sewers*
BS EN 1671: 1997 *Pressure sewerage systems outside buildings*
BS EN 1091: 1997 *Vacuum sewerage systems outside buildings*
BS 8000-14: 1989 *Workmanship on building sites. Code of practice for below ground drainage*
BS EN 12566: 2000 *Small wastewater treatment systems for up to 50 PT*
BS EN 805: 2000 *Water supply. Requirements for systems and components outside buildings*
BS EN 806-2:2005 *Specifications for installations inside buildings conveying water for human consumption. Design*
BS 7671: 2001 *Requirements for electrical installations. IEE Wiring Regulations.* 16th edition
BS 1722: *Fences* (Various Parts)
BS 7818: 1995 *Specification for pedestrian restraint systems in metal*

11 Modern methods of construction

The term 'modern methods of construction' (known as MMC) covers a broad range of construction types ranging from complete housing systems built in factories through to new site-based technologies. Older terms such as 'system building', 'off-site assembly', 'industrialised construction' and 'modular construction' are still used by many builders and designers.

A simple classification of modern methods by built form is:
- volumetric construction,
- panellised systems,
- hybrid construction,
- sub-assemblies and components,
- site-based modern methods of construction.

The first four categories are usually manufactured in a factory; the term 'site-based modern methods' covers systems that do not fall neatly into the first four categories. This chapter describes the form and nature of all these construction types to provide background for reference. There is more guidance in BRE's pocketbook, *Modern methods of house construction: a surveyor's guide* and LPS 2020-1.

Ross K. *Modern methods of house construction: a surveyor's guide*. FB11. Bracknell, IHS BRE Press, 2005
Loss Prevention Certification Board (LPCB). *Standard for innovative systems, elements and components for residential buildings*. LPS 2020-1. Available from www.RedBookLive.com

Volumetric construction

Volumetric construction involves the production of three-dimensional units in a factory (Figure 11.1). The units are transported to site where they are stacked onto prepared foundations to form a building. A typical house is made from four units, whereas flats might comprise one or (more usually) two units depending on size.

All the necessary internal finishes, services and, potentially, the furnishings can be installed at the factory, with the complete entity able to be transported to site and assembled. Some external finishes can also be factory fitted, but usually some work has to be done on site in order to make good the join between units. Figure 11.2 shows modular units being assembled on-site. In this example, the external finish is of brick slips, most of which have been factory-fitted. A conventional 100 mm brick

Figure 11.1 Factory production of volumetric units

Figure 11.2 Volumetric units being stacked on site (Courtesy of Guinness Trust)

outer leaf could have been built on site after the modules had been stacked as an alternative to brick slips.

To date most volumetric construction has been in the hotel, student and key-worker accommodation, health care and fast food sectors, where repetition of built form (either at the building level, or at the room level) is common. However, this method of construction is also now being used for housing. Figure 11.3 shows a pair of semi-detached dwellings comprising four volumetric modules plus panellised roof.

Volumetric units can also be added to existing buildings to create new facilities.

Figure 11.3 A pair of semi-detached dwellings manufactured by Britspace

> **Key Point**
> Volumetric units need to be strengthened to increase their rigidity and prevent damage during transportation and craning into position.

A requirement of volumetric construction is that, in allowing the units to be transported and craned into position without being damaged, they need to be strengthened sufficiently to increase their rigidity. In some instances, temporary bracing is attached to the modules for transportation, but in other cases, additional or larger members are incorporated into the structure, effectively making the units stronger than they need to be for in-service requirements.

Volumetric construction can be in almost any material. Examples are known in steel, timber and concrete. If the frames are made of steel they are usually cold-rolled, light-gauge, galvanised C sections with a typical steel thickness range of 1–3.2 mm. With domestic properties, the whole structure can be made using light-gauge, cold-rolled steel. However, some manufacturers use thicker hot-rolled sections for corner posts and edge beams to give the extra strength required for transportation and lifting.

Due to the physical limitations of road transport, volumetric units are usually less than 4 m in width and, although unit lengths of 16 m are possible, lengths within the range of 8–12 m are more typical.

Panellised systems

Flat panel units are produced in a factory using purpose-made jigs or machinery to ensure dimensional accuracy and are transported to site for assembly (Figures 11.4 and 11.5).

Panels can be of a variety of materials and constructions, ranging from framed panels in either timber or steel, to concrete and composite panels such as SIPs (structural insulated panels). Framed panels can be of two basic types: 'open' or 'closed'. The meaning of these terms has changed over time. During the period that non-traditional housing was being constructed, an 'open' system was one that could accept components from any source, whereas

Figure 11.4 Flat panel units being produced in a factory (Courtesy of Space4)

Figure 11.5 Panel units being assembled on site

> **Key Point**
> Fixing insulation to the external face of the frame creates a 'warm frame' construction that is effective at reducing thermal bridging.

Insulation to external walls is normally applied to the outside of the frame on site (although it can be factory-installed), sometimes supplemented by more insulation between the studs (Figure 11.6).

The practice of fixing insulation to the external face of the frame creates a 'warm frame' construction that is effective at reducing thermal bridging across steel members and also allows services to be installed between the steel studs.

Steel-framed panel units are also now being fabricated on site using 'portable factories'. These facilities take rolls of galvanised steel stock and extrude the profiles required to make the panel on site. The rolling machines are quite sophisticated in that they can be programmed to cut the sections to length and pre-punch holes for rivets and other fixings accurately. This means that the resulting panels are dimensionally more accurate than those produced on site using 'stick' construction (ie where the building is constructed from lengths of stock material and accuracy depends on the skill of the fabricator).

a 'closed' system was one that could accept components from only a few sources that produced components specific to the system. In the context of modern methods of construction, the terms have the following interpretations.

- *Open panel system*. This is a framing system (metal or timber) delivered to site before insulation, services, etc. are fitted.
- *Closed panel system*. This is a complex panel system that can have services, windows, doors, internal wall finishes and external claddings fitted at the factory.

Steel panels

Steel-framed panels are most commonly 'open' and do not include insulation, lining boards, etc. Such panels are also referred to as 'sub-frames'.

Timber panels

Conventional timber-frame panels normally arrive on site with sheathing board fixed but with no insulation, cladding, etc. With modern systems insulation, service conduits, linings and window frames can all be factory-fitted. In the finished building it can be difficult distinguishing between 'modern' and conventional timber frames.

Concrete panels

Large concrete panels have been used since the 1950s in non-traditional construction, especially in flats and high-rise construction. Panels are usually two-dimensional (Figure 11.7), but can be more complicated, for example L-shaped, which makes them self-stable when positioned.

More sophisticated concrete panels are also being produced that incorporate factory-fitted cladding

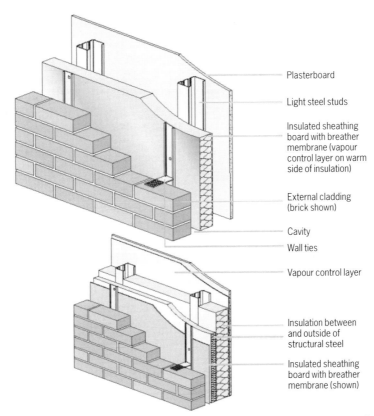

Plasterboard

Light steel studs

Insulated sheathing board with breather membrane (vapour control layer on warm side of insulation)

External cladding (brick shown)

Cavity

Wall ties

Vapour control layer

Insulation between and outside of structural steel

Insulated sheathing board with breather membrane (shown)

Figure 11.6 Typical steel frame constructions. © Crown copyright reproduced from *Limiting thermal bridging and air leakage: robust construction details for dwellings and similar buildings* 2001 with the permission of the Department for Communities and Local Government

Figure 11.7 Flat concrete panels being erected on site

and windows. One such system incorporates cladding of bricks cut lengthways to simulate a brick wall of stretcher bond.

Structural insulated panels (SIPs)

SIPs are essentially a sandwich construction comprising two layers of sheet material bonded to a foam insulation core (Figure 11.8). They have no internal studs within the panels and rely on the bond between the foam and the two layers of sheet material to form a load-bearing unit (though there may be studs at corners and around some openings). SIP panels are used in the same way as timber- or steel-framed panels. One advantage of SIP panels is that the insulation layer is more continuous than in normal framed panels. This leads to better thermal performance for a given thickness of panel because

Key Point
An advantage of SIP panels is that the insulation layer is more continuous than in normal framed panels.

Figure 11.8 Detail of a typical SIP showing the construction and one method of joining adjacent panels (Courtesy of Kingspan Tek)

Figure 11.9 A house being built using SIPs (Courtesy of Kingspan Tek)

of the absence of thermal bridges associated with studs. In the US where SIP construction has been used for a number of years, panels vary in thickness from 100 mm to around 300 mm.

The two layers of sheet material can be of a variety of materials; oriented strand board and cement-based boards are fairly common. The rigid

Bregulla J & Enjily V. *An introduction to building with structural insulated panels (SIPS)*. Information Paper IP 13/04. Bracknell, IHS BRE Press, 2004

foam core will usually be expanded polystyrene, polyurethane or polyisocyanurate.

Various devices, such as special cramps for use during assembly or cam-lock fittings built into the panels, are used to pull the panels together to assist with creating an airtight join.

BRE's *Information Paper IP 13/04* gives an introduction to building with SIPs.

Composite panels

There are a number of modern systems that produce walls of composite construction (strictly speaking, any panel that is made from a combination of different materials is a composite panel, though most are not referred to as such). Some, such as insulating formwork, are site-based methods, whereas others are factory-made panels. One type of 'modern' composite system has a foam insulation core within a wire space frame, all of which is encased in fine aggregate concrete. The systems range from:

● flat reinforced insulation panels that are erected and a render finish applied on site,
to:
● a system that comprises a number of standard panels along with a system of connecting brackets and channels to facilitate accurate assembly on site.

Hybrid construction

This method of construction is also known as semi-volumetric as it combines both volumetric and panellised approaches within the same building. Volumetric units can be used for the highly serviced areas such as kitchens and bathrooms units or pods, with the remainder of the dwelling being constructed with panels (Figure 11.10). Examples of hybrid construction have been built in a variety of materials, and it is feasible to use different construction materials for the different parts. For example, steel frame can be used for the volumetric element and timber frame for the panellised element. However, in such circumstances, care is needed in the design to cater for differences in thermal and moisture movements of the different materials.

Figure 11.10 Hybrid construction

Subassemblies and components

This section covers items that are not full 'systems', but which use factory-made components either within manufactured buildings or within otherwise traditionally built buildings. Components such as door sets, windows, stair strings and other manufactured components are commonly used in all forms of construction and therefore do not fall within the definition of modern methods of construction. The main items relevant to this chapter are innovative floor and roof constructions and modern composite joists.

Floor construction

A number of innovations have been introduced in floor technology for houses. Solid timber joists are being replaced by engineered products such as timber I beams (Figure 11.11) and lattice joists that are lighter and stiffer than solid timber. Although some techniques, such as steel lattices, have been used for many years, there have been considerable improvements in manufacturing processes and quality control. Joists up to ~ 0.5 m deep and up to 12 m long can be produced that are both lightweight and rigid, making it practical to span much larger distances without the need for intermediate structural support. Also, because of the relatively

Figure 11.11 Long spans resulting from the use of deep timber I beams (Courtesy of Excel Industries Ltd)

open structure that can be achieved, it is much easier to accommodate services within the floor.

Floor cassettes are prefabricated framed units that are delivered to site ready-assembled in the same way as wall panels (Figure 11.12). Openings for stair wells will normally be formed in the factory. The joists can be of timber, light-gauge steel or a composite.

The cassettes can be designed for long spans that require no support other than around the perimeter

Figure 11.12 A light-gauge steel floor cassette being installed

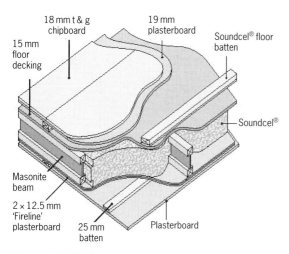

18 mm t & g chipboard

19 mm plasterboard

Soundcel® floor batten

15 mm floor decking

Soundcel®

Masonite beam

2 × 12.5 mm 'Fireline' plasterboard

25 mm batten

Plasterboard

Figure 11.13 A floor cassette designed specifically for good acoustic performance (Courtesy of Excel Industries Ltd)

Figure 11.14 A roof cassette panel being craned into place

(thus allowing open-plan accommodation to be produced), so may be much thicker to install than normal floors.

Floor cassettes can vary in the complexity of construction, ranging from simple box construction to systems designed for specific purposes such as separating floors with good acoustic properties. Figure 11.13 illustrates a floor cassette designed specifically for good acoustic performance.

Roof construction

Products and components for roof construction can be made up from materials that are similar to those for floor construction. The main difference is the need to incorporate thermal insulation. Roof cassettes comprise panels that span from eaves to ridge (Figure 11.14). They often require no intermediate structural support, although in some cases purlins are used. A range of designs and materials is employed. The panels may be of SIP-type construction or have internal studs, and panels can be hinged at the ridge. Both faces of the panels are normally finished with a sheet material, and

insulation is included within the structure of the panel. Tiling battens can also be factory-fitted. Roof panels are useful for room-in-the-roof constructions where the absence of structural timbers is beneficial. In some cases, the whole roof construction can be prefabricated on the ground and lifted into place.

Site-based modern methods of construction

This section relates to site-based assembly methods and the use of traditional components in an innovative way. Examples of construction forms that are generally accepted as 'modern methods of construction' are Tunnelform (cast-in-situ concrete using heated steel moulds), aircrete planks and thin joint blockwork. The next section describes other innovations in 'traditional' materials and techniques. These include the use of brick slips, insulating formwork and single-leaf masonry. These methods can be mixed with other forms of construction.

Tunnelform

Tunnelform construction is by no means new and has been used extensively in Holland and other parts of Europe for over 20 years. A substantial

---

Key Point

Roof panels are useful for room-in-the-roof constructions because structural timbers are absent.

Figure 11.15 Open-ended bays formed by Tunnelform

Figure 11.16 Thin layer mortar masonry
(Courtesy of H+H Celcon)

proportion of residential buildings are still built there using Tunnelform.

In this form of construction, 'L'-shaped steel shutters are used to cast concrete 'tunnels'. The moulds are heated overnight to accelerate the cure of the concrete and allow the moulds to be removed and re-used on a 24-hour cycle. Reinforcement and service conduits can be placed within the moulds as necessary before pouring the concrete, and openings for stairwells and interconnecting doors can also be formed. The resulting structure is a series of open-ended bays with concrete walls and ceilings. The open ends of the bays are closed with a different system, often a panel system, but could be anything including cavity masonry in a similar way to more conventional cross-wall construction. The bays can be sub-divided internally to make more than one room.

Aircrete (aerated concrete) products

There are two types of product that have recently been introduced into UK house building, namely thin joint blockwork and aerated concrete planks. As the name implies, thin joint blockwork is constructed using a thin (approx. 4 mm) bed of special mortar (Figure 11.16). The mortar is mixed on site with a standard plasterer's whisk attachment in an electric drill to produce a smooth, free-flowing adhesive. The blocks are larger than normal blocks, and are dimensionally more accurate. More information on thin layer mortar masonry can be

found in BRE's *Good Building Guide 58* and *Digest 438*.

Thin joint blockwork can be used as a direct substitute for conventional blockwork or in conjunction with aerated concrete planks, ie large reinforced planks of aerated concrete up to approx. 6 m in length, in a range of widths (up to approx. 600 mm) and thicknesses (200–300 mm). The planks can be used to form floor and roof elements of buildings. The whole structural envelope can be made of aircrete (Figure 11.17): the external walls will be of thin joint blockwork, internal walls will be of large thin joint blockwork, and floors and roof will be of aerated concrete plank construction.

de Vekey R. *Thin layer mortar masonry.* Good Building Guide 58. Bracknell, IHS BRE Press, 2003
de Vekey RC. *AAC 'aircrete' blocks and masonry.* Digest 468 (2 Parts). Bracknell, IHS BRE Press, 2002
BRE. *Aircrete: thin joint mortar systems.* Digest 432. Bracknell, IHS BRE Press, 1998

Figure 11.17 Inside a house being constructed with aircrete products

Figure 11.18 Insulating formwork under construction

Other innovations in traditional materials and techniques

Insulating concrete formwork

Insulating concrete formwork derives its name from the fact that an insulation material (usually expanded polystyrene) is used as permanent shuttering for a cast-in-situ structural concrete wall. A number of systems exist, some of which are based on two sheets of insulating material tied together, whereas others are in the form of large hollow building blocks.

This type of construction has been popular in the self-build sector because of the ease with which an energy-efficient, airtight construction can be produced with minimal construction skills being required on the part of the self-builder, and because the mould materials are easily man-handled without the need for lifting equipment. Reinforcement can be introduced locally to add strength, for example over openings to create integral lintels as the concrete is poured. Figure 11.18 shows an insulating formwork building under construction.

Brick slips

An increasingly common feature of off-site manufactured housing is the use of brick slips that have been fixed in panels on the external walls of the property to give the appearance of a 'traditional' brick outer leaf. Brick slips are produced in two main ways, either by cutting the faces off a normal brick (each brick yielding up to two brick slips) or by extrusion (ie the wet clay is pressed through a die and cut using wire in the same way that wire cut bricks are produced). The slips vary in thickness according to the manufacturer, but 15–20 mm is typical. L-shaped slips are produced for corner details.

The slips are usually adhesively bonded to a substrate that can be either some form of backing sheet (eg a galvanised steel sheet with a pvc coating) or a profiled rigid foam that also provides the thermal insulation to the structure (Figure 11.19). Figure 11.20 shows another system where a profiled extruded brick slip is clipped into a metal profile and therefore does not rely on an adhesive bond to retain the slip. Because of the unique profile of the brick slips in this system they are somewhat thicker than other slips and corner slips are produced by a bonded mitre joint.

Figure 11.19 Brick slips bonded to polystyrene insulation before pointing

Figure 11.20 Brick slip system designed for mechanical fixing

12 Access to buildings

In recent years there has been an increasing push to improve the accessibility of buildings for all and especially for disabled people. This has come about with the realisation that:

- in addition to disabled people, elderly people, pregnant women, small children, injured or other temporarily disabled people will benefit from an enhanced accessibility of the built environment and
- accessible design, which will make buildings more functional, will be to the benefit of all.

At the cross-point of these deliberations, numerous, often overlapping, concepts have been introduced: universal design, design for all, barrier-free design all claim different perspectives on an otherwise common goal, ie to create high quality and non-discriminatory building environments.

Improved access is being driven by changes to building regulations and the Disability Discrimination Act (1995). Relevant guidance has been developed in the form of BS 8300 and the supporting documents of the national building regulations:

- **England and Wales:** AD M,
- **Scotland:** Sections 3 and 4 of the Technical Handbooks, Domestic and Non-Domestic,
- **Northern Ireland:** Technical Booklet R.

In Scotland, a decision was made in 1999 to remove the specific building standard related to disabled access and instead to distribute the access provisions amongst other standards; this approach remains with the advent of the new Building Scotland Act (2003).

> **Key Point**
> The built environment can be designed to anticipate and overcome restrictions that prevent disabled people making full use of premises and their surroundings.

An accessible environment is one which a disabled person can enter and make use of independently or with help from a partner or assistant.

This chapter sets out basic requirements for accessible buildings. Designers should, however, refer specifically to the guidance in BS 8300 or the building regulations. There will also be some specialist areas concerned with the provision of assistive technology or non-standard buildings that are not specifically covered by this guidance. The Joseph Rowntree Foundation has produced specific guidance on lifetime homes (Carroll et al 1999), this is one branch of housing that sets standards for improved access for all. Other terms such as barrier-free homes will also be encountered by those seeking to achieve accessible housing.

HMSO. *Disability Discrimination Act 1995 (c. 50).* Amended 2005. London, The Stationery Office, 2005
British Standards Institution. BS 8300: 2001 *Design of buildings and their approaches to meet the needs of disabled people. Code of practice*

Designers need to remember that there are differences in the requirements between domestic and non-domestic buildings, although the principles set out in this guidance cover elements of both. The requirements of building regulations should be met in full.

Disability Discrimination Act, 1995

The Disability Discrimination Act, 1995 (DDA) made it unlawful to discriminate against disabled persons in connection with employment, the provision of goods, facilities and services or the disposal or management of premises; to make provision about the employment of disabled persons; and to establish a National Disability Council [8th November 1995]. The National Disability Council has been replaced by the Disability Rights Commission. The DDA, although not strictly dealing with buildings, has had a significant impact on the design of buildings, particularly in the adaptation of buildings to improve access. This often presents a challenge to building owners in adapting existing buildings to a good standard of accessibility.

The DDA 1995 in fact aims to end the discrimination that many disabled people face. It is human rights legislation that gives disabled people rights in the following areas:

● employment,
● education,
● access to goods, facilities and services,
● buying or renting land or property.

The DDA also allows the government to set minimum standards so that disabled people can use public transport easily.

The DDA defines a disabled person as someone who has a physical or mental impairment that has a substantial and long-term adverse effect on his or her ability to carry out normal day-to-day activities. The DDA 2005 amendment extends the definition of disability, removing the requirement that a mental illness should be clinically well recognised.

Carroll C, Cowans J & Darton D (Eds). *Meeting Part M and designing lifetime homes.* York, Joseph Rowntree Foundation, 1999

Service provision is covered by Part III of the DDA and has had the greatest impact on the built environment. Since October 2004 service providers have had to make reasonable adjustments to the physical features of their premises to overcome physical barriers to access. Examples of adaptations that could be made are:

● putting in a ramp to replace steps,
● providing a larger, well-defined sign for people with a visual impairment,
● improving access to toilet or washing facilities.

The DDA allows the Government to issue regulations (usually in the form of statutory instruments) giving more detailed legal requirements covering a variety of areas. These regulations amount to legislation. Although there is a regulation related to the services and premises (1996), there is no specific guidance to the DDA with regard to the removal of physical barriers from buildings in order to enable service provision. It should be noted, however, that the DDA requires reasonable provision. In the event of a legal case being advanced due to alleged discrimination the court will decide upon the provisions for access made.

The Disability Rights Commission (DRC) is the statutory body set up by the government to help secure civil rights for disabled people. The Government and the DRC have produced a number of Codes of Practice, explaining legal rights and requirements under the DDA 1995. These codes are practical guidance, particularly for disabled people, employers, service providers and education institutions, rather than definitive statements of the law. However, courts and tribunals must take them into account.

Car parking

This is important for disabled people as car transport may be the only form of transport that is viable for some people or under certain circumstances.

Sufficient allocated parking spaces for disabled people should be provided, the number of which depends on the type of building. For example, a designated parking place at a workplace is necessary

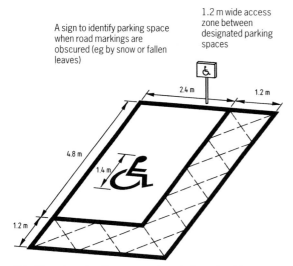

A sign to identify parking space when road markings are obscured (eg by snow or fallen leaves)

1.2 m wide access zone between designated parking spaces

2.4 m

1.2 m

4.8 m

1.4 m

1.2 m

Note: Dimensions of parking space are to centre lines of markings

Figure 12.1 Markings for designated off-street parking spaces. Reproduced from BS 8300 (Figure 3) by permission of BSI. Details of how to obtain British Standards are given on page 306

for every disabled employee, plus a sufficient number for visitors (one space or 2% of capacity). For churches at least two parking places are recommended (BS 8300).

Off-street parking spaces of dimensions 4.8 × 2.4 m should take the car; 1.2 m space should be provided between spaces and 1.2 m at the boot space (Figure 12.1).

Access routes

The number of barriers, restrictions or other hazards on the approach to and from a building should be as few as possible. Low-level bollards and chain-linked posts, for example, are particularly hazardous to blind and visually impaired people. For disabled people who need space when moving about, the provision of narrow approaches creates difficulties. The more frequently a specific route is used, the more often people will need to pass each other, and the wider the access route required. Uneven surfaces

British Standards Institution. BS 8300: 2001 *Design of buildings and their approaches to meet the needs of disabled people. Code of practice*

of loose materials and large gaps between paving materials cause problems for wheelchair users, people with impaired vision and people who are unsteady on their feet.

The longer or steeper an access route is, the greater a barrier it is to access for many building users.

Drop-off points on a roadway, or car parking designated for use by disabled people, require a short, direct and level route of no more than 45 m to the building.

Where the accessible route does not incorporate ramped access, the distance can be increased as long as resting points are provided at no more than 50 m intervals.

Width requirements
To permit adequate space for persons to pass, provide a footpath width of at least 1500 mm. Where traffic on an access route will be busy, consider a minimum width of 1800 mm. The footpath can be reduced locally to 1000 mm or more around unavoidable obstructions such as existing buildings or trees.

Lesser access route widths may be permissible in lightly trafficked areas. In such cases, use passing places to allow wheelchair users to pass safely.

Surface requirements
The surface of any accessible route should be uniform and of a material and finish that provides a level of traction that will minimise the possibility of slipping. This should take into account both anticipated use and environmental conditions. Where there are differing materials used on the route, they require similar frictional characteristics.

Surfaces should be firm and permit ease in manoeuvring; do not use cobbles or loose-laid materials like gravel. See also chapter 10 for additional information on pavings, and chapter 6 for more guidance on slipperiness.

Provide a visual contrast between the surface of an accessible route and any adjacent ground.

The surface of an accessible route, whether composed of modular paving units, tarmac or another suitable material should present no variation in surface profile greater than 5 mm; this is to limit projections that may offer trip hazards and avoid depressions that may result in standing water.

To minimise the possibility of tripping or entrapment, the following is recommended:

- a difference in level between units of no more than 5 mm,
- recessed filled joints no more than 10 mm wide and 5 mm deep, and
- unfilled joints no more than 5 mm wide.

Ramps

If the slope of an access route is steeper than 1 in 20, it is classified as a ramp and should comply with the requirements in Table 12.1.

A minimum ramp width of 1200 mm is required. Provide landings at the foot and head of a ramp with any intermediate landings at least 1500 mm long and clear of any swing or other obstruction. Unless it is under cover, design a slight cross-fall gradient not exceeding 1 in 50 to drain surface water from the landing.

Adequate handrails are required; recommendations are given in Box 12.1.

To prevent falls where there is a difference in level between the edge of a ramp and the adjacent floor, include precautions such as an upstand at least 100 mm high to any open side of the ramp with appropriate guarding. The provision of handrails is a separate issue from the provision of protective barriers.

An external ramp flight should have one of the following:

- a landscaped margin, level with the edge or the ramp for a distance of 600 mm before any grading; or
- a kerbed upstand, at least 100 mm high, on any open side of a flight, together with appropriate guarding.

Table 12.1 Pedestrian ramp requirements	
Going of a flight	Maximum gradient
10 m	1 in 20
5 m	1 in 15
Not exceeding 2 m	1 in 12

For goings between 2 m and 10 m, it is acceptable to interpolate between the maximum gradients, ie 1 in 14 for a 4 m going or 1 in 19 for a 9 m going

BRE undertook research into access ramps for the Office of the Deputy Prime Minister (ODPM) during 2001–2003. The research project involved a range of ramp issues, all of which were related to improving the design and safe use of disabled access ramps. This was in response to the greater emphasis in the UK construction industry on the provision of access to buildings for disabled people following the introduction of the Disability Discrimination Act and subsequently the changes in the building regulations and the introduction of BS 8300.

The ramp requirements of UK legislation are broadly similar to those of Australia and the USA, two countries that are well advanced in the field of accessibility.

Ramps are a fundamental part of building accessibility and their design must ensure that they are easy to use by the whole population (all of whom may have slightly differing needs) in a range of weather conditions.

It is clear that safe access and use of ramps is dependent on a number of design features, in particular the following:

- construction and materials,
- gradient,
- ramp width and surface,
- landings,
- handrail provision (ie number, positioning, size, shape),
- kerb guards,
- signage and lighting.

The incidence of ramp-related accidents in the UK is low and most injuries are of less than moderate harm. This may well indicate that current guidance is correct and that the resultant good ramp design is removing, or reducing, various risks.

To ensure that the requirements of the guidance in the ADs meet those of ramp users, they were compared with the results of work where experimental ramps were trialled by people with a range of abilities. A series of recommendations for the safe design of ramps was drawn up by BRE and is summarised in Box 12.2.

The use of an upstand kerb alone in open landscaping is not recommended as it may present a potential trip hazard.

The unobstructed length of a landing on a ramp should be not less than 1.5 m to allow users sufficient space to arrest forward movement. However, this issue is not relevant to a landing serving only the top of a flight, which may, therefore, have a reduced length of at least 1.2 m.

Stairs and steps

Ambulant disabled people generally find it easier to negotiate a flight of stairs than a ramp, although the presence of handrails for support is essential. For steps to be comfortable to use then the rise and going need to reflect stride length while keeping dimensional limits. Excessively high risers may result in excessive strain being placed on the knee or hip of ambulant disabled people. Steps without protective nosings are preferred. If a nosing does exist, a rise of between 150 mm and 170 mm is preferred, and the going preferably 300 mm with an overlap of no more than 25 mm.

No flight in a stepped access route should contain more than 12 risers and, as far as possible, the numbers of risers in successive flights should be uniform. A level landing should be provided at the top and bottom of each flight of steps.

Protective barriers are used where there is a sudden change in level and the possibility of severe injury from a fall. Where a significant drop occurs or where the location of any change of level increases the risk of injury, guarding should be provided.

Any drop at a change in direction on an access route can be a hazard, particularly to a person using a wheelchair or to someone with a visual impairment.

Entrance doors

The following provisions represent good practice for entrance doors:

- a canopy or other means of protecting people from the weather,
- an unobstructed space to the opening face, next to the leading edge, of at least 300 mm (unless the door is automatically controlled),
- any ironmongery or fittings provided to operate the door that contrast visually with the surrounding surfaces and that are of a form that can be used by a person with limited manual dexterity,
- if fitted with a closing device, be operable with an opening force of no more than 30 N (for the first 30° of opening) and 22.5 N (for the remainder of the swing) when measured at the leading edge of the leaf.

Effective clear door-opening widths are given in Table 12.2.

Box 12.2 Recommendations for safe ramp design

- Ramp surfaces must be slip-resistant and have colour contrasting between ramps and landings. In addition, tactile indication of the change in level would be beneficial.
- The risks, in wet and inclement weather, of using a 1:12 ramp may suggest that the ramp gradient should be set at 1:14 where the ramp is not covered and sufficient room exists to accommodate the ramp.
- The minimum surface width could be increased to 1600 mm to accommodate a higher percentage of wheelchairs.
- Landing lengths may need to be increased to 1600 mm to facilitate the opening and negotiating of doors.
- The height of safety kerbs should be increased to 150–170 mm and their use encouraged by fitting handrails and second rails.
- The handrail diameter should be reduced from 46 mm to 32 mm. This will facilitate safe use by a wider user group. Promoting oval and circular handrail styles would be beneficial.
- Clearer guidance on signage should be given to indicate the presence of a ramp, including pictorial signage, colour contrast and positioning signage at eye level, rather than at height.

Table 12.2 Effective clear door opening widths

	Preferred effective clear width (mm)
Straight on (without a turn or oblique approach)	800
At right angles from an access route at least 1500 mm wide	800
At right angles from an access route at least 1200 mm wide	825
At right angles from an access route at least 900 mm wide	850

Other design issues are as follows.
- Provide a threshold (internal and external) of no more than 15 mm upstand, chamfered or rounded if greater than 5 mm.
- Give weather protection by use of a canopy or recessed entrance.
- Provide visual clarity:
  - ❏ contrast with surroundings,
  - ❏ well lit and clearly signed,
  - ❏ not of high polished material (eg stainless steel),
  - ❏ give a clear view of the interior of the building or entrance lobby (using vision panels where possible).
- Door furniture — use handles that:
  - ❏ have a lever action,
  - ❏ are not cold to the touch,
  - ❏ are operable with one hand,
  - ❏ have visual contrast with the door.

Consistent location and design of the door furniture throughout a building is necessary.

Guidance on designing level external thresholds is given in BRE's *Good Building Guide 47.*

Powered doors

A powered entrance door may be provided to ensure a building is accessible. This may be preferable in many buildings as it will increase both general amenity and the range of people that may enter the building unaided. Powered doors are generally used in non-domestic buildings, however, they are used in housing to assist access by disabled people. BRE's guide *Domestic automatic doors and windows for use by elderly and disabled people, Good Building Guide 48* and *Information Paper 2/02* give further guidance.

Powered doors are commonly used at entrances to a building where the public have access. They can be controlled by either:
- a manually activated push-pad, or other activating device, or
- an automatic sensor, such as a motion detector or pressure mat.

In addition to the general requirements for accessible entrances, powered doors require safety features to ensure safe passage.

Guidance on safety aspects of automatic doors is given in BS 7036: Parts 1–5 which requires prominent signage to identify this type of door.

Inside buildings

To enable a building to be used and enjoyed, it must be accessible to everyone where people can move throughout a building and use the facilities present without unnecessary assistance.

Lobbies

The recommended width of a lobby is the same as for corridors. However, for a lobby where either of the doors can be secured against access, its width should be no less than 1500 mm wide to permit a person in a wheelchair to turn around.

Stirling C. *Level external thresholds: reducing moisture penetration and thermal bridging.* Good Building Guide 47. Bracknell, IHS BRE Press, 2001

BRE
Garvin SL. *Domestic automatic doors and windows for use by elderly and disabled people: a guide for specifiers.* BR 334. 1997
Garvin SL. *Installing domestic automatic door controls.* Good Building Guide 48. 2001
Garvin SL. *Whole life performance of domestic automatic door controls.* Information Paper IP2/02. 2002

British Standards Institution. BS 7036: 1996 *Code of practice for safety at powered doors for pedestrian use* (5 Parts)

Corridors

Corridors and passageways within a non-domestic building need to be wide enough to allow two-way traffic and provide adequate space for manoeuvring at junctions and when passing through doorways. To permit this, all corridors require an unobstructed width of at least 1200 mm.

Unless a corridor has a continuous unobstructed width of at least 1800 mm, localised widening of corridors to accommodate an area of 1800 mm in both length and width, free of obstructions should be provided at any changes of direction, junctions and any point within a corridor where people may congregate.

Projections into a corridor should be avoided wherever possible.

Surfaces

Floor surfaces in a building need to be uniform, permit ease in manoeuvring and of a material and finish that minimises the possibility of slipping. Where there are differing materials as part of a circulation route, they require similar frictional characteristics to reduce the potential for a fall.

Floors in a building should generally be level, other than to provide transition between different levels. Gradients between 1 in 50 and 1 in 20 may disorient some building users.

Internal doors

Internal doors in buildings are recommended to:
● have a door leaf that provides a clear opening width in accordance with Table 12.3,
● have, in all leafs, a clear glazed panel, of at least 150 mm width, giving a zone of visibility from a height of not more than 500 mm to at least 1500 mm above finished floor level, which may be interrupted by a solid element between 800 mm and 1150 mm above floor level if required,
● when of manual operation, provide an unobstructed space to the opening face of the door, next to the leading edge, of at least 300 mm,
● have ironmongery that contrasts visually with the surrounding surfaces,

Table 12.3 Opening widths for door leafs

Unobstructed space adjacent to door	Effective clear width of doorway
1500 mm or more in front of each face of the door	800 mm
1200 mm or more in front of each face of the door	825 mm
900 mm or more in front of each face of the door	850 mm

● be capable of being used by a person with limited manual dexterity, eg be opened using a lever or pull handles (Figure 12.2),
● if principally of glass, have guarding and manifestation.

Door-closing

To ensure that closing devices do not present a barrier, the opening force is of no more than 30 N (for the first 30° of opening) and 22.5 N (for remainder of swing) when measured at the leading edge of the leaf. These forces may be increased by approximately 2 N when measurement is made at the point that ironmongery is positioned, up to

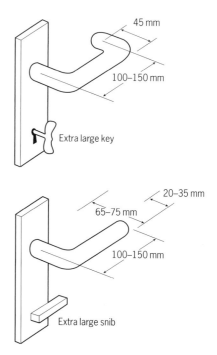

Figure 12.2 D (top) and L (bottom) type handles for doors

60 mm in from the door edge to simulate normal operation. Manual door closers are covered by BS EN 1154.

Vertical circulation

This is normally achieved between storeys by the provision of a passenger lift meeting the recommendations of BS EN 81-70.

Vertical circulation within a storey

Changes of level within a storey involve either of the following:
- a ramp,
- a lifting device.

Sanitary accommodation

> **Key Point**
> It is important that buildings are provided with sanitary accommodation that meets the needs of the occupants or users in terms of both availability and accessibility.

Sanitary facilities that can be reached and used by building occupants is an essential basic requirement.

If it permits future alteration of fitting type or position to enhance access for a person with a disability, without alteration of wall or door positions, it will greatly enhance the flexibility and sustainability of the building.

The number of sanitary facilities for males and females are calculated from the maximum number of persons the building is likely to accommodate. With the exception of small buildings, all buildings should provide separate facilities for males and females.

British Standards Institution

BS EN 1154: 1997 *Building hardware. Controlled door closing devices. Requirements and test methods*
BS EN 81-70: 2003 *Safety rules for the construction and installation of lifts. Particular applications for passenger and goods passenger lifts. Accessibility to lifts for persons including persons with disability*

> **Key Point**
> Within a house, there should be accommodation for at least one WC or waterless closet, one washbasin and one bath or shower, designed to permit use by all.

Wheelchair accessible toilets include the following:
- sufficient space to enter, manoeuvre and use facilities in privacy, both unassisted or with a companion, including an unobstructed space sufficient to turn a wheelchair,
- a consistent layout and relationship between WC, flanking wall and wash basin,
- a height of WC seat above floor level of 480 mm,
- an assistance alarm, with a visual indicator and audible tone distinguishable from a fire alarm provided both within and outside the compartment, that can be operated or reset from the seated WC position, and
- assistive aids in the form of fixed and folding grab rails. Any rails are to securely fixed to walls and capable of withstanding loads, applied in the normal direction of use.

A maximum horizontal distance of 45 m to be travelled to a wheelchair accessible toilet is advised.

Communication in buildings

Appropriate levels and forms of natural and artificial lighting and the reduction of glare and shadow will assist any person, visually impaired or not, to navigate through a building. It is a prerequisite to successful implementation of other elements of visual amenity such as visual contrast and signage.

People with a hearing impairment require assistance normally in the form of an electronic hearing enhancement system. There are three forms of system commonly in use:
- induction loop,
- infra-red, and
- radio systems.

The installation of such a system enhances sound communicated to the user, either through a standard hearing aid, or through an additional receiver supplied as part of the system.

Location and anticipated use will determine the most appropriate system for a given situation. The presence and type of hearing enhancement system requires appropriate signage for users.

Any fixed communication device fitted within a building, such as a door entry or lift intercom system requires a facility to assist users with a hearing impairment. This may take the form of an inductive coupler compatible with the 'T'-setting on a hearing aid, together with a visual indicator that a call has been made and received.

Keypads follow the 12-button telephone convention, with an embossed locater to the central '5' digit.

Permanent instruction on the operation of any device should be provided adjacent to any call point. This is particularly important where a system uses a keypad and multiple-digit entry for activation. In managed buildings, use an additional assistance button that can be easily operated by a single action.

The use of colour and contrast in fixed building elements will assist all building users, particularly those with a visual impairment, in wayfinding, use of facilities and also in the avoidance of hazards.

Factors such as colour and the surface finish of materials affect how a difference in contrast is perceived. Annex G of BS 8300 gives general advice on this issue.

Due to the inconsistency of natural lighting levels in external locations, tactile warnings are required in situations such as road crossings and stepped changes in level.

The provision of informative signage in a form that is consistent, clear and easy to understand will allow the building to be navigated with ease.

Access auditing

Access auditing gives an indication of the accessibility of an existing building at one point in time (CAE 2004). Such audits are a useful starting point in assessing the current state of accessibility and usability of existing buildings. Buildings that are designed from the outset for disabled people are likely to be more flexible and make it easier for

employers and service providers to meet the requirements of the DDA.

Access audits can be undertaken in a variety of ways and there is no standard for this process. The process normally involves the collection of data on the building through a series of checklists. The data gathered is compared with good practice as defined in BS 8300 or other documents. Improvements can be made to the accessibility of the building by identifying inaccessible areas and features.

Access plans and strategies are the means of ensuring that the information gathered and recommendations made in the access audit are effectively used. It is preferable to carry out further audits at regular intervals to ensure that access provisions continue.

Access statements are used in the design of new buildings to define and incorporate the principles of accessible design. They can be used in support of building regulation applications. Access statements have been used in the planning process in applications for large, high-profile schemes. This allows the access approach to be determined and allows communication between all parties in a construction contract. The statement provides an audit trail to determine whether or not particular aspects have been taken into account. It should not be a static document, but will evolve with time over a project. It should be kept as part of a health and safety file by the building owner.

One of the main problems when considering the accessibility of a given location is to be capable of measuring it, define some kind of methodology or criteria to be able to compare two different buildings or locations. Accessibility can be assessed in several ways, but measurement would enable the comparison of different designs of new buildings or performance of existing buildings.

A European project on accessibility, POLIS, is developing a means of measuring accessibility (www.polis-ubd.net) (Sakkas 2004). This is done in

British Standards Institution. BS 8300: 2001 *Design of buildings and their approaches to meet the needs of disabled people. Code of practice*
Centre for Accessible Environments (CAE). *Designing for accessibility.* 2004 edition. London, CAE

the context of universal building design (UBD). The methodology is based on buildings that supply services and have service paths between each. This type of approach could be a valuable contribution to the accessibility of new buildings; however, it would need to be used as a voluntary approach to reinforce the standards and building regulations in the UK.

Assistive technology

The variable needs of people as they progress through their lives in terms of their abilities to carry out everyday functions, perform their jobs and undertake their chosen leisure pursuits has a significant impact on the physical environment they occupy and on the tools they require to undertake tasks. One of these tools is 'assistive technology' (AT). AT is the generic name given to technologies used in education, rehabilitation, employment, service provision and independent living to help train, assist and support people with disabilities to obtain the maximum quality of life possible commensurate with the level of their disability.

In the United States of America, the Assistive Technology Act of 1998 includes the following definition:

'assistive technology device means any item, piece of equipment, or product system, whether acquired commercially, modified, or customized, that is used to increase, maintain, or improve functional capabilities of individuals with disabilities'.

In recent years, the term 'smart homes' has come into common usage. It is the popular name for the integration of 'telematics' into the electrical installation of a home. Many products in common home usage contain microchips and can be controlled by means of independent controllers, eg television, video recorders and central heating controllers. In 'smart homes' technology a large number of sensors and controllers are connected together to interact, thus making the home seem 'intelligent'. This ability to automate and control functions centrally is of great assistance to disabled and older people. However, it also has attractions for non-disabled people by providing increased comfort and energy saving.

Essentially, AT is technology that assists a person to undertake a task that without that technology would be difficult or otherwise impossible to execute. Thus, at a basic level a hammer is a form of AT and so is a computer that allows a person to send a message instantly to any other part of the world. In terms of providing assistance to disabled persons, AT is encountered in four main areas, as follows:

- *Access:* help in gaining access to and moving within a building level or between levels of a building, for example, automatic opening of doors, audio-visual indication of floor level and direction of travel of an elevator, 'hands-free' operation of facilities such as flushing toilets.
- *Communication:* a means of providing the user communications internally and externally, eg adaptations of computer interface devices and software, text phones for the deaf, speech devices for the blind or visually impaired.
- *EADL* (Electronic Aid to Daily Living): a means of giving the user the ability to interact with and manipulate one or more electronic appliances, eg television, CD player, lights, blinds, ventilation, etc.
- *Furniture and equipment:* covers a wide range of activities and tasks such as sport and leisure, personal care, mobility, seating and beds, moving and handling and household, kitchen and bathroom equipment.

The level of incorporation of AT into buildings is dependent on the type of facility and the type of people that facility is intended to serve. For example, it is likely a home for a disabled person would include more assistive technology than a work environment as the home has to cater for more human functions than an office environment does; but both would contain common features and technologies that would provide a certain level of access and assistance.

Sakkas N. *Design support tools and policy initiatives in support of universal design in buildings: the POLIS Project.* Designing for the 21st century III: an international conference on universal design, 7–12 December 2004, Rio de Janeiro, Brazil

Clearly, it is not practicable to include technologies that would serve all types of disabled people in all building archetypes at the outset. However, it should be possible to ensure that the introduction of specialised technologies to suit a particular special need can be achieved relatively easily. A good example would be when an employee becomes disabled and is not able to undertake his job function without the use of an assistive technology. It makes financial sense to be able to provide that technology as the cost of absence, recruitment and training a replacement can be much more expensive than acquiring the AT; and, perhaps more importantly, that person's sense of self-worth

and quality of life is greatly enhanced by assisting them to remain in employment.

Assistive technologies come in many varieties from simple 'dumb' tools, like ergonomically designed cutlery to 'smart' tools, such as microprocessor-controlled communication systems. The past 2–3 decades have seen an explosion in the availability of Information and Communications Technologies (ICT). Research has shown that there are literally tens of thousands of AT products in the marketplace that broadly fit into the four main AT areas noted above. They can be categorised as given in Box 12.3.

Box 12.3 Available assistive technologies (AT)

Architectural elements
Internal environment, safety and security, external environment, vertical lift, lighting, signage, ramps, handrails.

Blind and low vision
Computers, education aids, health care, information storage, kitchen aids, labelling, magnification, office equipment, orientation and mobility, reading, recreation, sensors, telephones, time, tools, travel, typing, weather, writing (Braille).

Communications
Alternative and augmentative communications (AAC), headwands, mouthsticks, signalling systems, telephones, typing, writing.

Computers
Software, hardware, accessories.

Controls
Environmental controls, control switches, remote switching.

Deaf and hard of hearing
Amplification, hearing aids, driving, recreational electronics, signal switches, speech training, telephones, time.

Deaf/blind
Braille telephone devices, tactile paging, tactile measurement, vibrating pedestrian signal.

Education
Classroom equipment, instructional materials.

Health, safety & security
Remote monitoring, radio-frequency identification [RFID], automatic fire and smoke detection, carbon monoxide detection, intruder detection, movement detectors, anti-scalding devices, entry systems).

Home management
Food preparation, housekeeping, furniture, utensils, automatic doors and windows.

Mute/speech impediment
Text to speech, audio visual.

Orthotics
Extensions, splints, supports.

Personal care
Feeding, carrying, drinking, holding, grooming/hygiene, transfer, toileting, dispensing aids, bathing, handle padding, clothing, dressing, health care, child care, stretching.

Prosthetics
Lower extremity, general, upper extremity.

Recreation
Crafts, reading, sewing, sports, toys, music, gardening.

Seating
Seating systems, cushions, therapeutic seats, beds.

Therapeutic aids
Thermal, water, pressure, massage, sensory integration, evaluation, positioning.

Transportation
Mass transit, transit facilities, vehicles, vehicle accessories.

Walking
Canes, crutches, walking frames.

Wheeled mobility
Manual, sport, powered wheel chairs, chair alternatives, chair accessories, carts, transporters, stretchers.

Workplace
Assessment, training, work stations, office equipment, tools.

Regulations and standards

Building regulations

Improved access to buildings has been spurred on by changes to the Disability Discrimination Act (1995) and to national building regulations:

- **England and Wales:** AD M,
- **Scotland:** Sections 3 and 4 of the Technical Handbooks, Domestic and Non-Domestic,
- **Northern Ireland:** Technical Booklet R.

In Scotland, the specific building standard related to disabled access was removed in 1999 when access provisions became an integral part of other standards whenever relevant.

Standards

BS 8300 provides guidance in support of the building regulations.

HMSO. *Disability Discrimination Act 1995 (c. 50).* Amended 2005. London, The Stationery Office, 2005
British Standards Institution. BS 8300: 2001 *Design of buildings and their approaches to meet the needs of disabled people. Code of practice*

13 References and further reading

The following list of references and further reading is not exhaustive but it will provide a useful resource. The list is correct at the time of publication but regulations, standards and other information are being reissued frequently. Readers should check with the publishers for the latest revisions. Contact details for the more frequently occurring publishers/distributors are included to assist in locating their publications. A discussion of the national building regulations is included at the end of the chapter 1.

Association of Geotechnical Specialists (AGS)
AGS publications are available from www.ags.org.uk

BRE
BRE and BRE Trust publications are available from IHS BRE Press: Tel 01344 328038; email brepress@ihsatp.com; www.ihsbrepress.com

For BREEAM products see www.breeam.org

For lists of approved products and services see www.RedBookLive.com

For lists of certified Environmental Profiles see www.GreenBookLive.com

Brick Development Association (BDA)
BDA publications are available from www.brick.org.uk/

British Standards Institution (BSI)
BSI publications are available from www.bsonline.bsi-global.com; email: cservices@bsi-global.com

Comité Européen des Assurances (CEA)
CEA publications are available from www.italique.lib@wanadoo.fr

Chartered Institution of Building Services Engineers (CIBSE)
CIBSE publications are available from www.cibse.org

CIRIA
CIRIA publications are available from www.ciria.org.uk

Health and Safety Executive (HSE)
HSE publications are available from www.hsebooks.com

Institution of Civil Engineers (ICE)
ICE publications are available from www.thomastelford.com/books/

Loss Prevention Certification Board (LPCB)
LPCB publications are available from www.redbooklive.com

NHBC
NHBC publication are available from www.nhbcbuilder.co.uk

The Stationery Office (TSO)
TSO publications are available from www.tsoshop.co.uk

TRADA Technology Ltd
TRADA publications are available from www.trada.co.uk

Chapter 1: Introduction

BRE

Digests

BRE. *Low-rise building foundations: the influence of trees on clay soils.* Digest 298. Revised 1999

McGrath C & Anderson M. *Waste minimisation on a construction site.* Digest 447. 2000

Good Building Guides

BRE. *Construction and demolition waste.* Good Building Guide 57. 2 Parts. 2003

BRE. *Climate change: impact on building design and construction.* Good Building Guide 63. 2004

Good Repair Guide

BRE. *Damage to buildings caused by trees.* Good Repair Guide 2. 1996

Information Papers

Roys M & Wright M. *Proprietary nosings for non-domestic stairs.* IP 15/03. 2003

Pollution Control Guides

Kukadia V et al. *Controlling particles, vapour and noise pollution from construction sites.* Pollution Control Guides. 5 Parts. AP 60. 2003

Books, Guides and Reports

Anderson J & Howard N. *The Green Guide to Housing Specification: an environmental profiling system for building materials and components used in housing.* BR 390. 2003

Anderson J & Shiers DE. *The Green Guide to specification: an environmental profiling system for building materials and components used in housing.* 3rd edition. Oxford, Blackwell Publishing, 2002

Bourke K, Ramdas V, Singh S et al. *Achieving whole life value in infrastructure and buildings.* BR 476. 2005

BRE. *Specifying dwellings with enhanced sound insulation: a guide.* BR 406. 2000

BRE. *Part L explained: the BRE guide.* BR 489. 2006

BRE. *Site organisation and management pack* (covering site communications, site waste and offsite construction). AP 212

BRE. *Sustainability and green issues pack* (covering BRE guidance on environmental impacts and benchmarking, sustainability lessons and whole life costing and whole life value). AP 213.

BRE. *Waste materials and recycling pack* (covering waste, reclamation and recycling and waste minimisation). AP 214

BRE. *Energy use and efficiency pack* (covering thermal bridging, condensing boilers, energy efficiency, solar collectors and thermal mass principles). AP 218

BRE. *Wind, floods and climate pack* (covering performance of materials in hot and cold climates, microclimates, wind loads and wind actions). AP 224

BRE & Cyril Sweett. *Putting a price on sustainability.* FB10. 2005

BRE & Net Composites. *Green Guide to composites: an environmental profiling system for composite materials and products.* BR 475. 2004

Brownhill D & Rao S. *A sustainability checklist for developments: A common framework for developers and local authorities.* BR 436. 2002

Graves HM & Phillipson MC. *Potential implications of climate change in the built environment.* FB2. 2000

Hadi M, Rao S, Sargant H & Rathouse K. *Working with the community: a good practice guide for the construction industry.* BR 472. 2004

Howard N, Edwards S & Anderson J. *BRE methodology for environmental profiles of construction materials, components and buildings.* BR 370. 1999

Hurley JW, McGrath C, Fletcher SL & Bowes HM. *Deconstruction and reuse of construction materials.* BR 418. 2001

Kukadia V, Upton S & Hall D. *Control of dust from construction and demolition activities.* BR 456. 2003

Pascoe T & Bartlett P. *Making crime our business: a crime audit guide for Registered Social Landlords.* BR 383. 2000

Rao S et al. *EcoHomes: the environmental rating for homes.* BR 389. 2003

Raw GJ, Aizlewood CE & Hamilton RM. *Building regulation, health and safety.* BR 417. 2001

Yates A, Baldwin R, Howard N & Rao S. *BREEAM '98 for offices.* BR 350. 1998

Websites

www.breeam.org
www.GreenBookLive.com
www.RedBookLive.com

British Standards Institution (BSI)

BS 5228: *Noise and vibration control on construction and open sites* (5 Parts)

BS 5837: 2005 *Trees in relation to construction. Recommendations.*

BS 6206: 1981 *Specification for impact performance requirements for flat safety glass and safety plastics for use in buildings*

BS 7671: 2001 *Requirements for electrical installations. IEE wiring Regulations.* 16th edition

BS 8104: 1992 *Code of practice for assessing exposure of walls to wind-driven rain*

BS EN ISO 14020: 2001 *Environmental labels and declarations. General principles*

BS EN ISO 14040: 1997 *Environmental management. Life cycle assessment. Principles and framework*

BSI also provide the CEDREC CD ROM containing summaries of UK and EC legislation including air pollution, dangerous substances, noise pollution, water discharge, wildlife and countryside.

Chartered Institution of Building Services Engineers (CIBSE)

CIBSE. *Testing buildings for air leakage.* Technical Memoranda TM23. 2000

CIBSE. *Energy efficiency in buildings: CIBSE Guide F.* 2004

CIRIA

CIRIA. *The essential green guide.* SP158. 2005

Connolly, S & Charles P (eds). *Environmental good practice: site guide.* C650. 2005

Connolly, S & Charles P (eds). *Environmental good practice: pocket book.* C651. 2005

Coventry S , Shorter B & Kingsley M. *Demonstrating waste minimisation benefits in construction.* C536. 2001

Garvin S, Reid J & Scott M. *Standards for the repair of buildings following flooding.* C623. 2005

Vivian S, Williams N & Rogers W. *Climate change risks in building: an introduction.* C638. 2005

Health and Safety Executive (HSE)

Ferguson I. *Dust and noise in the construction process.* CRR73. London, HSE, 1995. Available as a pdf

HSE. *Health and safety in construction.* HSG150. 2001

HSE. *A guide to managing health and safety in construction.* 1995

HSE. *Designing for health and safety in construction.* 1995

The Stationery Office (TSO)

DCLG. *Building Regulations Approved Documents Complete Set (Paper Copy).* Various dates. Also available from www.odpm.gov.uk

Lacy RE. *Climate and building in Britain.* BRE Report. London, HMSO, 1977

National Audit Office. *Improving health and safety in the construction industry.* HC 531. 2003/2004. Available as a pdf from www.nao.co.uk.

National Audit Office/CABE. *Getting value for money from construction projects through design: how auditors can help.* 2004

Other

Architectural Energy Corporation, Boulder, USA www.archenergy.com/library/general/performance

Association of British Insurers. *Flood resilient homes.* London, ABI, 2004

Association of Chief Police Officers (ACPO). *Secured by design.* London, ACPO Crime Prevention Initiatives. 2004 www.securedbydesign.com

Carroll B & Turpin T. *Environmental impact assessment handbook: a practical guide for planners, developers and communities.* London, Thomas Telford, 2002

CCF. *Whole life costing: A client's guide.* London, Confederation of Construction Clients. 1999

Department for Culture, Media and Sport (DCMS). *Better public buildings.* 2000. Available as a pdf from www.culture.gov.uk

Flood Repairs Forum. *Repairing flooded buildings: an insurance guide to investigation and repair.* EP 69. Bracknell, IHS BRE Press, 2006

Hawken P, Lovins, AB & Lovins LH. *Natural capitalism: creating the next industrial revolution.* Snowmass, Colorado, Rocky Mountain Institute, 1999

NHBC. *NHBC Standards.* Chapter 4.2: Building near trees. Available from www.nhbcbuilder.co.uk

Office of Government Commerce (OGC). *Achieving sustainability in construction procurement.* 2000. Available as a pdf from www.ogc.gov.uk

Romm J & Browning WD. *Greening the building and the bottom line.* Snowmass, Colorado, Rocky Mountain Institute, 1994

Sustainable Construction Task Group (SCTG). *Reputation, risk and reward.* Report prepared by BRE and Environment Agency. 2002

The Architecture Foundation. *Creative spaces: a toolkit for participatory urban design.* London, The Architecture Foundation, 2000

UK Climate Impacts Programme (UKCIP). *Climate change scenarios for the United Kingdom.* The UKCIP02 Briefing Report. Norwich, The Tyndall Centre for Climate Change Research, 2002. www.tyndall.ac.uk

UK Climate Impacts Programme (UKCIP). *Climate adaptation: risk, uncertainty and decision-making.* UKCIP Technical Report. Norwich, The Tyndall Centre for Climate Change Research, 2003. www.tyndall.ac.uk

Chapter 2: Site investigation and preparation

Association of Geotechnical Specialists (AGS)

AGS. AGS *Code of conduct for site investigation.* 1998

AGS. AGS *Guidelines for good practice in site investigation.* 1998

AGS. AGS *guide to the selection of geotechnical soil laboratory testing.* 1998

AGS. *Guidelines for combined geoenvironmental and geotechnical investigation.* 2005

BRE

Digests

BRE. *Concrete in aggressive ground.* Special Digest 1. 3rd edition. 2005

BRE. *Soils and foundations.* Digests 63–67. Reprinted 1985

BRE. *Fill: Classification and load carrying characteristics.* Digest 274. Revised 1991

BRE. *Fill: Site investigation, ground improvement and foundation design.* Digest 275. Minor revisions, 1992

BRE. *Site investigation for low-rise building: desk studies.* Digest 318. 1987

BRE. *Site investigation for low-rise building: the walk-over survey.* Digest 348. 1989

BRE. *Site investigation for low-rise building: trial pits.* Digest 381. 1993

BRE. *Slurry trench cut-off walls to contain contamination.* Digest 395. 1994

BRE. *Site investigation for low-rise building: direct investigations.* Digest 411. 1995

BRE. *Optimising ground investigation.* Digest 472. 2002

Good Building Guides

BRE. *Buildings and radon.* Good Building Guide 25. 1996

BRE. *Minimising noise from domestic fan systems.* Good Building Guide 26. 1996

BRE. *Simple foundations for low-rise housing.* Good Building Guide 39

    Part 1: *Site investigation.* 2000

    Part 2: *'Rule of thumb design'.* 2001

    Part 3: *Groundworks: getting it right.* 2002

Charles JA. *Building on brownfield sites.* Good Building Guide 59

    Part 1: *identifying the hazards.* 2003

    Part 2: *reducing the risks.* 2004

Information Paper

Charles JA & Watts KS. *Building on fill: collapse compression on inundation.* IP 5/97. 1997

Books, Guides and Reports

BRE. *Radon: guidance on protective measures for new dwellings.* BR 211. 1999 edition

BRE. *Radon: guidance on protective measures for new dwellings in Scotland.* BR 376. 1999

BRE. *Specifying vibro stone columns.* BR 391. 2000

BRE. *Radon: guidance on protective measures for new dwellings in Northern Ireland.* BR 413. 2001

BRE. *Thermal insulation: avoiding risks.* BR 262. 2002 edition

BRE. *Specifying dynamic compaction.* BR 458. 2003

BRE. *Cover systems for land regeneration: Thickness of cover systems for contaminated land.* BR 465 (booklet and CD ROM). 2004

Charles JA. *Geotechnics for building professionals.* BR 473. 2005

Charles JA, Chown RC, Watts KS & Fordyce G. *Brownfield sites: ground-related risks for buildings.* BR 447. 2002

Charles JA & Watts KS. *Building on fill: geotechnical aspects.* 2nd edition. BR 424. 2001

Harrison HW & Trotman PM. *BRE Building Elements. Foundations, basements and external works. Performance, diagnosis, maintenance, repair and the avoidance of defects.* BR 440. 2002

Hartless R. *Construction of new buildings on gas-contaminated land.* BR 212. 1991

Johnson R. *Protective measures for housing on gas-contaminated land.* BR 414. 2001

Paul V. *Performance of building materials in contaminated land.* BR 255. 1994

Scivyer CR & Gregory TJ. *Radon in the workplace.* BR 293. 1995

Skinner HD, Charles JA & Tedd P. *Brownfield sites: An integrated ground engineering strategy.* BR 485. 2005

British Standards Institution

BS 1377: 1990 *Methods of test for soils for civil engineering purposes*

    Part 1: *General requirements and sample preparation*

    Part 2: *Classification tests*

    Part 3: *Chemical and electro-chemical tests*

    Part 4: *Compaction-related tests*

    Part 5: *Compressibility, permeability and durability tests*

    Part 6: *Consolidation and permeability tests in hydraulic cells and with pore pressure measurement*

    Part 7: *Shear strength tests (total stress)*

    Part 8: *Shear strength tests (effective stress)*

    Part 9: *In-situ tests*

BS 5930: 1999 *Code of practice for site investigations*

BS 8004: 1986 *Code of practice for foundations*

BS 8006: 1995 *Code of practice for strengthened/reinforced soils and other fills*

BS 8103-1: 1995 *Structural design of low-rise buildings. Code of practice for stability, site investigation, foundations and ground floor slabs for housing*

BS 10175: 2001 *Investigation of potentially contaminated sites. Code of practice*

BS EN 1997-1: 2004 *Eurocode 7. Geotechnical design. General rules*

DD ENV 1997-2: 2000 *Eurocode 7. Geotechnical design. Design assisted by laboratory testing*

CIRIA

CIRIA publications are available from www.ciria.org.uk

Charles JA & Watts KS. *Treated ground: engineering properties and performance.* C572. 2002

Leach BA & Goodger HK. *Building on derelict ground.* SP078. 1991

Mitchell J & Jardine FM. *A guide to ground treatment.* C573. 2002

Institution of Civil Engineers (ICE)

Institution of Civil Engineers Geotechnical Engineering Group. *Specification for ground treatment.* 1987

Site Investigation Steering Group (SISG). *Site investigation in construction (4 Volumes).* 1993

NHBC

National House-Building Council. *NHBC Standards*

Chapter 4.1 Land quality: managing land conditions

Chapter 4.6 Vibratory ground improvement

The Stationery Office

DCLG. Planning Policy Guidance Notes

    14   *Development on unstable land*

    25   *Development and flood risk*

Defra and Environment Agency. *Assessment of risks to human health from land contamination: an overview of the development of soil guideline values and related research.* CLEA Report CLR7. 2002

Defra and Environment Agency. *Priority contaminants for the assessment of land.* CLEA Report CLR8. 2002

Defra and Environment Agency. *Contaminants in soil: collation of toxicological data and intake values for humans.* CLEA Report CLR9. 2002

Defra and Environment Agency. *The Contaminated Land Exposure Assessment Model (CLEA): technical basis and algorithms.* CLEA Report CLR10. 2002

Defra and Environment Agency. *Model procedure for the management of land contamination.* CLEA Report CLR11. 2004

Environment Agency. *Risks of contaminated land to buildings, building materials and services: a literature review.* R & D Report P331. 2000

Environment Agency. *Assessment and management of risks to buildings, building materials and services from land contamination.* R & D Report P5-035/TR/01. 2001

Garvin S, Hartless R, Smith M et al. *Risks of contaminated land to buildings, building materials and services; a literature review.* Environment Agency 2000. Available as a pdf from www.environment-agency.gov.uk

HMSO. *Health and Safety at Work etc Act 1974.* Chapter 37

HMSO. *Environmental Protection Act 1990*

HMSO. *The Construction (Design and Management) Regulations 1994. Statutory Instrument 1994 No 3140.*

HMSO. *The Waste Management Licensing Regulations 1994. Statutory Instrument 1994 No 1056*

HMSO. *Party Wall etc Act 1996.* Chapter 40

HMSO. *The Groundwater Regulations 1998. Statutory Instrument 1998 No 2746*

Chapter 3: Site environment and orientation

BRE

Digests

BRE. *The assessment of wind loads.* Digest 346
  Part 1: *Background and method.* Revised 1992
  Part 2: *Classification of structures.* 1989
  Part 3: *Wind climate in the United Kingdom.* 1992
  Part 4: *Terrain and building factors and gust peak factors.* Revised 1992
  Part 5: *Assessment of wind speed over topography.* 1989
  Part 6: *Loading coefficients for typical buildings.* 1989
  Part 7: *Wind speeds for serviceability and fatigue assessments.* 1989
  Part 8: *The assessment of wind loads.* 1990
BRE. *Wind around tall buildings.* Digest 390. 1994
BRE. *Wind actions on buildings and structures.* Digest 406. 1995
BRE. *Roof loads due to local drifting of snow.* Digest 439. 1999

Information Papers

Littlefair PJ. *Average daylight factor: a simple basis for daylight design.* IP 15/88. 1988
Littlefair P. *Site layout planning for daylight.* IP 5/92. 1992
Littlefair PJ. *Measuring daylight.* IP 23/93. 1993
Littlefair PJ & Aizlewood ME. *Daylight in atrium buildings.* IP 3/98. 1998
Littlefair P. *Developments in innovative daylighting.* IP 9/00. 2000

Books, Guides and Reports

Bell J & Burt W. *Designing buildings for daylight.* BR 288. 1996
BRE. *The Government's Standard Assessment Procedure for energy rating of dwellings.* 2005 edition. Available from http://projects.bre.co.uk/sap2005/
BRE. *Part L explained: the BRE guide.* BR 489. 2006
Crisp VHC, Littlefair PJ, Cooper I & McKennan G. *Daylighting as a passive solar energy option: an assessment of its potential in non-domestic buildings.* BR 129. 1988
Littlefair P. *Summertime performance of windows with shading devices.* FB9. 2005

Littlefair PJ. *Designing with innovative daylighting.* BR 305. 1996
Littlefair PJ. *Site layout planning for daylight and sunlight: a guide to good practice.* BR 209. 1998
Littlefair PJ. *Solar shading of buildings.* BR 364. 1999
Littlefair PJ, Santamouris M, Alvarez S et al. *Environmental site layout planning: solar access, microclimate and passive cooling in urban areas.* BR 380. 2000

British Standards Institution (BSI)

BS 6399-2: 1997 *Loading for buildings. Code of practice for wind loads*
BS 6399-3: 1988 *Loading for buildings. Code of practice for imposed roof loads*
BS 8104: 1992 *Code of practice for assessing exposure of walls to wind-driven rain*
BS 8206-2: 1992 *Lighting for buildings. Code of Practice for daylighting*

BS EN 12056-3: 2000 *Gravity drainage systems inside buildings. Roof drainage, layout and calculation*

Chartered Institution of Building Services Engineers (CIBSE)

Daylighting and window design. Lighting Guide LG10. 1999
CIBSE Guide A. Environmental design. 2006
Designing for improved solar shading control. TRM 37. 2006

The Stationery Office (TSO)
Department of Education and Employment (Architects and Building Branch)

Guidelines for environmental design in schools. Building Bulletin 87. London, The Stationery Office, 1987
Ventilation of school buildings. Building Bulletin 101. London, The Stationery Office, 2006

Other

DCLG. *Housing Health and Safety Rating System: the Guidance (Version 1).* 2000
Energy Saving Trust. *Reducing overheating: a designer's guide.* Energy Efficiency Best Practice in Housing CE129. London, Energy Saving Trust, 2005
HSE. *Workplace (Health, Safety and Welfare) Regulations 1992. Approved Code of Practice and Guidance.* L24

Chapter 4: Foundations and basements

BRE

Digests

BRE. *Concrete in aggressive ground.* Special Digest 1. 3rd edition. 2005

BRE. *Low-rise buildings on shrinkable clay soils.* Part 1. Digest 240. New edition 1993

BRE. *Low-rise buildings on shrinkable clay soils.* Part 2. Digest 241. Minor revisions 1990

BRE. *Low-rise buildings on shrinkable clay soils.* Part 3. Digest 242. 1980

BRE. *Low-rise building foundations: the influence of trees in clay soils.* Digest 298. New edition 1999

BRE. *Mini-piling for low-rise buildings.* Digest 313. 1986

BRE. *Choosing piles for new construction.* Digest 315. 1986

BRE. *Site investigation for low-rise building: desk studies.* Digest 318. 1987

BRE. *Site investigation for low-rise building: procurement.* Digest 322. 1987

BRE. *Site investigation for low-rise building: the walk-over survey.* Digest 348. 1989

BRE. *Underpinning.* Digest 352. Revised 1993

BRE. *Why do buildings crack?* Digest 361. 1991

BRE. *Site investigation for low-rise building: trial pits.* Digest 381. 1993

BRE. *Site investigation for low-rise building: soil description.* Digest 383. 1993

BRE. *Monitoring building and ground movement by precise levelling.* Digest 386. 1993

BRE. *Damage to structures from ground-borne vibration.* Digest 403. 1995

BRE. *Site investigation for low-rise building: direct investigations.* Digest 411. 1995

BRE. *Desiccation in clay soils.* Digest 412. 1996

BRE. *Low-rise buildings on fill.* Digest 427

    Part 1: *Classification and load-carrying characteristics.* 1997

    Part 2: *Site investigation, ground movement and foundation design.* 1998

    Part 3: *Engineered fill.* 1998

BRE. *Low-rise building foundations on soft ground.* Digest 471. 2002

BRE. *Optimising ground investigation.* Digest 472. 2002

BRE. *Tilt of low-rise buildings with particular reference to progressive foundation movement.* Digest 475. 2003

Reynolds TN. *Timber piles and foundations.* Digest 479. 2003

Good Building Guides

BRE. *Damp proofing basements.* Good Building Guide 3. 1990

BRE. *Simple foundations for low-rise housing.* Good Building Guide 39

    Part 1: *Site investigation.* 2000

    Part 2: *'Rule of thumb design'.* 2001

    Part 3: *Groundworks: getting it right.* 2002

BRE. *Foundations for low-rise building extensions.* Good Building Guide 53. 2002

Good Repair Guides

BRE. *Damage to buildings caused by trees.* Good Repair Guide 2. 1996

BRE. *Treating dampness in basements.* Good Repair Guide 23. 1999

Information Papers

Anderson BR. *U-values for basements.* IP 14/94. 1994

Charles JA & Burford D. *The effect of a rise in water table on the settlement of opencast mining backfill.* IP 15/85. 1985

Charles JA & Watts KS. *Preloading uncompacted fills.* IP 16/86. 1986

Charles JA & Watts KS. *Building on fill: collapse compression on inundation.* IP 5/97. 1997

Crilly MS & Chandler RJ. *A method of determining the state of desiccation in clay soils.* IP 4/93. 1993

St John HD, Hunt RJ & Charles JA. *The use of 'vibro' ground improvement techniques in the United Kingdom.* IP 5/89. 1989

Books, Guides and Reports

Bonshor RB & Bonshor LL. *Cracking in buildings.* BR 292. 1995

BRE. *Specifying vibro stone columns.* BR 391. 2000

BRE. *Specifying dynamic compaction.* BR 458. 2003

Charles JA. *Geotechnics for building professionals.* BR 473. 2005

Charles JA & Watts KS. *Building on fill: geotechnical aspects.* BR 424. 2nd edition, 2001

Charles JA, Chown RC, Watts KS & Fordyce G. *Brownfield sites: ground-related risks for buildings.* BR 447. 2002

Harrison HW & Trotman PM. *BRE Building Elements. Foundations, basements and external works: Performance, diagnosis, maintenance, repair and the avoidance of defects.* BR 440. 2002

Paul V. *Performance of building materials in contaminated land.* BR 255. 1994

Skinner HD, Charles JA & Tedd P. *Brownfield sites: an integrated ground engineering strategy.* BR 485. 2005

British Standards Institution (BSI)

BS 1377: 1990 *Methods of test for soils for civil engineering purposes*
- Part 1 *General requirements and sample preparation*
- Part 2 *Classification tests*
- Part 3 *Chemical and electro-chemical tests*
- Part 4 *Compaction-related tests*
- Part 5 *Compressibility, permeability and durability tests*
- Part 6 *Consolidation and permeability tests in hydraulic cells and with pore pressure measurement*
- Part 7 *Shear strength tests (total stress)*
- Part 8 *Shear strength tests (effective stress)*
- Part 9 *In-situ tests*

BS 5837: 2005 *Trees in relation to construction. Recommendations*

BS 5588-1: 1990 *Fire precautions in the design, construction and use of buildings. Code of practice for residential buildings*

BS 5930: 1999 *Code of practice for site investigations*

BS 6031: 1981 *Code of practice for earthworks*

BS 6262-4: 2005 *Glazing for buildings. Code of practice for safety related to human impact*

BS 8002: 1994 *Code of practice for earth retaining structures*

BS 8004: 1986 *Code of practice for foundations*

BS 8006: 1995 *Code of practice for strengthened/reinforced soils and other fills*

BS 8007: 1987 *Code of practice for design of concrete structures for retaining aqueous liquids*

BS 8081: 1989 *Code of practice for ground anchorages*

BS 8102: 1990 *Code of practice for protection of structures against water from the ground*

BS 8103-1: 1995 *Structural design of low-rise buildings. Code of practice for stability, site investigation, foundations and ground floor slabs for housing*

BS 8110: *Structural use of concrete*
- Part 1: 1997 *Code of practice for design and construction*
- Part 2: 1985 *Code of practice for special circumstances*

BS 10175: 2001 *Investigation of potentially contaminated sites. Code of practice*

BS EN 1536: 2000 *Execution of special geotechnical work. Bored piles*

BS EN 1537: 2000 *Execution of special geotechnical work. Ground anchors*

BS EN 1538: 2000 *Execution of special geotechnical works. Diaphragm walls*

BS EN 1997-1: 2004 *Eurocode 7. Geotechnical design. General rules*

BS EN 12063: 1999 *Execution of special geotechnical work. Sheet pile walls*

BS EN 12600: 2002 *Glass in building. Pendulum test. Impact test method and classification for flat glass*

BS EN 12699: 2001 *Execution of special geotechnical work. Displacement piles*

BS EN 12715: 2000 *Execution of special geotechnical work. Grouting*

BS EN 12716: 2001 *Execution of special geotechnical works. Jet grouting*

BS EN 13564-1: 2002 *Anti-flooding devices for buildings. Requirements*

BS EN 14199: 2005 *Execution of special geotechnical works. Micropiles*

BS EN 14679: 2005 *Execution of special geotechnical works. Deep mixing*

BS EN 14731: 2005 *Execution of special geotechnical works. Group treatment by deep vibration*

BS EN ISO 14688-1: *Geotechnical investigation and testing. Identification and classification of soil*
- Part 1: 2002 *Identification and description*
- Part 2: 2004 *Principles for a classification*

BS EN ISO 14689-1: 2003 *Geotechnical investigation and testing. Identification and classification of rock. Identification and description*

BS EN ISO 22476-2:2005 *Geotechnical investigation and testing. Field testing. Dynamic probing*

BS EN ISO 22476-3: 2005 *Geotechnical investigation and testing. Field testing. Standard penetration test*

DD ENV 1997-2: 2000 *Eurocode 7. Geotechnical design. Design assisted by laboratory testing*

CIRIA

Charles JA & Watts KS. *Treated ground: engineering properties and performance.* C572. 2002

Johnson RA. *Water resisting basements: a guide. Safeguarding new and existing basements against water and dampness.* R139. 1995

CIRIA (cont'd)

Leach BA & Goodger HK. *Building on derelict ground.* SP078. 1991

Mitchell J & Jardine FM. *A guide to ground treatment.* C573. 2002

Health and Safety Executive (HSE)

HSE. *Health and safety in excavations: be safe and shore.* HSG185. 1999

NHBC

National House-Building Council. *NHBC Standards*

Chapter 4.2 Building near trees

Chapter 4.4 Strip and trench fill foundations

Chapter 4.5 Raft, pile, pier and beam foundations

Chapter 4.6 Vibratory ground movement

The Stationery Office (TSO)

HMSO. *Party Wall etc. Act 1996*

HMSO. *Water Supply (Water Fittings) Regulations 1999. Statutory Instrument 1999 No 1148.* 1999

HMSO. *The Water Supply (Water Quality) (Scotland) Regulations 2000*

HMSO. *Private Water Supply Regulations 2000*

HMSO. *The Water Supply (Water Quality) Regulations 2001*

HMSO. *The Water Supply (Water Quality) (Amendment) Regulations (Northern Ireland) 2003*

HMSO. *Water Authorities Enforcement Regulations*

HSE. *Workplace (Health, Safety and Welfare) Regulations 1992. Approved Code of Practice and Guidance.* L24

Other

Atkinson MF. *Structural foundations manual for low-rise buildings.* 2nd edition. Abingdon, Spon Press, 2003

Butcher AP, Powell JJM & Skinner HD (eds). *Reuse of foundations for urban sites: proceedings of international conference.* EP73. Bracknell, IHS BRE Press, 2006

Butcher AP, Powell JJM & Skinner HD (eds). *Reuse of foundations for urban sites: a best practice handbook.* EP75. Bracknell, IHS BRE Press, 2006

HAPM. *Feedback from data 1991–1994.* TechNote 7. HAPM, 1997

The Basement Information Centre. *The Building Regulations 2000. Approved Document: Basements for dwellings.* Camberley, Surrey, TBIC, 2004. Available from www.tbic.org.uk

Chapter 5: External walls, windows and doors

BRE

Digests

BRE. *Damp proof courses.* Digest 380. 1993

de Vekey RC. *Clay bricks and clay brick masonry.* Digest 441 (2 Parts). 2000

Garvin SL. *Insulating glazing units.* Digest 453. 2000

Quillin KC. *Corrosion of steel in concrete: service life design and prediction.* Digest 455. 2001

Good Building Guides

BRE. *Choosing external rendering.* Good Building Guide 18. 1994

BRE. *Building brickwork or blockwork retaining walls.* Good Building Guide 27. 1996

BRE. *Connecting walls and floors.* Good Building Guide 29. 1997

BRE. *Building damp-free cavity walls.* Good Building Guide 33. 1999

BRE. *Building without cold spots.* Good Building Guide 35. 1999

BRE. *Installing wall ties.* Good Building Guide 41. 2000

BRE. *Insulating masonry cavity walls.* Good Building Guide 44. 2001

   Part 1: *techniques and materials*

   Part 2: *principal risks and guidance*

de Vekey R. *Building masonry with lime-based bedding mortars.* Good Building Guide 66. 2005

Harrison H. *Plastering and internal rendering.* Good Building Guide 65 (2 Parts). 2005

Stirling C. *Level external thresholds: reducing moisture penetration and thermal bridging.* Good Building Guide 47. 2001

Stirling C. *Timber frame construction: an introduction.* Good Building Guide 60. 2003

Information Paper

Littlefair PJ. *Daylighting design for display-screen equipment.* Information Paper IP 10/95. 1995

Webber GMB & Aizlewood CE. *Falls from domestic windows.* IP17/93. 1993

Books, Guides and Reports

BRE and CIRIA. *Sound control for homes.* BR 238. 1993

BRE. *Quiet homes: a guide to good practice and reducing the risk of poor sound insulation between dwellings.* BR 358. 1998

BRE. *Specifying dwellings with enhanced sound insulation: a guide.* BR 406. 2000

BRE. *Thermal insulation: avoiding risks.* BR 262. 2002 edition

BRE. *Part L explained: the BRE guide.* BR 489. 2006

BRE. *Simplified Building Energy Model* (SBEM). Part L software. Download available from www.ncm.bre.co.uk. Update released 2006

Garvin SL & Blois-Brooke TRE. *Double-glazing units: a BRE guide to improved durability.* BR 280. 1995

Harrison HW & de Vekey RC. *BRE Building Elements. Walls, windows and doors. Performance, diagnosis, maintenance, repair and the avoidance of defects.* BR 352. 1998

Kelly D, Garvin S & Murray I. *Sloping glazing: understanding the risks.* BR 471. 2004

Littlefair PJ. *Solar shading of buildings.* BR 364. 1999

Littlefair P. *Summertime solar performance of windows with shading devices.* FB 9. 2005

Rennie D & Parand F. *Environmental design guide for naturally ventilated and daylit offices.* BR 345. 1998

Ross K. *Modern methods of house construction: a surveyor's guide.* FB 11. 2005

Trotman P, Sanders C & Harrison H. *Understanding dampness.* BR 466. 2004

Website
www.projects.bre.co.uk/hemphomes/
www.RedBookLive.com

Brick Development Association (BDA)

Haseltine BA & Tutt N. *The design of brickwork retaining walls.* DG2. Revised 1991

Haseltine BA & Tutt N. *External walls: design for wind loads.* DG4. Revised 1984

Curtin WG, Shaw G, Beck JK & Bray WA. *Design of brick diaphragm walls.* DG11. Revised 1982

Korff J. *Design of free-standing walls.* DG12. 1984

Morton J. *The design of laterally loaded walls.* DG16. 1986

Curtin WG, Shaw G, Beck JK & Howard J. *Design of post-tensioned brickwork.* DG20. 1989

British Standards Institution (BSI)

BS 747: 2000 *Reinforced bitumen sheets for roofing. Specification*
Note: This Standard will be replaced by BS EN 13707.

BS 1202: *Specification for nails*
Part 1: 2002 *Steel nails*
Part 2: 1974 *Copper nails*
Part 3: 1974 *Aluminium nails*

BS 2523: 1966 *Specification for lead-based priming paints*

BS 5250: 2002 *Code of practice for control of condensation in buildings*

BS 5262: 1991 *Code of practice for external renderings*

BS 5268: *Structural use of timber*
Part 2: 2002 *Code of practice for permissible stress design, materials and workmanship*
Part 5: 1989 *Code of practice for the preservative treatment of structural timber*

BS 5534: 2003 *Code of practice for slating and tiling (including shingles)*

BS 5617: 1985 *Specification for urea-formaldehyde (UF) foam systems suitable for thermal insulation of cavity walls with masonry or concrete inner and outer leaves*

BS 5618: 1985 *Code of practice for thermal insulation of cavity walls (with masonry or concrete inner and outer leaves) by filling with urea-formaldehyde (UF) foam systems*

BS 5628-3: 2001 *Code of practice for use of masonry. Materials and components, design and workmanship*

BS 6399: *Loading for buildings*
Part 1: 1996 *Code of practice for dead and imposed loads*
Part 2: 1997 *Code of practice for wind loads*
Part 3: 1988 *Code of practice for imposed roof loads*

BS 8000-10: 1995 *Workmanship on building sites. Code of practice for plastering and rendering*

BS 8104: 1992 *Code of practice for assessing exposure of walls to wind-driven rain*

BS 8206-2: 1992 *Lighting for buildings. Code of practice for daylighting*

BS 8213-1: 2004 *Windows doors and rooflights. Design for safety in use and during cleaning of windows, including door-height windows and roof windows. Code of practice*

BS 8215: 1991 *Code of practice for design and installation of damp-proof courses in masonry construction*

BS 8233: 1999 *Sound insulation and noise reduction for buildings. Code of practice*

BS 8298: 1994 *Code of practice for design and installation of natural stone cladding and lining*

BS EN 998-2: 2003 *Specification for mortar for masonry. Masonry mortar*

Chartered Institution of Building Services Engineers (CIBSE)

Daylighting and window design. Lighting Guide LG10. 1999

CIBSE Guide A. Environmental design. 2006

CIRIA

Gillinson B. *Wall technology.* Volumes A–G. SP087. 1992

Ledbetter SR, Hurley S & Sheehan A. *Sealant joints in the external envelope of buildings: a guide to design, specification and construction.* R178. 1998

Keiller A, Walker A, Ledbetter S & Wolmuth W. *Guidance on glazing at height.* C632. 2005

Health and Safety Executive (HSE)

HSE publications are available from www.hsebooks.com

HSE. *Workplace (Health, Safety and Welfare) Regulations 1992. Approved Code of Practice and Guidance.* L24

HSE. *Health and safety (Display Screen Equipment) Regulations 1992*

Loss Prevention Certification Board (LPCB)

LPCB. *Test and evaluation requirements for the LPCB approval and listing of fire doorsets, lift landing doors and shutters.* LPS 1056-6.1

LPCB. *Requirements and tests for LPCB approval of wall and floor penetration and linear gap seals.* LPS 1132-4.1

LPCB. *Requirements and testing of burglary resistance building components, strongpoints and security enclosures.* LPS 1175-5.3

LPCB. *Requirements and tests for external thermal insulated cladding systems with rendered finishes (ETICS) or rainscreen cladding systems (RSC) applied to the face of a building.* LPS 1181-4

The Stationery Office (TSO)

DTLR and Defra. *Limiting thermal bridging and air leakage: robust construction details for dwellings and similar buildings.* 2001

DfES. *Acoustic design of schools.* Building Bulletin 93 (BB 93). Revised 2003. Available as a pdf from www.teachernet.gov.uk or as a publication from The Stationery Office

Tinker A & Littlewood J. *Families in flats.* 1990

TRADA Technology Ltd

Hislop P. *High performance wood windows.* 2nd edition. TBL65. 2004

Hislop PJ. *External timber cladding.* DG3. 2000

TRADA Technology. *Timber frame construction.* 3rd edition. TBL58. 2001

TRADA Technology. *Introduction to timber framed construction.* Wood Information Sheet WIS 0-3. 2003

TRADA Technology. *Timber frame building: materials specification.* Wood Information Sheet WIS 0-5. 2003

Other

Battle T. The commercial offices handbook. London, RIBA Publishing, 2003

Beckett HE & Godfrey JA. *Windows: performance, design and installation.* New York, Van Nostrand Reinhold, 1974

DCLG. *Planning and noise.* Planning Policy Guidance 24. 1994. Available at www.odpm.gov.uk

Gorgolewski MT, Grubb PJ & Lawson RM. *Building design using cold formed steel sections: Light steel framing in residential construction.* Ascot, Berkshire, Steel Construction Institute, 2001

Keefe L. *Earth building: methods and materials, repair and conservation.* London, Routledge, 2005

Ministry of Housing and Local Government. *Safety in the home.* Design Bulletin 13. London, HMSO, 1967

NHS Estates. *Acoustics: design considerations.* Health Technical Memorandum (HTM) 2045. 1996

NHS Estates. *Building components: partitions.* Health Technical Memorandum (HTM) 56. 1997

Robust Details. *Robust Details Handbook.* 2nd edition. Part E, Resistance to the passage of sound. Milton Keynes, Robust Details Ltd. Updated 2005. www.robustdetails.com

Sinnot R. *Safety and security in the home.* London, Collins, 1985

Von Burg E. Home accidents and design. *Build International* 1971: **October:** 268–270

Walker P, Keable R, Martin J & Maniatidis V. *Rammed earth: design and construction guidelines.* EP62. Bracknell, IHS BRE Press, 2005

Chapter 6: Floors and ceilings

BRE

Digests

BRE. *Selecting wood-based panel products.* Digest 323. 1992

BRE. *Sound insulation of separating walls and floors.* Part 2: *Floors.* Digest 333. 1988

BRE. *Design of timber floors to prevent decay.* Digest 364. 1993

BRE. *Plywood.* Digest 394. 1994

Good Building Guides

BRE. *Joist hangers.* Good Building Guide 21. Revised 1996

BRE. *Domestic floors: construction, insulation and damp-proofing.* Good Building Guide 28, Part 1. 1997

BRE. *Connecting walls and floors.* Good Building Guide 29. 1997

    Part 1: *A practical guide*

    Part 2: *Design and performance*

BRE. *Insulating ground floors.* Good Building Guide 45. 2001

Information Papers

Anderson BR. *The U-value of ground floors: application to building regulations.* IP 3/90. 1990

Dinwoodie JM. *Wood chipboard: recommendations for use.* IP 3/85. 1985

Jeary AP. *Determining the probable dynamic response of suspended floors.* IP 17/83. 1983

Roys M & Wright M. *Proprietary nosings for non-domestic stairs.* IP 15/03. 2003

Defect Action Sheet

Suspended timber floors: chipboard flooring — specification. Defect Action Sheet 31. 1983

Books, Guides and Reports

BRE. *BRE guide to radon remedial measures in existing dwellings.* BR 250. 1993

BRE. *Quiet homes: a guide to good practice and reducing the risk of poor sound insulation between dwellings.* BR 358. 1998

BRE. *Specifying dwellings with enhanced sound insulation: a guide.* BR 406. 2000

BRE. *Simplified Building Energy Model* (SBEM). Part L software. Download available from www.ncm.bre.co.uk. Update released 2006

BRE and CIRIA. *Sound control for homes.* BR 238. 1993

Harrison HW & Trotman PM. *BRE Building Elements: Foundations, basements and external works: Performance, diagnosis, maintenance, repair and the avoidance of defects.* BR 440. 2002

Pye PW & Harrison HW. *BRE Building Elements: Floors and flooring: Performance, diagnosis, maintenance, repair and the avoidance of defects.* 2003 edition. BR 460. 2003

Website

www.RedBookLive.com

British Standards Institution (BSI)

BS 476: *Fire tests on building materials and structures*

    Part 20: 1987 *Method for determination of the fire resistance of elements of construction (general principles)*

    Part 21: 1987 *Methods for determination of the fire resistance of loadbearing elements of construction*

BS 648: 1964 *Schedule of weights of building materials*

BS 1447: 1988 *Specification for mastic asphalt (limestone fine aggregate) for roads, footways and pavings in building*

BS 5268: *Structural use of timber* (various parts)

BS 5287: 1988 *Specification for assessment and labelling of textile floor coverings tested to BS 4790*

BS 5385: *Wall and floor tiling*

    Part 3: 1989 *Code of practice for the design and installation of ceramic floor tiles and mosaics*

    Part 4: 1992 *Code of practice for tiling and mosaics in specific conditions*

    Part 5: 1994 *Code of practice for the design and installation of terrazzo tile and slab, natural stone and composition block floorings*

BS 5395: *Stairs, ladders and walkways*

    Part 1: 2000 *Code of practice for the design, construction and maintenance of straight stairs and winders*

    Part 2: 1984 *Code of practice for the design of helical and spiral stairs*

    Part 3: 1985 *Code of practice for the design of industrial type stairs, permanent ladders and walkways*

BS 5442-1: 1989 *Classification of adhesives for construction. Classification of adhesives for use with flooring materials*

BS 5588-5: 2004 *Fire precautions in the design, construction and the use of buildings. Code of practice for firefighting stairs and lifts*

BS 5950: *Structural use of steelwork in building* (9 Parts)

    Part 1: 2000 *Code of practice for design. Rolled and welded sections*

    Part 2: 2001 *Specification for materials, fabrication and erection. Rolled and welded sections*

    Part 3.1: 1990 *Design in composite construction. Code of practice for design of simple and continuous composite beams*

    Part 4: 1994 *Code of practice for design of composite slabs with profiled steel sheeting*

    Part 5: 1998 *Code of practice for design of cold formed thin gauge sections*

    Part 6: 1995 *Code of practice for design of light gauge profiled steel sheeting*

British Standards Institution (cont'd)

BS 5950: Structural use of steelwork in building (cont'd)

Part 7: 1992 *Specification for materials and workmanship: cold formed sections*

Part 8: 2003 *Code of practice for fire resistant design*

Part 9: 1994 *Code of practice for stressed skin design*

BS 6399-1: 1996 *Loading for buildings. Code of practice for dead and imposed loads*

BS 6925: 1988 *Specification for mastic asphalt for building and civil engineering (limestone aggregate)*

BS 7263-1: 2001 *Precast concrete flags, kerbs, channels, edgings and quadrants. Precast, unreinforced concrete paving flags and complementary fittings. Requirements and test methods*

BS 7671: 2001 *Requirements for electrical installations. IEE Wiring Regulations. 16th edition*

BS 7976: 2002 *Pendulum testers*

Part 1: *Specification*

Part 2: *Method of operation*

Part 3: *Method of calibration*

BS 8000 *Workmanship on building sites*

Part 2: 1990 *Code of practice for concrete work*

Section 2.1 *Mixing and transporting concrete*

Section 2.2 *Sitework with in situ and precast concrete*

Part 9: 2003 *Cementitious levelling screeds and wearing screeds. Code of practice*

Part 11: 1989 *Code of practice for wall and floor tiling.*

Section 1: *Ceramic tiles, terrazzo tiles and mosaics*

BS 8110: Structural use of concrete

Part 1: 1997 *Code of practice for design and construction*

Part 2: 1985 *Code of practice for special circumstances*

BS 8201: 1987 *Code of practice for flooring of timber, timber products and wood based panel products. Tables 1–11*

BS 8203: 2001 *Code of practice for installation of resilient floor coverings*

BS 8204: *Screeds, bases and in situ floorings.*

Part 1: 2003 *Concrete bases and cement sand levelling screeds to receive floorings. Code of practice*

Part 2: 2003 *Concrete wearing surfaces. Code of practice*

Part 4: 2004 *Cementitious terrazzo wearing surfaces. Code of practice*

Part 5: 2004 *Mastic asphalt underlays and wearing surfaces. Code of practice*

BS 8233: 1999 *Sound insulation and noise reduction for buildings. Code of practice*

BS EN 204: 2001 *Classification of thermoplastic wood adhesives for non-structural applications*

BS EN 300: 1997 *Oriented strand boards (OSB). Definitions, classification and specifications*

BS EN 312: 2003 *Particleboards. Specifications*

BS EN 548: 2004 *Resilient floor coverings. Specification for plain and decorative linoleum*

BS EN 622 *Fibreboards. Specifications.*

Part 3: 2004 *Requirements for medium boards*

Part 4: 1997 *Requirements for softboards*

Part 5: 1997 *Requirements for dry process boards (MDF)*

BS EN 634 *Cement-bonded particle boards. Specification.*

Part 1: 1995 *General requirements*

Part 2: 1997 *Requirements for OPC bonded particleboards for use in dry, humid and exterior conditions*

BS EN 636: 2003 *Plywood. Specifications*

BS EN 649: 1997 *Resilient floor coverings. Homogeneous and heterogeneous polyvinyl chloride floor coverings. Specification*

BS EN 654: 1997 *Resilient floor coverings. Semi-flexible polyvinyl chloride tiles. Specification*

BS EN 1817: 1998 *Resilient floor coverings. Specification for homogeneous and heterogeneous smooth rubber floor coverings*

BS EN 13810-1: 2002 *Wood-based panels. Floating floors. Performance specifications and requirements*

BS EN ISO 12944: 1998 *Paints and varnishes. Corrosion protection of steel structures by protective paint systems* (8 Parts)

CIRIA

CIRIA publications are available from www.ciria.org.uk

Carpenter J, Lazarus D & Perkins C. *Safer surfaces to walk on: reducing the risk of slipping. C652. 2006*

The Stationery Office (TSO)

TSO publications are available from www.tsoshop.co.uk

Barker P, Berrick J & Wilson R. *Building sight.* RNIB and The Stationery Office. 1995

DTLR and Defra. *Limiting thermal bridging and air leakage: robust construction details for dwellings and similar buildings.* 2001

Great Britain Home Office. *Fire Precautions Act 1971. Guide to fire precautions in premises used as hotels and boarding houses which require a fire certificate.* 1991

Greenwood T. *Guide to fire precautions in existing places of entertainment and like premises.* 6th impression. 2000

TRADA Technology Ltd
Kaczmar P. *Sealing timber floors: a best practice guide.* Report 2/2001. 2001. 28 pp
TRADA Technology. *Decorative timber flooring .* Wood Information Sheet WIS 1-46. 2004

Other
Commission of the European Communities, Directorate General for Science. Research and Development. Guidelines for ventilation requirements in buildings. *European Concerted Action: Indoor air quality and its impact on Man. Report No 11.* Luxembourg, Office for Publications of the European Communities, 1992
Davies VJ & Tomasin K. *Construction safety handbook.* 2nd edition. London, Thomas Telford, 1996
European Union of Agrément (UEAtc). Directive for the assessment of floorings. Method of Assessment and Test No 2(25). Paris, UEAtc, 1970

Chapter 7: Separating and compartment walls and partitions

BRE
Digests
BRE. *Improving the sound insulation of separating walls and floors.* Part 1: *Walls.* Digest 293. 1985
BRE. *Selecting wood based panel products.* Digest 323. 1992
BRE. *Sound insulation of separating walls and floors.* Part 1: *Walls.* Digest 333. 1988

Books, Guides and Reports
BRE and CIRIA. *Sound control for homes.* BR 238. 1993
BRE. *Quiet homes: a guide to good practice and reducing the risk of poor sound insulation between dwellings.* BR 358. 1998
BRE. *Specifying dwellings with enhanced sound insulation: a guide.* BR 406. 2000

Website
www.RedBookLive.com

British Standards Institution (BSI)
BS 8233: 1999 *Sound insulation and noise reduction for buildings. Code of practice*

The Stationery Office (TSO)
DTLR and Defra. *Limiting thermal bridging and air leakage: robust construction details for dwellings and similar buildings.* 2001
HMSO. *Party Wall etc. Act 1996*

TRADA Technology Ltd
Pitts G. *Acoustic performance of party floors and walls in timber framed buildings.* Report 1/2000. 2000. 36 pp
TRADA Technology. *Timber frame construction.* 3rd edition. TBL58. 2001

Chapter 8: Roofs

BRE
Digests
Blackmore P. *Slate and tile roofs: avoiding damage from aircraft wake vortices.* Digest 467. 2002
Blackmore P. *Wind loads on roof-based photovoltaic systems.* Digest 489. 2004
BRE. *Stability under wind load of loose-laid external roof insulation boards.* Digest 295. 1985
BRE. *Wind scour of gravel ballast on roofs.* Digest 311. 1986
BRE. *Flat roof design: the technical options.* Digest 312. 1986
BRE. *Flat roof design: thermal insulation.* Digest 324. 1987
BRE. *Swimming pool roofs: minimising the risk of condensation using warm-deck roofing.* Digest 336. 1988
BRE. *Flat roof design: waterproof membranes.* Digest 372. 1992
BRE. *Wind actions on buildings and structures.* Digest 406. 1995
BRE. *Flat roof design: bituminous waterproof membranes.* Digest 419. 1996
BRE. *Protecting buildings against lightning.* Digest 428. 1998
BRE. *Wind loadings on buildings.* Part 1: *brief guidance for using BS 6399-2: 1987.* Digest 436. 1999
Currie D. *Roof loads due to local drifting of snow.* Digest 439. 1999
Saunders G. *Safety considerations in designing roofs.* Digest 493. 2005
Saunders GK. *Reducing the effects of climate change by roof design.* Digest 486. 2004

BRE (cont'd)

Good Building Guides

BRE. *Erecting, fixing and strapping trussed rafter roofs.* Good Building Guide 16. 1993

BRE. *Ventilating thatched roofs.* Good Building Guide 32. 1999

BRE. *Building a new felted flat roof.* Good Building Guide 36. 1999

BRE. *Site-cut pitched roofs.* Good Building Guide 52. 2002
  Part 1: *Design*
  Part 2: *Construction*

BRE. *Disposing of rainwater.* Good Building Guide 38. 2000

Harrison H. *Tiling and slating pitched roofs.* Good Building Guide 52. 2005
  Part 1: *Design criteria, underlays and battens*
  Part 2: *Plain and profiled clay and concrete tiles*
  Part 3: *Natural and manmade slates*

Stirling C. *Insulating roofs at rafter level: sarking insulation.* Good Building Guide 37. 2000

Stirling C. *Insulated profiled metal roofing.* Good Building Guide 43. 2000

Stirling C. *Ventilated and unventilated cold pitched roofs.* Good Building Guide 51. 2002

Stirling C. *Insulated profiled metal roofing.* Good Building Guide 43. 2000

Information Papers

Beech JC & Saunders GK. *The movement of foam plastics insulants in warm deck flat roofs.* IP6/84. 1984

Beech JC & Saunders GK. *Thermal performance of lightweight inverted warm deck flat roofs.* IP2/89. 1989

Beech JC & Uberoi S. *Ventilating cold deck flat roofs.* IP13/87. 1987

Hopkins C. *Rain noise from glazed and lightweight roofing.* IP2/06. 2006

Jenkins MR & Saunders GK. *Bituminous roofing membranes: performance in use.* IP7/95. 1995

Menzies JB & Gulvanessian H. *Eurocode 1: The code for structural loading (2 Parts).* IP 13/98. 1998

Sanders C. *Modelling and controlling interstitial condensation in buildings.* IP 2/05. 2005

Saunders GK. *Designing roofs with safety in mind.* IP 7/04. 2004

Ward TI. *Assessing the effects of thermal bridging at junctions and around openings.* IP 1/06. 2006

Books, Guides and Reports

Anderson B. *Conventions for U-value calculations.* 2nd edition. BR 443. 2006

BRE. *Thermal insulation: avoiding risks.* BR 262. 2002 edition

BRE/CIRIA. *Sound control for homes.* BR 238. 1993

Harrison HW. *BRE Building Elements. Roofs and roofing: Performance, diagnosis, maintenance, repair and the avoidance of defects.* BR 302. 2000 edition

Kelly D, Garvin S & Murray I. *Sloping glazing: understanding the risks.* BR 471. 2004

Ridal J, Reid J & Garvin S. *Highly glazed buildings: assessing and managing the risks.* BR 482. 2005

Website

www.RedBookLive.com

British Standards Institution (BSI)

BS 476: *Fire tests on building materials and structures*
  Part 3: 2004 *Classification and method of test for external fire exposure to roofs*
  Part 20: 1987 *Method for determination of the fire resistance of elements of construction (general principles)*
  Part 21: 1987 *Methods for determination of the fire resistance of loadbearing elements of construction*
  Part 22: 1987 *Methods for determination of the fire resistance of non-loadbearing elements of construction*
  Part 23: 1987 *Methods for determination of the contribution of components to the fire resistance of a structure*

BS 747: 2000 *Reinforced bitumen sheets for roofing. Specification*
Note: This Standard will be replaced by BS EN 13707

BS 3083: 1988 *Specification for hot-dip zinc coated and hot-dip aluminium/zinc coated corrugated steel sheets for general purposes*

BS 5250: 2002 *Code of practice for the control of condensation in buildings*

BS 5268-3: 1998 *Structural use of timber. Code of practice for trussed rafter roofs*

BS 5427-1: 1996 *Code of practice for the use of profiled sheet for roof and wall cladding on buildings. Design*

BS 5516: 2004 *Patent glazing and sloping glazing for buildings*
  Part 1: *Code of practice for design and installation of sloping and vertical patent glazing*
  Part 2: *Code of practice for sloping glazing*

BS 5534: 2003 *Code of practice for slating and tiling (including shingles)*
BS 5589: 1989 *Code of practice for preservation of timber*
BS 5803: 1985 *Thermal insulation for use in pitched roof spaces in dwellings*
   Part 2: *Specification for man-made mineral fibre thermal insulation in pelleted or granular form for application by blowing*
   Part 3: *Specification for cellulose fibre thermal insulation for application by blowing*
   Part 5: *Specification for installation of man-made mineral fibre and cellulose fibre insulation*
BS 6229: 2003 *Flat roofs with continuously supported coverings. Code of practice*
BS 6399: *Loading for buildings*
   Part 1: 1996 *Code of practice for dead and imposed loads*
   Part 2: 1997 *Code of Practice for wind loads*
   Part 3: 1988 *Code of practice for imposed roof loads*
BS 6651: 1999 *Code of practice for protection of structures against lightning*
BS 6925: 1988 *Specification for mastic asphalt for building and civil engineering (limestone aggregate)*
BS 8000: *Workmanship on building sites*
   Part 4: 1989 *Code of practice for waterproofing*
   Part 7: 1990 *Code of practice for glazing*
   Part 6: 1990 *Code of practice for slating and tiling of roofs and claddings*
BS 8103-3:1996 *Structural design of low-rise buildings. Code of practice for timber floors and roofs for housing*
BS 8217: 2005 *Reinforced bitumen membranes for roofing. Code of practice*
BS 8218: 1998 *Code of practice for mastic asphalt roofing*

BS EN 485:1994 *Aluminium and aluminium alloys. Sheet, strip and plate* (4 Parts)
BS EN 490: 2004 *Concrete roofing tiles and fittings for roof covering and wall cladding. Product specifications*
BS EN 988: 1997 *Zinc and zinc alloys. Specification for rolled flat products for building*
BS EN 1304: 2005 *Clay roofing tiles and fittings. Product definitions and specifications*
BS EN 12056-3: 2000 *Gravity drainage systems inside buildings. Roof drainage, layout and calculation*
BS EN 12588: 1999 *Lead and lead alloys. Rolled lead sheet for building purposes*

BS EN 13707: 2004 *Flexible sheets for waterproofing. Reinforced bitumen sheets for roof waterproofing. Definitions and characteristics*
BS EN 13859-1: 2005 *Flexible sheets for waterproofing. Definitions and characteristics of underlays. Underlays of discontinuous roofing*
BS EN 13956: 2005 *Flexible sheets for waterproofing. Plastic and rubber sheets for roof waterproofing. Definitions and characteristics*

Chartered Institution of Building Services Engineers (CIBSE)
CIBSE. *CIBSE Guide A. Environmental design.* 2006

Health and Safety Executive (HSE)
HSE. *The Construction (Design and Management) (CDM) Regulations 1994.* Available as a pdf from www.hse.gov.uk/pubns
HSE. *Health and safety in roof work.* HSG33. 1998

Loss Prevention Certification Board (LPCB)
LPCB. *Requirements and tests for built-up cladding and sandwich panel systems for use as part of the external element of buildings.* LPS 1181-4.1

The Stationery Office (TSO)
DTLR and Defra. *Limiting thermal bridging and air leakage: robust construction details for dwellings and similar buildings.* 2001
HMSO. *The Reporting of Injuries, Diseases and Dangerous Occurrences Regulations 1995.* Statutory Instrument 1995 No. 3163. London, The Stationery Office, 1995

TRADA Technology Ltd
TRADA Technology. *Trussed rafters.* Wood Information Sheet WIS 1-29. 1991
TRADA Technology. *Principles of pitched roof construction.* Wood Information Sheet WIS 1-10. 1993
TRADA Technology. *Room in the roof construction for new houses.* Wood Information Sheet WIS 0-12. 2003
TRADA Technology. *Span tables for solid timber members in floors, ceilings and roofs (excluding trussed rafter roofs) for dwellings.* DA 1/2004. 2005

Other
English Nature. *Bats in buildings.* Free leaflet IN15.2. Peterborough, English Nature

English Nature and Defra. *Bats, buildings and barn owls: a guide to safeguarding protecting species when renovating traditional buildings.* 2004. Available as a pdf from www.english-nature.org.uk or www.defra.gov.uk

Grant G. *Green roofs and façades.* Bracknell, IHS BRE Press, 2006

Hydraulics Research Ltd. *Performance of syphonic drainage systems for roof gutters.* Report SR 463. Hydraulics Research Ltd, Wallingford

Chapter 9: Building services

BRE

Digests
BRE. *Natural ventilation in non-domestic buildings.* Digest 399.1994

BRE. *Installing BMS to meet electromagnetic compatibility requirements.* Digest 424.1997

Littlefair P. *Selecting lighting controls.* Digest 498. 2006

Good Building Guides
BRE. *Lighting.* Good Building Guide 61. 2004
  Part 1: *General principles*
  Part 2: *Domestic and exterior*
  Part 3: *Non-domestic*

Jaggs M & Scivyer C. *Achieving airtightness* (3 Parts). Good Building Guide 67. 2006

Good Repair Guide
BRE. *Improving energy efficiency. Part 2: Boilers and heating systems, draughtstripping.* Good Repair Guide 26. 1999

Information Papers
Aizlewood M. *Interior lighting calculations: a guide to computer programs.* IP 16/98. 1998

Baker P, McEvoy M & Southall R. *Improving air quality in homes with supply air windows.* IP 6/03. 2003

BRE. *Preventing hot water scalding in bathrooms: using TMVs.* IP 14/03. 2003

Griggs JC & Hall J. *Hard water scale in hot water storage cylinders.* IP 13/93. 1993

Kukadia V, Ajiboye P & White M. *Ventilation and indoor air quality in schools.* IP 6/05. 2005

Lauchlan C, Griggs J & Escharameia M. *Drainage design for buildings with reduced water use.* IP 1/04. 2004

Littlefair P. *Photoelectric control of lighting: design, setup and installation issues.* IP 2/99. 1999

Littlefair P. *Dwellings and energy efficient lighting: new regulation Part L.* IP 5/02. 2002

Loe D & Perry M. *Hospitals in the best light: an introduction to hospital lighting.* IP 14/00. 2000

Perry MJ. *New ways of predicting discomfort glare.* IP 24/93. 1993

Shouler MC, Griggs JC & Hall J. *Water conservation.* IP 15/98. 1998

Slater AI. *Lighting uniformity: subjective studies.* IP 15/93. 1993

Slater AI, Bordass WT, Heasman TA. *People and lighting controls.* IP 6/96. 1996

Stephen RK. *Domestic mechanical ventilation: guidelines for designers and installers.* IP 18/88. 1988

Stephen RK, Parkins LM and Wooliscroft M. *Passive stack ventilation systems: design and installation.* IP 13/94. 1994

Webber GMB. *Emergency wayfinding lighting systems.* IP 1/93. 1993

Webber GMB & Aizlewood CE. *Emergency wayfinding lighting systems in smoke.* IP 17/94. 1994

Webber GMB, Wright MS & Cook GK. *Emergency lighting and wayfinding for visually impaired people.* IP 9/97. 1997

White MK, Griggs JC & Sutcliffe S. *Self-sealing waste valves for domestic use: an assessment.* IP 5/05. 2005

Books, Guides and Reports
BRE. *Part L explained: the BRE guide.* BR 489. 2006

Bromley K, Perry M, Webb G & Ross K. *Smart homes, a briefing guide for housing associations.* 2003. Available at: www.bre.co.uk/pdf/smarthomesbriefing.pdf

Coward SKD, Llewellyn J, Raw GJ et al. *Indoor air quality in homes in England.* BR 433. 2001

Dimitroulopoulou C, Crump D, Coward SKD et al. *Ventilation, air tightness and indoor air quality in new homes.* BR 477. 2005

Griggs JC, Pitts NJ, Hall J & Shouler MC. *Water conservation: Design, installation and maintenance requirements for the use of low flush WCs flushed syphonically or by valves.* BR 328. 1997

Harrison HW & Trotman PM. *BRE Building Elements. Building services: Performance, diagnosis, maintenance, repair and the avoidance of defects.* BR 404. 2000

Littlefair P. *Energy efficient lighting: Part L of the Building Regulations explained.* BR 430. 2001

Malhotra HL. *Fire safety in buildings.* BR 96. 1987

Paul V. *Performance of building materials in contaminated land.* BR 255. 1994

Trotman P, Sanders C & Harrison H. *Understanding dampness.* BR 466. 2004

Webb BC & Barton R. *Airtightness in commercial and public buildings.* BR 488. 2002

Website
www.RedBookLive.com

British Standards Institution (BSI)

BS 416-1: 1990 *Discharge and ventilating pipes and fittings, sand-cast or spun in cast iron. Specification for spigot and socket systems*

BS 2871-1: 1971 *Specification for copper and copper alloys. Tubes. Copper tubes for water, gas and sanitation*

BS 4305-1: 1989 and EN 198: 1987 *Baths for domestic purposes made of acrylic material. Specification for finished baths*

BS 5250: 2002 *Code of practice for the control of condensation in buildings*

BS 5266-1: 2005 *Emergency lighting. Code of practice for the emergency lighting of premises*

BS 5306-2: 1990 *Fire extinguishing installations and equipment on premises. Specification for sprinkler systems*

BS 5504-1: 1977, EN 34: 1992 *Wall hung WC pan. Wall hung WC pan with close coupled cistern. Connecting dimensions*

BS 5504-2:1977, EN 38:1992 *Wall hung WC pan. Wall hung WC pan with independent water supply. Connecting dimensions*

BS 5506-3: 1977 *Specification for wash basins. Wash basins (one or three tap holes). Materials, quality, design and construction*

BS 5588 *Fire precautions in the design, construction and use of buildings* (Various parts)

BS 5839 *Fire detection and alarm systems for buildings*
  Part 1: 2002 *Code of practice for system design, installation, commissioning and maintenance*
  Part 5: 1988 *Specification for optical beam smoke detectors.* Note: This Standard has been superseded by BS EN 54-12
  Part 6: 2004 *Code of practice for the design, installation and maintenance of fire detection and fire alarm systems in dwellings*
  Part 8: 1998 *Code of practice for the design, installation, commissioning, and maintenance of voice alarm systems*

BS 5925: 1991 *Code of practice for ventilation principles and designing for natural ventilation*

BS 6340: 1983 Shower units:
  Part 1: *Guide on choice of shower units and their components for use in private dwellings*
  Part 5: *Specification for prefabricated shower trays made from acrylic material*
  Part 6: *Specification for prefabricated shower trays made from porcelain enamelled cast iron*
  Part 7: *Specification for prefabricated shower trays made from vitreous enamelled sheet steel*
  Part 8: *Specification for prefabricated shower trays made from glazed ceramic*

BS 6700: 1987 *Specification for design, installation, testing and maintenance of services supplying water for domestic use within buildings and their curtilages*

Current but partly replaced by BS EN 806-2:2005 *Specifications for installations inside buildings conveying water for human consumption. Part 2: Design*

BS 6701: 2004 *Telecommunications equipment and telecommunications cabling. Specification for installation, operation and maintenance*

BS 6891: 2005 *Installation of low pressure gas pipework of up to 35 mm (R1¼) in domestic premises (2nd family gas). Specification*

BS 7671: 2001 *Requirements for electrical installations. IEE Wiring Regulations.* 16th edition

BS 8207: 1985 *Code of practice for energy efficiency in buildings*

BS 8300: 2001 *Design of buildings and their approaches to meet the needs of disabled people. Code of practice*

BS 12101: *Smoke and heat control systems*
  Part 2: 2003 *Specification for natural smoke and heat exhaust ventilators*
  Part 3: 2002 *Specification for powered smoke and heat exhaust ventilators*
  Part 6: 2005 *Smoke and heat control systems. Specification for pressure differential systems. Kits*

BS EN 32: 1999 *Wall-hung wash basins. Connecting dimensions*

BS EN 33: 1999 *Pedestal W.C. pans with close-coupled flushing cistern. Connecting dimensions*

BS EN 35: 2000 *Specification for bidets*

BS EN 37: 1999 *Pedestal WC pans with independent water supply. Connecting dimensions*

BS EN 54 *Fire detection and fire alarm systems*
  Part 1: 1996 *Introduction*
  Part 2: 1998 *Control and indicating equipment*
  Part 3: 2001 *Fire alarm devices. Sounders*
  Part 4: 1998 *Power supply equipment*

British Standards Institution (cont'd)

BS EN 54 *Fire detection and fire alarm systems* (cont'd)

    Part 5: 2001 *Heat detectors. Point detectors*

    Part 7: 2001 *Smoke detectors. Point detectors using scattered light, transmitted light or ionization*

    Part 10: 2002 *Flame detectors. Point detectors*

    Part 11: 2001 *Manual call points*

    Part 12: 2002 *Smoke detectors. Line detectors using an optical light beam*

    Part 13: 2005 *Compatibility assessment of system components*

    Part 17: 2005 *Short-circuit isolators*

    Part 18: 2005 *Input/output devices*

    Part 20: 2006 *Aspirating smoke detectors*

    Part 21: 2006 *Alarm transmission and fault warning routing equipment*

BS EN 111: 2003 *Wall-hung hand rinse basins. Connecting dimensions*

BS EN 232: 2003 *Baths. Connecting dimensions*

BS EN 251: 2003 *Shower trays. Connecting dimensions*

BS EN 263: 2002 *Crosslinked cast acrylic sheets for baths and shower trays for domestic purposes*

BS EN 274: 2002 *Waste fittings for sanitary appliances.* (Parts 1–3)

BS EN 329: 1997 *Sanitary tapware. Waste fittings for shower trays. General technical specifications*

BS EN 806-2: 2005 *Specifications for installations inside buildings conveying water for human consumption. Design*

BS EN 997: 2003 *WC pans and WC suites with integral trap*

BS EN 1116: 2004 *Kitchen furniture. Co-ordinating sizes for kitchen furniture and kitchen appliances*

BS EN 12101-2: *Smoke and heat control systems*

    Part 2: 2003 *Specification for natural smoke and heat exhaust ventilators*

    Part 3: 2002 *Specification for powered smoke and heat exhaust ventilators*

BS EN 12109: 1999 *Vacuum drainage systems inside buildings*

BS EN 12050: 2001 *Wastewater lifting plants for buildings and sites. Principles of construction and testing* (Parts 1–4)

BS EN 12056: 2000 *Gravity drainage systems inside buildings*

    Part 1: *General and performance requirements*

    Part 2: *Sanitary pipework, layout and calculation*

    Part 3: *Roof drainage, layout and calculation*

    Part 4: *Wastewater lifting plants. Layout and calculation*

    Part 5: *Installation and testing, instructions for operation, maintenance and use*

BS EN 12464-1: 2002 *Light and lighting. Lighting of work places. Indoor work places*

BS EN 12763: 2000 *Fibre-cement pipes and fittings for discharge systems for buildings. Dimensions and technical terms of delivery*

BS EN 13829: 2001 *Thermal performance of buildings. Determination of air permeability of buildings. Fan pressurization method*

BS EN 14511: Parts 1–4: 2004 *Air conditioners, liquid chilling packages and heat pumps with electrically driven compressors for space heating and cooling*

BS EN 50131 *Alarm systems. Intrusion systems*

    Part 1: 1997 *General requirements*

    Part 5-3: 2005 *Requirements for interconnections equipment using radio frequency techniques*

    Part 6: 1998 *Power supplies*

    Part 7: 2003 *Application guidelines*

BS EN 50136 *Alarm systems. Alarm transmission systems and equipment*

    Part 1-1: 1998 *General requirements for alarm transmission systems*

    Part 2-1: 1998 *General requirements for alarm transmission equipment*

CP 342-2: 1974 *Code of practice for centralized hot water supply. Buildings other than individual dwellings*

DD 243: *Installation and configuration of intruder alarm systems designed to generate confirmed alarm conditions. Code of practice*

DD TS 50131: *Alarm systems Intrusion systems*

    Part 2-2: 2004 *Requirements for passive infrared detectors*

    Part 2-3: 2004 *Requirements for microwave detectors*

    Part 2-4: 2004 *Requirements for combined passive infrared and microwave detectors*

    Part 2-5: 2004 *Requirements for combined passive infrared and ultrasonic detectors*

    Part 2-6: 2004 *Requirements for opening contacts (magnetic)*

    Part 3: 2003 *Control and indicating equipment*

PD 6662: 2004 *Scheme for the application of European Standards for intruder and hold-up alarm systems*

Chartered Institution of Building Services Engineers (CIBSE)

CIBSE. *Lighting for hostile and hazardous environments.* Application Guide. 1983

CIBSE. *Emergency lighting.* Technical Memorandum TM12. 1986

CIBSE. *Lighting Guide: The visual environment for display screen use.* LG03. 1996

CIBSE. *Natural ventilation in non-domestic buildings.* CIBSE Applications Manual AM 10. 1997

CIBSE. *Testing buildings for air leakage.* Technical Memorandum TM23. 2000

CIBSE. *Lighting Guide: Surface reflectance and colour.* LG11. 2001

CIBSE. *Guide to fibre-optic and remote-source lighting.* ILE02. 2001

CIBSE. *Commissioning Code L: Lighting.* 2003

CIBSE. *Code for interior lighting.* 2004

CIBSE. *Lighting Guide: Office lighting.* LG07. 2005

CIBSE. *CIBSE Guide A. Environmental design.* 2006

Comité Européen des Assurances (CEA)

CEA. *Fire protection systems — Specifications for fire detection and fire alarm systems — Requirements and tests methods for multisensor detectors, which respond to smoke and heat, and smoke detectors with more than one smoke sensor.* CEA 4021

CEA. *Fire protection systems — Specifications for fire detection and fire alarm systems — Requirements and tests methods for aspirating smoke detectors.* CEA 4022

Loss Prevention Certification Board (LPCB)

LPCB. *Requirements for certificated fire detection and fire alarm firms.* LPS 1014

LPCB. *Requirements for component compatibility for fire detection and fire alarm systems.* LPS 1054

LPCB. *Requirements and testing procedures for radio linked fire detection and fire alarm equipment.* LPS 1257

LPCB. *Requirements and testing procedures for the LPCB approval and listing of carbon monoxide fire detectors using electrochemical cells.* LPS 1265

LPCB. *Testing procedures for the LPCB approval and listing of carbon monoxide/heat multisensor fire detectors using electrochemical cells.* LPS 1274

LPCB. *Requirements for LPCB approval and listing of alarm transmission systems and equipment.* LPS 1277

LPCB. *Point multisensor fire detectors using optical or ionization smoke sensors and electrochemical cell CO sensors and, optionally, heat sensors.* LPS 1279

LPCB. *Requirements for LPCB approval and listing of intruder alarm movement detectors.* LPS 1602

LPCB. *Requirements for LPCB approval and listing of intruder alarm control and indicating equipment.* LPS 1603

The Stationery Office (TSO)

DCLG. The Construction (Design and Management) Regulations 1994. Statutory Instrument No. 3140

DCLG. *The Energy Efficiency (Ballasts for fluorescent lighting) Regulations 2001.* Statutory Instrument 2001 No. 3316. 2001

DCLG. *Fire safety — risk assessment* (Series of titles, various dates). Also available from: www.firesafetyguides.communities.gov.uk

HMSO. *Environmental Protection Act 1990* (c. 43)

HMSO. *Health and Safety (Display Screen Equipment) Regulations 1992. Statutory Instrument 1992 No. 2792*

HMSO. The Waste Management Licencing Regulations 1994. Statutory Instrument 1994 No 1056

HMSO. Gas Safety (Installation and Use) Regulations 1998. Statutory Instrument 1998 No. 2451

HMSO. Water Supply (Water Fittings) Regulations 1999. Statutory Instrument 1999 No 1148. 1999

HMSO. Water Supply (Water Fittings) Regulations 1999. Statutory Instrument 1999 No 1148. 1999

HMSO. Private Water Supply Regulations 2000

HMSO. The Water Supply (Water Quality) (Scotland) Regulations 2000

HMSO. The Water Supply (Water Quality) Regulations 2001

HMSO. The Water Supply (Water Quality) (Amendment) Regulations (Northern Ireland) 2003

HSE. The Health and Safety at Work etc. Act 1974

HSE. *The Workplace (Health, Safety and Welfare) Regulations 1992.* Statutory Instrument 1992 No.3004

HSE. *Lighting at work.* HSG 38. 1998

Other

Association of Chief Police Officers (ACPO). *Strategy position statement on Police response to security systems.* London, ACPO Crime Prevention Initiatives

Bromley AKR. The Electromagnetic Compatibility Directive and its implications for equipment installers. *Electrical Contractor* 1990: 28–30

EU. *The Electromagnetic Compatibility (EMC) Directive* 89/336/EEC

European Parliament and the Council of the European Union. *Energy Performance of Buildings.* Directive 2002/91/EC. Available as a pdf from http://europa.eu.int

Rayment R. Energy from the wind. *Building Services Engineer* 1976: **44**(3)

Scottish Water Authorities. Water Byelaws 2000 (Scotland)

Wilkins AJ, Nimmo-Smith I, Slater AI & Bedocs L. Fluorescent lighting, headaches and eyestrain. *Lighting Research & Technology* 1989: **21**(1): 11–18

Chapter 10: External works

BRE

Digest
BRE. *Soakaway design.* Digest 365. 1991

Good Building Guides
BRE. *Building simple plan brick or blockwork freestanding walls.* Good Building Guide 14. 1994

BRE. *Building reinforced, diaphragm and wide plan freestanding walls.* Good Building Guide 19. 1994

BRE. *Building brickwork or blockwork retaining walls.* Good Building Guide 27. 1996

BRE. *Disposing of rainwater.* Good Building Guide 38. 2000

BRE. *Lighting.* Good Building Guide 61. 2004
    Part 1: General principles
    Part 2: Domestic and exterior
    Part 3: Non-domestic

Griggs J & Grant N. *Reed beds.* Good Building Guide 42. 2000
    Part 1: *application and specification*
    Part 2: *design, construction and maintenance*

Information Papers
Hall J & Griggs J . *Rats in drains.* IP 6/90. 1990

Lauchlan C, Griggs J & Escharameia M. *Drainage design for buildings with reduced water use.* IP 1/04. 2004

Littlefair P. *Photoelectric control of lighting: design, setup and installation issues.* IP 2/99. 1999

Littlefair P. *Dwellings and energy efficient lighting: new regulation Part L.* IP 5/02. 2002

Books, Guides and Reports
Grant N & Griggs J. *Reed beds for the treatment of domestic wastewater.* BR 420. 2001

Harrison HW & Trotman PM. *BRE Building Elements. Foundations, basements and external works: Performance, diagnosis, maintenance, repair and the avoidance of defects.* BR 440. 2002

Littlefair P. *Energy efficient lighting: Part L of the Building Regulations explained.* BR 430. 2001

Paul V. *Performance of building materials in contaminated land.* BR 255. 1994

Phelps RDS & Griggs J. *Mound filter systems for the treatment of domestic wastewater.* BR 478. 2005

Brick Development Association (BDA)

A range of design guides and file notes, eg:
Design of brick diaphragm walls. DG11. 1982
Design of free-standing walls. DG12. 1984
BDA guide to successful brickwork. 2nd edition. 2000
The design of brickwork retaining walls. DG2. 1991
External walls: design for wind loads. DG4. 1984
The design of flexible pavements surfaced with clay pavers. DG21. 1990

British Standards Institution (BSI)

BS 1196: 1989 *Specification for clayware field drain pipes and junctions*

BS 1722: *Fences* (Various Parts)

BS 3690: *Bitumens for building and civil engineering* (3 Parts)

BS 4729: 2005 *Clay and calcium silicate bricks of special shapes and sizes. Recommendations*

BS 4962: 1989 *Specification for plastics pipes and fittings for use as subsoil field drains*

BS 4987: *Coated macadam (asphalt concrete) for roads and other paved areas* (Various parts)

BS 5489: 2003 *Code of practice for the design of road lighting*
    Part 1: *Lighting of roads and public amenity areas*
    Part 2: *Lighting of tunnels*

BS 5642-2: 1983 *Sills and copings. Specification for copings of precast concrete, cast stone, clayware, slate and natural stone*

BS 5911: *Concrete pipes and ancillary concrete products* (Various Parts)

BS 6297: 1983 *Code of practice for design and installation of small sewage treatment works and cesspools*

BS 6717: 2001 *Precast, unreinforced concrete paving blocks. Requirements and test methods*

BS 7158: 2001 *Plastics inspection chambers for drains and sewers. Specification*

BS 7263-1: 2001 *Precast concrete flags, kerbs, channels, edgings and quadrants. Precast, unreinforced concrete paving flags and complementary fittings. Requirements and test methods*

BS 7533: *Pavements constructed with clay, natural stone or concrete pavers* (Various parts)

BS 7671: 2001 *Requirements for electrical installations. IEE Wiring Regulations.* 16th edition

BS 7818: 1995 *Specification for pedestrian restraint systems in metal*

BS 8000-14: 1989 *Workmanship on building sites. Code of practice for below ground drainage*

BS 8002: 1994 *Code of practice for earth retaining structures*

BS 8204-2: 2003 *Screeds, bases and in situ floorings. Concrete wearing surfaces. Code of practice*

BS 8220-1: 2000 *Guide for security of buildings against crime. Dwellings*

BS EN 752: *Drain and sewer systems outside buildings:*
    Part 1: 1996 *Generalities and definitions*
    Part 2: 1997 *Performance requirements*
    Part 3: 1997 *Planning.*
    Part 4: 1998 *Hydraulic design and environmental considerations*
    Part 6: 1998 *Pumping installations*

BS EN 805: 2000 *Water supply. Requirements for systems and components outside buildings*

BS EN 806-2: 2005 *Specifications for installations inside buildings conveying water for human consumption. Design*

BS EN 1091: 1997 *Vacuum sewerage systems outside buildings*

BS EN 1343: 2001 *Kerbs of natural stone for external paving. Requirements and test methods*

BS EN 1610: 1998 *Construction and testing of drains and sewers*

BS EN 1671: 1997 *Pressure sewerage systems outside buildings*

BS EN 12566: *Small wastewater treatment systems for up to 50 PT*
    Part 1: 2000 *Prefabricated septic tanks*
    Part 3: 2005 *Packaged and/or site assembled domestic. Wastewater treatment plants*

BS EN 13380: 2001 *General requirements for components used for renovation and repair of drain and sewer systems outside buildings*

BS EN 13566: 2002 *Plastics piping systems for renovation of underground non-pressure drainage and sewerage networks* (Parts 1–4)

prEN 12464-2: *Light and lighting. Lighting of work places. Outdoor work places*

Chartered Institution of Building Services Engineers (CIBSE)

CIBSE. *Lighting guide: The outdoor environment.* LG 06. 1992

CIBSE. *Lighting the environment: a guide to good urban lighting.* ILE01. 1995

CIBSE. *Guide to fibre-optic and remote-source lighting .* ILE02. 2001

CIBSE. *Lighting Guide: Office lighting.* LG07. London, CIBSE, 2005

CIBSE. *CIBSE Guide A. Environmental design.* 2006

CIRIA

CIRIA. *Septic tanks and small sewage treatment works.* TN 146. 1993

CIRIA. *On-site sewage dispersal options.* SP 144/L2. 1998

CIRIA. *Design of embedded retaining walls.* C580. 2001

Hall MJ, Hockin DL, Ellis JB. *Design of flood storage reservoirs.* B014. 1993

Martin P et al. *Sustainable urban drainage systems: design manual for Scotland and Northern Ireland.* C521. 2000

Martin P et al. *Sustainable urban drainage systems: design manual for England and Wales.* C522. 2000

Martin P et al. *Sustainable urban drainage systems: best practice manual for England, Scotland, Wales and Northern Ireland.* C523. 2001

Health and Safety Executive (HSE)

HSE. *The Construction (Design and Management) (CDM) Regulations 1994.* Available as a pdf from www.hse.gov.uk/pubns

HSE. *Electrical safety on construction sites.* HSG 141. 1995

HSE. *Avoiding danger from underground services.* HSG 47. 2000

The Stationery Office (TSO)

HMSO. *Health and Safety at Work etc Act 1974.* 1974

HMSO. *The Construction (Design and Management) Regulations 1994. Statutory Instrument 1994 No 3140.* 1994

HMSO. The Waste Management Licencing Regulations 1994. Statutory Instrument 1994 No 1056

HMSO. *Gas Safety (Installation and Use) Regulations 1998. Statutory Instrument 1998 No. 2451.*

HMSO. *The Groundwater Regulations 1998. Statutory Instrument 1998 No 2746*

HMSO. Water Supply (Water Fittings) Regulations 1999. Statutory Instrument 1999 No 1148. 1999

HMSO. The Water Supply (Water Quality) (Scotland) Regulations 2000

HMSO. Private Water Supply Regulations 2000

HMSO. The Water Supply (Water Quality) Regulations 2001

HMSO. The Water Supply (Water Quality) (Amendment) Regulations (Northern Ireland) 2003

Water Regulations Advisory Scheme (WRAS)

Reclaimed water systems. Information about installing, modifying or maintaining reclaimed water systems. 1999 Reclaimed water systems. Marking and identification of pipework for reclaimed (greywater) systems. 1999

Other

Arcelor. *Piling handbook.* 8th edition. Luxembourg, Arcelor RPS. 2005

British Water. *Flows and loads 2: Sizing criteria, treatment capacity for small wastewater treatment systems (package plant).* British Water, London, 2004, revised 2005

Horticultural Trades Association. *The national plant specifications.* Reading, Berkshire, Horticultural Trades Association, 2002

Joint Council for Landscape Industries. *Trees and shrubs for landscape planting.* London, The Landscape Institute, 1989

Institute of Lighting Engineers (ILE). *Guidance notes for the reduction of obtrusive light.* Rugby, ILE, 2005. Available as a pdf from www.ile.org.uk

Scottish Water Authorities. Water Byelaws 2000 (Scotland)

Chapter 11: Modern methods of construction

BRE

Digests

BRE. *Aircrete: thin joint mortar systems.* Digest 432. 1998

de Vekey RC. *AAC 'aircrete' blocks and masonry.* Digest 468 (2 Parts). 2002

Good Building Guides

de Vekey R. *Thin layer mortar masonry.* Good Building Guide 58. 2003

Stirling C. *Offsite construction: an introduction.* Good Building Guide 56. 2003

Stirling C. *Timber frame construction: an introduction.* Good Building Guide 60. 2003

Information Papers

Bagenholm C, Yates A & McAllister I. *Prefabricated housing in the UK.* IP 16/01. 2001

    Part 1: *A case study: Murray Grove, Hackney*

    Part 2: *A case study: CASPAR II, Leeds*

    Part 1: *A summary paper*

Bregulla J & Enjily V. *An introduction to building with structural insulated panels (SIPS).* IP 13/04. 2004

Books, Guides and Reports

BRE & TRADA. *Multi-storey timber frame buildings: a design guide.* BR 454. 2003

Ross K. *Modern methods of house construction: a surveyor's guide.* FB11. 2005

CIRIA

Gibb A & Pendlebury M. *The offsite project toolkit.* C631. 2005

Loss Prevention Certification Board (LPCB)

LPCB. *Standard for innovative systems, elements and components for residential buildings.* LPS 2020-1

TRADA Technology Ltd

Bainbridge R & Milner M. *Timber floors: improvements through process re-engineering.* Report 3/99. 1999. 28 pp

Pitts G. *Timber frame re-engineering for affordable housing.* Report 2/2000. 2000

Other

Gibb AGF. *Offsite fabrications: prefabrication, pre-assembly and modularisation.* Caithness, Whittles Publishing. 1999

Chapter 12: Access to buildings

BRE

Good Building Guides

Stirling C. *Level external thresholds: reducing moisture penetration and thermal bridging.* Good Building Guide 47. 2001

Garvin SL. *Installing domestic automatic door controls.* Good Building Guide 48. 2001

Garvin SL. *Installing domestic automatic window controls.* Good Building Guide 49. 2001

Information Papers

Garvin SL. *Whole life performance of domestic automatic door controls.* IP2/02. 2002

Garvin SL. *Whole life performance of domestic automatic window controls.* IP3/02. 2002

Books, Guides and Reports

Garvin SL. *Domestic automatic doors and windows for use by elderly and disabled people: a guide for specifiers.* BR 334. 1997

British Standards Institution (BSI)

BS 7036: 1996 *Code of practice for safety at powered doors for pedestrian use* (5 Parts)

BS 8300: 2001 *Design of buildings and their approaches to meet the needs of disabled people. Code of practice*

BS EN 81-70: 2003 *Safety rules for the construction and installation of lifts. Particular applications for passenger and goods passenger lifts. Accessibility to lifts for persons including persons with disability*

BS EN 1154: 1997 *Building hardware. Controlled door closing devices. Requirements and test methods*

CIRIA

Bright K, Flanagan S, Embleton J et al. *Buildings for all to use: improving the accessibility of public buildings and environments.* C610. 2004

The Stationery Office (TSO)

HMSO. *Disability Discrimination Act 1995 (c. 50).* Amended 2005. 2005

Other

Carroll C, Cowans J & Darton D (Eds). *Meeting Part M and designing lifetime homes.* York, Joseph Rowntree Foundation, 1999

Centre for Accessible Environments (CAE). *Designing for accessibility.* 2004 edition. London, CAE

Sakkas N. *Design support tools and policy initiatives in support of universal design in buildings: the POLIS Project.* Designing for the 21st century III: an international conference on universal design, 7–12 December 2004, Rio de Janeiro, Brazil

Thorpe S & Habinteg Housing Association. *Wheelchair housing design guide.* 2nd edition. Bracknell, IHS BRE Press. 2006

Appendix A: Writing a performance specification

Given that minimum requirements are met for building elements and components, as, for example, those prescribed in building regulations, there will remain a bewildering array of alternatives from which the designer needs to choose.

Using a performance specification is a method which has proved to be a useful discipline for making a rational choice, or for delegation of responsibility for parts of the design, although the

sponsor will need to be satisfied that adequate means of verification or test exist.

The actual procedure chosen will need to relate to the complexity of the part of the building selected. The steps which BRE has found useful in choosing from a range of competing flooring products are given in Box A1 as a possible model for drawing up a performance specification for other elements.

Box A1 How to choose a flooring

1 The first task of the specifier is precisely to demarcate the extent of the different areas of the building having different categories of performance for the flooring. Particular attention should be paid to substrates, whether these are subject to deflection or movement, the relative traffic densities and likelihood of impacts and the likely or existing ambient conditions in that area of the building. Also, how much thickness can be allowed for the finish.

2 Next, the specifier should decide what headings to use in the performance specification. The main headings used in this book are repeated here for convenience:
- Strength and stability,
- Dimensional stability,
- Energy conservation, thermal insulation, air penetration and ventilation,
- Dampness and condensation,
- Comfort and safety,
- Fire protection and resistance to high temperatures,
- Appearance and reflectivity,
- Noise, sound insulation and quietness,
- Durability and Quality Assurance,
- Inspection and maintenance.

3 Next, the specifier will need to set ranges of acceptable performance for each of the chosen categories. Some of the categories are easier to assess than others. For example, combustibility, slipperiness, surface hardness,

impact resistance and thermal characteristics will be easier to deal with than colour and pattern or appearance generally, and quietness.

4 Some attempt must be made to set the life-cycle cost plan, taking into account initial cost, maintenance requirements, ease of repair and ease of removal for replacement. Warranties too are relevant.

5 The relative importance of the various attributes must next be decided, in conjunction with the client. Often, it will be found that this is one of the most significant determinants of choice.

6 Some attempt must be made to set acceptable limits on site processes. For example, how long can the laying of moisture-susceptible flooring be delayed waiting for screeds to dry?

7 Assembly of the competing floorings is the next step, with listing of the relevant performance attributes. With traditional solutions, the risk of misrepresentation of likely performance is low, though newer solutions may justify third-party certification.

8 Finally, if the choice is not already immediately apparent, a tabulation of requirements against performance attributes of competing floorings will usually yield one or more acceptable compromises, albeit that an entirely objective choice will not usually be possible.

Appendix B: BREEAM assessment of acoustic criteria

An approach to the assessment of acoustic criteria in buildings not covered by Building Regulations is that adopted in BRE's Environmental Assessment Method (BREEAM). BREEAM considers a range of characteristics for a building or development: from energy consumption to aspects that affect the health and well-being of its occupants. This holistic approach to assessment includes the acoustic environment in buildings.

BREEAM credits for the acoustic environment are awarded if specified acoustic criteria are met. Acoustics credits obtained not only enable an objective assessment of the acoustic environment but, when considered with credits awarded for other aspects such as thermal comfort or ventilation, allow an assessment of overall performance against objective criteria.

BREEAM assessments can be carried out for offices, schools, industrial buildings and dwellings. For dwellings, Ecohomes credits can be awarded depending on the extent of sound insulation testing and the results of the tests.

Ecohomes credits can be awarded where a test programme comprising a greater number of pre-completion tests than that required by AD E of the Building Regulations (England & Wales) is conducted in completed buildings. The number of tests required is based on the number of dwellings and the number of groups or sub-groups in a development. Section 1 in AD E contains the definition of groups and sub-groups. An example of how credits can be awarded to a development is given in Box B1.

Ecohomes credits can be awarded for developments where separating floors or both separating floors and walls need to be tested. The testing regimes necessary for developments that are different from the example in Box B1 can be obtained from the Ecohomes website: www.breeam.org/ecohomes.html.

The acoustic environment in a building is only one aspect that has to be considered when designing a quality building. However, design choices such as whether to use natural or mechanical ventilation can affect what is practically achievable in terms of indoor ambient noise levels. Therefore, it is advisable that design teams responsible for creating good acoustic conditions in a building should:

- be clear about how the building will be used,
- be aware of the guidance that is available and appropriate,
- be aware of the potential consequences of design choices on the internal acoustic environment,
- include all relevant areas of expertise to enable a holistic approach to design and specification.

BRE. For BREEAM products, visit website: www.breeam.org

Box B1 Example: awarding Ecohomes credits to a new housing development

A new development consists of two rows of terraces each having 12 houses. There are 24 units in the same group or sub-group.

The number of tests required for Ecohomes credits and the performance standards that must be achieved to gain the different numbers of credits are shown in Table B1. Also shown in Table B1 is the number of tests and minimum performance standard required for compliance with AD E of the Building Regulations (England & Wales).

Table B1 shows that Ecohomes credits can be awarded even though the minimum performance standard for airborne sound insulation in AD E is not necessarily exceeded. The reason for this is that the greater number of tests required for 1 and 2 Ecohomes credits gives increased confidence that the minimum performance standard needed is likely to be achieved between all dwellings in the development.

Figure B1 shows that the numbers of tests required for Ecohomes credits for different numbers of units. It can be seen that more than 30 tests within the same group or sub-group are never required.

Table B1: Number of tests and performance standards required for Ecohomes credits

AD E		Ecohomes		
Number of wall tests needed	Airborne sound insulation $D_{nT,w} + C_{tr}$ (dB) (Minimum values)	Credits desired	Number of wall tests needed	Airborne sound insulation $D_{nT,w} + C_{tr}$ (dB) (Minimum values)
6	45	1	12	45
		2	16	45
		3	16	48
		4	16	50

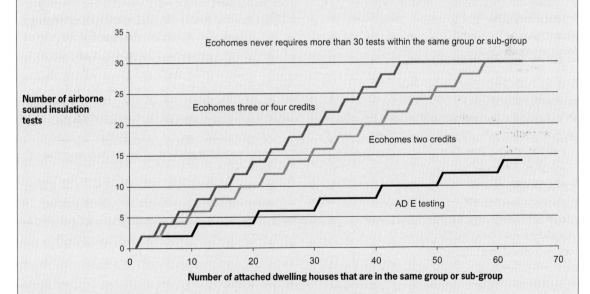

Figure B1 Comparison between Ecohomes and AD E for the number of airborne sound insulation tests required on the separating walls of attached dwelling houses that are in the same group or sub-group

Index